Quantitative Evaluation of Safety in Drug Development

Design, Analysis and Reporting

Chapman & Hall/CRC Biostatistics Series

Editor-in-Chief

Shein-Chung Chow, Ph.D., Professor, Department of Biostatistics and Bioinformatics,
Duke University School of Medicine, Durham, North Carolina

Series Editors

Byron Jones, Biometrical Fellow, Statistical Methodology, Integrated Information Sciences,
Novartis Pharma AG, Basel, Switzerland

Jen-pei Liu, Professor, Division of Biometry, Department of Agronomy,
National Taiwan University, Taipei, Taiwan

Karl E. Peace, Georgia Cancer Coalition, Distinguished Cancer Scholar, Senior Research Scientist
and Professor of Biostatistics, Jiann-Ping Hsu College of Public Health,
Georgia Southern University, Statesboro, Georgia

Bruce W. Turnbull, Professor, School of Operations Research and Industrial Engineering,
Cornell University, Ithaca, New York

Published Titles

Chapman & Hall/CRC Biostatistics Series

Quantitative Evaluation of Safety in Drug Development

Design, Analysis and Reporting

Edited by

Qi Jiang

Amgen Inc
Thousand Oaks, California, USA

H. Amy Xia

Amgen Inc
Thousand Oaks, California, USA

CRC Press is an imprint of the
Taylor & Francis Group, an **informa** business

A CHAPMAN & HALL BOOK

CRC Press
Taylor & Francis Group
6000 Broken Sound Parkway NW, Suite 300
Boca Raton, FL 33487-2742

First issued in paperback 2020

© 2015 by Taylor & Francis Group, LLC
CRC Press is an imprint of Taylor & Francis Group, an Informa business

No claim to original U.S. Government works

Version Date: 20141017

ISBN 13: 978-0-367-57600-4 (pbk)
ISBN 13: 978-1-4665-5545-7 (hbk)

Visit the Taylor & Francis Web site at
http://www.taylorandfrancis.com

and the CRC Press Web site at
http://www.crcpress.com

Contents

Section I Study Design

Section II Safety Monitoring

Section III   Evaluation/Analysis

Preface

Drug safety has always been important, but it has undergone increased scrutiny in recent years. This book is about quantitative evaluation of the safety of biopharmaceutical products. The book is meant to be comprehensive, covering design, monitoring, analysis, and reporting issues for both clinical trials and observational studies. In addition, since statistical methods for drug safety assessment are still evolving and new methods are under development, the book will cover developments in state-of-the-art methods for drug safety assessment. We hope our book will bring the most advanced knowledge and statistical methodology to the statistical, clinical, and safety community; share best practices; and stimulate further research and methodology development in the drug safety area.

The book presents key challenges that impact regulatory agencies, industry, and academia. It elaborates on Bayesian methods, presents approaches to creating effective safety graphics, and includes consideration for risk-benefit evaluation. The book is mainly for professionals with a quantitative background in fields such as statistics or epidemiology working in pharmaceutical, biotechnology, and device industries; regulatory agencies that govern these industries; and academic departments involved in activities related to clinical trials or observational studies in biopharmaceutical product development. The book is mainly suitable for research and could also be used as supplementary material for teaching; for example, for courses on pharmaceutical science, biostatistics, applied statistics, statistical computing, clinical trials, causal inference, and observational studies.

The chapters discuss quantitative approaches to safety evaluation and risk management in drug development. They cover broad topics such as study design, safety monitoring, and data evaluation/analysis. First, study design includes the Bayesian meta-experimental design for evaluating cardiovascular risk, non-inferiority study design and analysis for safety endpoints, and the program safety analysis plan. Second, safety monitoring is critical, and on this topic the book includes chapters on why a DMC safety report differs from a safety report written at the end of the trial, safety surveillance and signal detection processes, and Bayesian adaptive trials for drug safety. Finally, the book covers evaluation/analysis topics such as observational safety study design, analysis, and reporting; the observational medical outcomes partnership (OMOP); a roadmap for causal inference in safety analysis; safety graphics; Bayesian network meta-analysis for safety evaluation; regulatory issues in meta-analysis of safety data; Bayesian applications for drug safety evaluation; risk-benefit assessment approaches; detecting safety signals in subgroups, and an overview of safety evaluation and quantitative approaches during preclinical and early phases of drug development.

As editors of this book, we are pleased to see that such broad and critical safety topics have been addressed by well-established and experienced authors. We trust you will enjoy reading them and find them informative and useful in your drug safety evaluation as much as we do.

We are very grateful for our family members' tremendous support and patience during the completion of the book. We also wish to express our great appreciation for the terrific support received from David Grubbs, Laurie Schlags, and the production team at Taylor & Francis Group, LLC.

Qi Jiang, PhD
H. Amy Xia, PhD

Editors

Qi Jiang, PhD, is an executive director of Global Biostatistical Science at Amgen Inc., Thousand Oaks, California. In this role she is biostatistical therapeutic area head for bone-related therapies (including indications in osteoporosis and oncology), is head of the safety biostatistics department, and provides oversight to Amgen's biostatistical efforts in Asia. Before joining Amgen she worked at the Harvard School of Public Health, Merck, and Novartis. She has over 18 years of clinical trial experience in early and late clinical development phases across a broad spectrum of therapeutic areas and has authored numerous publications on method development, study design, and data analysis and reporting.

Dr. Jiang is a co-lead of American Statistical Association Biopharmaceutical Section Safety Working Group and the Quantitative Sciences in the Pharmaceutical Industry (QSPI) Benefit-Risk working group and a member of many initiatives including the Pharmaceutical Research and Manufacturers of America (PhRMA) Limited Duration Key Issue Team on Benefit-Risk Harmonization, the Drug Information Association (DIA) Next Steps Working Group (NSWG) on benefit-risk efforts, and a Clinical Trials Transformation Initiative (CTTI) team working on the implementation of the FDA (Food and Drug Administration) IND (Investigational New Drug) safety rule. Dr. Jiang is an associate editor of the journal *Statistics in Biopharmaceutical Research* (*SBR*), a referee for many other statistical journals and a coeditor for special issues of two statistical journals on safety and benefit-risk. She is a fellow of the American Statistical Association. In 2013, Dr. Jiang received a Healthcare Businesswomen's Association (HBA) Rising Star award. Her research interests include drug safety monitoring and analyses, benefit-risk assessment, clinical meaningfulness, dichotomization, meta-analyses, network meta-analysis, adaptive design, and noninferiority study design and analyses.

H. Amy Xia, PhD, is executive director in Global Biostatistical Science at Amgen Inc., Thousand Oaks, California. Dr. Xia has worked on designing, executing, and reporting clinical trials as well as observational studies for developing pharmaceutical and medical device products in a wide range of different disease areas in the past 18 years. Her research interests include safety biostatistics, signal detection, Bayesian design, monitoring and analysis of clinical trials, meta-analysis, and adaptive design. Dr. Xia has presented her work at a variety of professional conferences and published them in many peer-reviewed journals and books. She is a member of the PhRMA SPERT (Safety Planning, Evaluation, and Reporting Team), the Council for International Organizations of Medical Sciences (CIOMS) X Working Group, the Drug Information Association (DIA) Bayesian Scientific Working Group where she co-leads the safety subteam, and the Biostatistics Workgroup for CTTI (Clinical Trials Transformation Initiative)'s IND (Investigational New Drug) Safety Project. Dr. Xia has taught a short course entitled "Bayesian Methods for Drug Safety Evaluation and Signal Detection" at the DIA/FDA (Food and Drug Administration) Statistics Forum 2014. She holds an MS and a PhD in biostatistics from the University of Minnesota, and a medical degree from Peking University, China.

Contributors

Mary Banach
Department of Biostatistics
Vanderbilt University School of
  Medicine
Nashville, Tennessee

Jesse A. Berlin
Department of Epidemiology
Johnson & Johnson
Titusville, New Jersey

Jordan C. Brooks
Life Expectancy Project
San Francisco, California

Bradley P. Carlin
Division of Biostatistics
University of Minnesota
Minneapolis, Minnesota

Aloka G. Chakravarty
Division of Biometrics VII
Office of Biostatistics
Office of Translational Sciences
Center for Drug Evaluation and
  Research
U.S. Food and Drug Administration
Silver Spring, Maryland

Ming-Hui Chen
Department of Statistics
University of Connecticut
Storrs, Connecticut

Max Cherny
Clinical Programming
GlaxoSmithKline
King of Prussia, Pennsylvania

Christy Chuang-Stein
Statistical Research and Consulting
  Center
Pfizer
Kalamazoo, Michigan

Jason T. Connor
Berry Consultants
Austin, Texas

and

College of Medicine
University of Central Florida
Orlando, Florida

Brenda Crowe
Eli Lilly and Company
Indianapolis, Indiana

Susan P. Duke
Benefit Risk Evaluation
Global Clinical Safety and
  Pharmacovigilance
GlaxoSmithKline
Research Triangle Park, North
  Carolina

Alan S. Go
Division of Research
Kaiser Permanente
Oakland, California

Hisham Hamadeh
Amgen, Inc.
Thousand Oaks, California

Violeta Hennessey
Amgen, Inc.
Thousand Oaks, California

Hwanhee Hong
Department of Mental Health
Johns Hopkins Bloomberg School
   of Public Health
Baltimore, Maryland

Liping Huang
CSL Behring
King of Prussia, Pennsylvania

Joseph G. Ibrahim
Department of Biostatistics
University of North Carolina
   at Chapel Hill
Chapel Hill, North Carolina

Yoichi Ii
Pfizer
Tokyo, Japan

Qi Jiang
Amgen, Inc.
Thousand Oaks, California

Norisuke Kawai
Pfizer
Tokyo, Japan

Chunlei Ke
Amgen, Inc.
Thousand Oaks, California

Osamu Komiyama
Pfizer
Tokyo, Japan

Jürgen Kübler
CSL Behring
Marburg, Germany

Kazuhiko Kuribayashi
Pfizer
Tokyo, Japan

Mark Levenson
Division of Biometrics VII
Office of Biostatistics
Office of Translational Sciences
Center for Drug Evaluation and
   Research
U.S. Food and Drug Administration
Silver Spring, Maryland

Thomas Liu
Amgen, Inc.
Thousand Oaks, California

David Madigan
Department of Statistics
Columbia University
New York, New York

and

Observational Medical Outcomes
   Partnership
Foundation for the National
   Institutes of Health
Bethesda, Maryland

Mary Nilsson
Eli Lilly and Company
Indianapolis, Indiana

Karen L. Price
Eli Lilly and Company
Indianapolis, Indiana

Hong Qiu
Department of Epidemiology
Janssen Research and Development
Titusville, New Jersey

Patrick Ryan
Janssen Research and Development
Titusville, New Jersey

and

Observational Medical Outcomes
   Partnership
Foundation for the National
   Institutes of Health
Bethesda, Maryland

Mark Schactman
Statistics Collaborative, Inc.
Washington, DC

Martijn Schuemie
Janssen Research and Development
Titusville, New Jersey

and

Observational Medical Outcomes
   Partnership
Foundation for the National
   Institutes of Health
Bethesda, Maryland

Seta Shahin
Amgen, Inc.
Thousand Oaks, California

Atsuko Shibata
Amgen, Inc.
Thousand Oaks, California

Daniel E. Singer
Department of Medicine
Massachusetts General Hospital
Boston, Massachusetts

Steven Snapinn
Amgen, Inc.
Thousand Oaks, California

Paul E. Stang
Department of Epidemiology
Janssen Research and Development
Titusville, New Jersey

John Sullivan
Global Regulatory Affairs and Safety
Washington, DC

Mark J. van der Laan
Division of Biostatistics
University of California, Berkeley
Berkeley, California

José M. Vega
Merck Research Laboratories
North Wales, Pennsylvania

Wei Wang
Eli Lilly Canada, Inc.
Toronto, Ontario, Canada

Janet Wittes
Statistics Collaborative, Inc.
Washington, DC

H. Amy Xia
Amgen, Inc.
Thousand Oaks, California

Section I

Study Design

1

Incorporating Quantitative Safety Evaluation into Risk Management

Jürgen Kübler

CONTENTS

1.1 Introduction

Risk management is a concept of growing attention not only in the pharmaceutical industry but also in other industries. In general terms, risk management can be defined as "The identification, assessment, and prioritization of risks followed by coordinated and economical application of resources to minimize, monitor, and control the probability and/or impact of unfortunate events" (Hubbard 2009, p. 10). of risk assessment and risk minimization according to the U.S. Food and Drug Administration (FDA) (U.S. Food and Drug Administration 2005a). The FDA underlines further, "risk management is an iterative process of (1) assessing a product's benefit-risk balance, (2) developing and implementing tools to minimize its risks while preserving its benefits, (3) evaluating tool effectiveness and reassessing the benefit-risk balance, and (4) making adjustments, as appropriate, to the risk minimization tools to further improve the benefit-risk balance." Similarly, the European Medicine Agency (EMA) defines a risk management system as a "set of pharmacovigilance activities and interventions designed to identify, "characterize, prevent or minimize" risks relating to medicinal products including the assessment of the effectiveness of those activities and interventions" (European Medicine Agency 2012).

As with other new concepts, safety risk management is a rapidly evolving field and in fact marks a substantial change to previous practice with the FDA guidance becoming effective in 2005. The traditional approach to safety risk management had been to a large extent a reactive one. While rigorous planning and execution were applied to provide firm evidence of effectiveness of new medicines, safety analyses were considered to be mostly exploratory in nature and therefore best addressed by using descriptive statistics (International Conference on Harmonisation 1998). Once the drug was approved to the market, regulators and industry focused on signal detection and observation of the drug under real-world conditions including postmarketing safety studies. New safety information led to updates of labels, occasionally letters to health-care professionals, and ultimately to product restrictions or drug withdrawals.

The notion of risk management as an *iterative process* speaks to a much more proactive approach with a continuous generation of new information and its assessment. More precisely, risk management consists of an ongoing cycle of risk assessment, interventions, assessment of the effectiveness of these interventions, and modifications of actions throughout the life cycle of the drug. This new approach lends itself to the need for and application of quantitative approaches at almost each step of the cycle.

In this chapter, we will point out challenges and opportunities for quantitative approaches throughout the iterative process of risk management. Most of the specific techniques are not specific to risk management but are described elsewhere in the book.

The chapter is organized along the risk management process and will discuss the topics of risk identification, risk characterization, benefit–risk balance, and risk minimization. Dependence on other regulatory documents with relevance to safety will be pointed out throughout all sections.

1.2 Risk Identification

The first step in the risk management process is the identification of potential risks. Historically, this process was mainly based on biological and medical knowledge and judgment. While some risks were identified or postulated based on knowledge of the mechanism of action, preclinical findings for observations from drugs of the same class, the identification of new risks during drug development was and to a large extent still is based on the medical evaluation of single cases. This approach is also reflected in the requirements of the size of a clinical development project. A typical development program in a not life-threatening chronic indication is usually expected to include about 1500 patients treated for 6 months including 100 patients followed up for a year (ICH E1). This provides an 80% chance to observe at least one event within 6 months of treatment if the true rate is 1 in 1000 patients.

Quantitative approaches to risk identification nowadays go well beyond just the sample size calculations for the chance to observe a single event. Safety evaluations in preclinical development provide first insights into potential risks that warrant further evaluation during clinical development (cf. Chapter 17). As mentioned earlier, there is a limited chance to fully characterize a safety event of interest based on a clinical endpoint during clinical development. This has led to an increasing demand for surrogate endpoints and biomarkers for safety (U.S. Food and Drug Administration 2005b) like Hy's law cases (U.S. Food and Drug Administration 2007) and QT/QTc prolongation (ICH E14), to just name two well-known examples. The use of these endpoints comes with a whole range of opportunities and challenges. They not only extend the tools for signal detection but also lend themselves to quantitative evaluations. Routine use of these endpoints is perceived to strengthen the signal detection process for monitoring drug safety (Chapter 6). It also allows to more systematically conduct signal detection in subgroups of patients (Chapter 16). The increasing number of signal detection runs raises the risk of false-positive alarms, that is, methods to cope with multiplicity in signal detection should be exploited (Chapter 14). In addition, the validity of safety surrogate endpoints and biomarkers needs to be taken into account when discussing potential actions based on signals generated based on these endpoints.

Finally, we would like to briefly mention the large number of quantitative approaches that are utilized for *data mining* of spontaneous reporting systems in the postmarketing phase (Stephenson and Hauben 2007). In contrast to clinical trial data during development, spontaneous reporting systems hold data from a large number of reported adverse drug reactions with a limited amounts of information per case. That is, these approaches only make use of the reported adverse drug reaction term and cannot systematically use information from other data domains such as laboratory results. Most approaches apply quantitative methods at the Medical Dictionary for Regulatory Activities (MedDRA) preferred term level. Recent research, however, indicates that the choice of endpoint definition, for example, as defined by narrow and broad versions of the respective standardized MedDRA queries (SMQs) (CIOMS 2004), has an impact on the performance characteristic of the methods (Berlin et al. 2012). Although mostly beyond the scope of this book, signals generated from these sources may influence any drug development project by providing information on drugs of the same or a similar class.

1.3 Risk Characterization

Due to the limitations of the size of the clinical database, the evaluation of safety data and, therefore, risk characterization have long not been in the focus of statistical considerations. The evaluation of safety data is only briefly addressed by ICH E9. Although this document states that "later phase controlled trials represent an important means of exploring in an unbiased manner," there is little guidance as to how this can/should be achieved. While recommending a comprehensive collection of safety data and variables from different data domains, there is no link provided to specific safety questions that might be of interest. The general statement of the multiplicity of safety and the recommendation that "in most trials the safety and tolerability implications are best addressed by applying descriptive statistical methods to the data" may have contributed to the practice of providing large amounts of summary tables by data domain, that is, looking at individual data endpoints. The shortcomings of this approach are well recognized (O'Neill 2008; Crowe et al. 2009).

Significant progress has been made over the past decades. It starts off with the observation that not all potential safety questions are equally important. While the assessment of safety is a multidimensional problem by its nature, there are a number of aspects that can be taken into consideration. Based on the cumulative experience over the past decades, there are a number of safety questions that have to be addressed for a relatively large number of compounds and even classes of drugs (U.S. Food and Drug Administration 2005b). The challenges for the quantitative assessment of these risks can be exemplified with the example of the potential risk of QT prolongation. The first step in the process is to understand and to clearly articulate the underlying medical question or concern. At a first glance, this seems to be obvious in this case as the risk is called QT prolongation. A closer look at the relevant guidelines (International Conference on Harmonisation 2005) reveals that the QT prolongation is considered as a marker for a potentially serious clinical condition, namely, cardiac arrhythmias and more specifically torsade de pointes (TdP) and possibly other ventricular tachyarrhythmias. The reason why a marker is used instead of the clinical outcome leads us back to the discussion of the size of a clinical development program. TdP are very rare and, therefore, the risk of this event cannot be fully characterized within a typical drug development program. So, the question becomes what else can be done. In general terms, a good surrogate maker that allows for the prediction of the risk of the clinical output is needed. In this particular example, QT prolongation is a generally accepted marker although "the degree of QT prolongation is recognized as an imperfect biomarker for proarrhythmic risk." QT prolongation at this point is a medical concept that needs to be translated in a meaningful endpoint definition before conducting a quantitative assessment. This is reflected by the intensive discussion of various methods to a *corrected* QT prolongation that accounts for the fact that the interpretation of the raw QT values that describe the time from the beginning of the QRS complex to the end of the T wave on the surface electrocardiogram needs to take the heart rate into consideration. Once consensus on the relevant endpoint is achieved, the appropriate quantitative approaches can be utilized.

A solid understanding of what constitutes a clinically meaningful effect on this biomarker is required for the interpretation of the results. A detailed discussion on how this is addressed for the assessment of QT prolongation is beyond the scope of this book chapter. However, the general principles as discussed earlier can be applied for other risks as well. A discussion of various study designs for the analysis of safety endpoints can be found in Chapters 3 through 5.

The example of QT prolongation is particularly interesting for two reasons. First, there are some drugs that consistently induce a QT prolongation and are not associated with an increased risk of cardiac arrhythmias, which underlines the assessment of QT prolongation as an *imperfect biomarker*. Second, drugs with some level of QT prolongation might still be granted the market approval after thorough evaluation of the benefit–risk balance. Further monitoring and assessment in the postmarketing phase might be warranted in this case. Again, the questions needs to be asked what is to be achieved at this stage needs to be asked. At a first glance, the approach could be to further evaluate QT prolongation as this would now be considered an identified risk of the drug treatment. It is worthwhile, though, to go back to the original question. Therefore, the question arises if we are truly interested in the further evaluation of a biomarker or if we are now in the position to tackle the original question, namely, the assessment and quantification of the risk of cardiac arrhythmias. Even if we go for the latter, we may still be interested to consider QT prolongation as a secondary endpoint. Now, another challenge arises. This biomarker is based on ECG measurements that can be very reliably obtained in the context of a carefully designed and conducted clinical trial as the so-called thorough QTc study. Specifically in observational postmarketing studies, ECG measurements will not routinely be available in most instances, and, therefore, an unbiased estimation of the risk of QT prolongation is hard to achieve. At the same time, a much larger population is exposed to the new treatment that may allow for an assessment of the clinical endpoint, now. That is, instead of QT prolongation, the risk to be further evaluated may be cardiac arrhythmia at the time of marketing authorization. Again, the question what exactly is to be investigated depends on the underlying medical question and what is feasible to be explored with the data that are available or can be generated at the current stage. A more in-depth discussion of observational study design, analysis, and reporting can be found in Chapters 8 and 9.

The earlier case of QT prolongations is a special example of an adverse event of special interest (AESI). AESI is defined as an event (serious or nonserious) that is "one of scientific and medical concern specific to the sponsor's product or program" (CIOMS VI 2005). The Council for International Organizations of Medical Sciences (CIOMS) VI working group provides thorough discussion of AESIs with focus on signal detection and risk management during the conduct of clinical trials. Highly sensitive searches are often recommended to ensure that all potential cases will be medically evaluated. When solely based on adverse events, broad SMQs or their company-specific search terms if no SMQs are available are often recommended. These broad searches, however, are not the best possible choice for quantitative approaches as they are prone to lead to biased risk estimates. Therefore, so-called narrow searches and/or case adjudication are generally preferable for the quantitative assessment of risks (Crowe et al. 2009).

The SMQs consist of different adverse event terms with the objective to describe a common medical concept. In an attempt to increase the sensitivity of the assessment, it is quite common to combine search terms for medically closely related concepts, that is, to use a composite endpoint for the assessment of risks. A typical example of a composite safety endpoint is the *major adverse cardiovascular events* (MACE) criterion. This endpoint consists of myocardial infarction, stroke, and cardiovascular death. While this approach is very appealing at a first glance, great caution should be applied when using these composite endpoints. As in the case of MACE, the individual components of the composite do not necessarily have the same medical importance. Therefore, the results of the individual components of the composite should always be reported in addition to the overall result. This is particularly important if equivalence or noninferiority of the overall risk is to be claimed, and then it would also be interesting to know whether a treatment reduces the risk of stroke at the expense of an increased risk of cardiac mortality.

Individual clinical trials are usually not adequately powered to assess safety risks. Therefore, the assessment of safety should be based on all evidence that is available. While the shortcomings of simple pooling of all studies are well understood and described, meta-analyses of safety data are challenging as well (Berlin et al. 2013). Meta-analyses of safety have been frequently used in the past decade and are challenging from a regulatory point of view as well (cf. Chapter 13). Of course, the considerations discussed earlier should be applied for integrated summaries of safety as well. In addition to these, the consistency of approaches across trials is of utmost importance. While this sounds obvious, many practical challenges need to be addressed. Even in the situation that an existing SMQ has been used for the assessment of a safety risk, changes during the course of the development plan, for example, different versions of MedDRA used for coding in the individual studies, have to be taken into consideration (Crowe et al. 2009). Whenever possible, the corresponding analyses and approaches should be planned and documented well in advance and the same standards should be applied across all trials. Ideally, this also facilitates the conduct of meta-analyses (cf. Chapters 2 and 4).

Both integrated summaries and results of individual trials are used to develop the labeling information at the time of submissions. In general, unique medical concepts are to be used for the labeling information. That is, the same considerations as the ones given earlier should be applied. It should be noted that a frequency estimation is required to determine risk categories for adverse events, that is, reference rates for placebo and/or comparator drugs are not reported in the labeling (European Commission 2009). This raises the question as to how an unbiased estimate of a relative frequency can be derived from a clinical development program with studies in different patient and risk groups. Further research may be warranted in this space.

Finally, it should be noted that a more in-depth evaluation of potential bias should be taken into account in special circumstances. Differential dropouts, the potential for unblinding of treatment due to serious adverse events and other sources of information, may limit the validity of a quantitative risk assessment (Hammad et al. 2011).

1.4 Benefit–Risk Balance

It is noteworthy that safety of a drug treatment cannot be discussed in isolation but ultimately needs to be put in context to its benefits. Ultimately, a decision has to be made if a specific drug does have an acceptable risk profile. The CIOMS working group IV defines acceptable risk as the level of risk a population or group is prepared to accept in exchange for perceived benefits. Clearly, the threshold for an acceptable risk will be much higher for a treatment of an acute and life-threatening condition compared with a drug used for prevention. Generally speaking, the amount of the unmet medical need that is addressed by the treatment under consideration needs to be taken into account as it relates to therapeutic alternatives. Likewise, similar considerations apply at the other end of the spectrum, that is, for unacceptable risks. The line between these two areas, however, is not obvious, especially when taking uncertainty both for the assessment of benefits and for the assessment of risks into considerations. Furthermore, there often is no single accepted way to summarize both risks and benefits in a single number. That is, multiple risks usually have to be taken into account. Likewise, benefits are usually not limited to an assessment of the primary endpoint used for assessing the efficacy of a drug.

Unfortunately, there currently is no generally accepted formal approach as to how to assess the benefit–risk balance in a quantitative manner. Medical judgment has been the method of choice by regulatory bodies in the past. More recently, the need to switch from *implicit* to more *explicit* decisions has been demanded by Eichler et al. (2009) in Europe. This, however, calls for more quantitative approaches to risk assessment. Major initiatives were found to systematically evaluate these approaches, for example, the IMI-Protect project. Chapter 15 discusses benefit–risk assessment in depth. Noteworthy, the FDA continues to not support these approaches for the time being (U.S. Food and Drug Administration 2013). There is consensus, though, on the value of the use of a formal framework for benefit–risk evaluation.

1.5 Risk Minimization

Once the drug receives market approval, regulatory authorities may request additional measures to minimize risks. The EMA (2012) explicitly mentions the need for an assessment of the effectiveness of these activities. Additional guidance in this area is expected to be provided in Module XVI of the Good Pharmacovigilance Practice regulations. Due to the wide range of possible risk minimization actions, it is unlikely that general recommendation for the assessment of effectiveness of these measures will be available. However, the topic lends itself to the use of quantitative approaches as well. This is underlined by the fact that EMA mentions "studies looking at the effectiveness of risk minimisation measures" in this context. Likewise, the FDA (U.S. Food and Drug Administration 2009) emphasizes the need for the assessment of risk minimization actions and provides a discussion of different scenarios and approaches. Further research seems to be warranted to address the manifold challenges in this developing field (Banerjee et al. 2014).

References

Banerjee AK, Zomerdijk IM, Wooder S, Ingate S, Mayall SJ. 2014. Post-approval evaluation of effectiveness of risk minimisation: Methods, challenges and interpretation. *Drug Safety* 37: 33–42.

Berlin C, Blanch V, Lewis DJ, Maladorno DD, Michel C, Petrin M, Sarp S, Close P. 2012. Are all quantitative postmarketing signal detection methods equal? Performance characteristics of logistic regression and multi-item gamma Poisson Shrinker. *Pharmacoepidemiology and Drug Safety* 21: 622–630.

Berlin JA, Crowe BJ, Whalen E, Xia HA, Koro CE, Kuebler J. 2013. Meta-analysis of clinical trial safety data in a drug development program: Answers to frequently asked questions. *Clinical Trials* 10(1): 20–31.

Council for International Organizations of Medical Sciences (CIOMS) Working Group VI. 2005. Management of safety information from clinical trials. CIOMS, Geneva, Switzerland.

Council for International Organizations of Medical Sciences (CIOMS) Working Group on Standardised MedDRA Queries (SMQs). 2004. Development and rational use of standardized MedDRA queries (SMQs): Retrieving adverse drug reactions with MedDRA. CIOMS, Geneva, Switzerland.

Crowe BJ, Xia HA, Berlin JA, Watson DJ, Shi H, Lin SL, Kuebler J. et al. 2009. Recommendations for safety planning, data collection, evaluation and reporting during drug, biologic and vaccine development: A report of the safety planning, evaluation, and reporting team. *Clinical Trials* 6(5): 430–440.

Eichler H-G, Abadie E, Raine JM, Salmonson T. 2009. Safe drugs and the cost of good intentions. *New England Journal of Medicine* 360(14): 1378–1380.

European Commission. 2009. A guideline on summary of product characteristics (SmPC). Revision 2. Available from: http://ec.europa.eu/health/files/eudralex/vol-2/c/smpc_guideline_rev2_en.pdf. Accessed August 5, 2014.

European Medicine Agency. 2012. Guideline on good pharmacovigilance practices (GVP). Module V—Risk management systems. Available from: http://www.ema.europa.eu/docs/en_GB/document_library/Scientific_guideline/2012/06/WC500129134.pdf. Accessed August 5, 2014.

Hammad TA, Pinheiro SP, Neyarapally GA. 2011. Secondary use of randomized controlled trials to evaluate drug safety: A review of methodological considerations. *Clinical Trials* 8: 559–570.

Hubbard, D. 2009. *The Failure of Risk Management: Why It's Broken and How to Fix It.* John Wiley & Sons.

IMI-Protect. 2014. The pharmacoepidemiological research on outcomes of therapeutics by a European consortium. Available from: http://www.imi-protect.eu. Accessed August 5, 2014.

International Conference on Harmonisation. 1998. E9: Statistical principles for clinical trials. Available from: http://www.ich.org/fileadmin/Public_Web_Site/ICH_Products/Guidelines/Efficacy/E9/Step4/E9_Guideline.pdf. Accessed August 5, 2014.

International Conference on Harmonisation. 2005. E14: The clinical evaluation of QT/QTc interval prolongation and proarrhythmic potential for non-antiarrhythmic drugs. Available from: http://www.ich.org/fileadmin/Public_Web_Site/ICH_Products/Guidelines/Efficacy/E14/E14_Guideline.pdf. Accessed August 5, 2014.

O'Neill RT. 2008. A perspective on characterizing benefits and risks derived from clinical trials: Can we do more? *Drug Information Journal* 42: 235–245.

Stephenson WP, Hauben M. 2007. Data mining for signals in spontaneous reporting databases: Proceed with caution. *Pharmacoepidemiology and Drug Safety* 16: 359–365.

U.S. Food and Drug Administration. 2005a. Guidance for industry: Development and use of risk minimization action plans. Available from: http://www.fda.gov/downloads/RegulatoryInformation/guidances/ucm126830.pdf. Accessed August 5, 2014.

U.S. Food and Drug Administration. 2005b. Guidance for industry: Premarketing risk assessment. Available from: http://www.fda.gov/downloads/regulatoryinformation/guidances/ucm126958.pdf. Accessed August 5, 2014.

U.S. Food and Drug Administration. 2007. Guidance for industry: Drug-induced liver injury: Premarketing clinical evaluation. Available from: http://www.fda.gov/downloads/drugs/guidancecomplianceregulatoryinformation/guidances/ucm174090.pdf. Accessed August 5, 2014.

U.S. Food and Drug Administration. 2009. Guidance for industry format and content of proposed risk evaluation and mitigation strategies (REMS), REMS assessments, and proposed REMS modifications. Available from: http://www.fda.gov/downloads/Drugs/Guidances/UCM184128.pdf. Accessed August 5, 2014.

U.S. Food and Drug Administration. 2013. Structured approach to benefit-risk assessment in drug regulatory decision-making draft PDUFA V implementation plan. Available from: http://www.fda.gov/downloads/forindustry/userfees/prescriptiondruguserfee/ucm329758.pdf. Accessed August 5, 2014.

2

Bayesian Meta-Experimental Design for Evaluating Cardiovascular Risk

Joseph G. Ibrahim, Ming-Hui Chen, H. Amy Xia,
Thomas Liu, and Violeta Hennessey

CONTENTS

Recent guidance from the US Food and Drug Administration (FDA) for the evaluation of new therapies in the treatment of Type 2 diabetes mellitus (T2DM) calls for a program-wide meta-analysis of cardiovascular (CV) outcomes. In this context, we consider a Bayesian meta-analysis approach using survival regression models to assess whether the size of a clinical development program is adequate to evaluate a particular safety endpoint. The Bayesian meta-analysis trial design is general and flexible, which allows us to control the type I error and power. In addition, within the Bayesian framework, historical meta-survival data can be easily incorporated into the statistical design, which leads to a substantial reduction in the sample size compared to a frequentist design. In this chapter, we provide an in-depth discussion of meta-regression models, the general methodology for Bayesian meta-analysis design, prior elicitation based on historical data, and computational algorithms. The design of a phase 2/3 development program including a noninferiority clinical trial for CV risk assessment in T2DM studies is presented to illustrate the methodology.

2.1 Introduction

A typical drug development program may consist of multiple clinical studies with different study objectives, endpoints, and possibly different patient populations. Recent literature for single trial Bayesian sample size determination includes Wang and Gelfand (2002), Spiegelhalter et al. (2004), Inoue et al. (2005), De Santis (2007), M'Lan et al. (2008), and Chen et al. (2011). Recently, Ibrahim et al. (2012) developed a Bayesian meta-analysis sample size determination method for planning a phase 2/3 antidiabetes drug development program. Their approach focuses on sample size determination and power at a drug development program level by taking into account between-study heterogeneity within a meta-analysis framework. Sutton et al. (2007) proposed a hybrid frequentist–Bayesian approach for sample size determination for a future randomized clinical trial (RCT) using the results of meta-analyses reported in the literature and suggested that the power can be highly dependent on the statistical model used for meta-analysis and even very large studies may have little impact on a meta-analysis when there is considerable between-study heterogeneity. This raises a critical issue regarding how to appropriately account for between-study heterogeneity in a statistical model for meta-analysis.

In this chapter, we discuss a new Bayesian meta-analysis approach using survival models to assess whether the size of a clinical development program is adequate to evaluate a particular safety endpoint. The fitting and sampling priors of Wang and Gelfand (2002) are considered for designing a Bayesian meta-analysis clinical trial with a focus on controlling the type I error and power. The historical survival data are incorporated via the power priors of Ibrahim and Chen (2000). Various properties of the Bayesian meta-analysis design along with various computational algorithms are discussed in detail. The Bayesian methodology is applied to the design of a phase 2/3 development program including a noninferiority clinical trial for CV risk assessment in T2DM studies.

The rest of the chapter is organized as follows. Section 2.2 presents the federal guidelines for designs of clinical trials and meta-analysis assessments of CV risks. Section 2.3 describes a motivating case study for designing a phase

2/3 development program of a new T2DM therapy. In Section 2.4, we present both log-linear random effects and fixed effects regression models for meta-survival data. The hypotheses for "noninferiority" testing and a general Bayesian methodology for the meta-analysis clinical trial design are discussed in Section 2.5. Section 2.6 specifies the sampling prior and formulates the fitting prior using historical data. The posterior computations, simulation-based computational algorithms for generating the prior predictive data, and the computation of the type I error and power are presented in Section 2.7. In Section 2.8, we apply the Bayesian methodology to design a meta-analysis trial for evaluating CV risks. We conclude the chapter with some discussion and directions of future research on Bayesian meta-analysis design in Section 2.9.

2.2 Federal Guidelines for Bayesian Design and CV Risk Assessments

On February 5, 2010, the FDA released "Guidance for the Use of Bayesian Statistics in Medical Device Clinical Trials" (www.fda.gov/MedicalDevices/DeviceRegulationandGuidance/GuidanceDocuments/ucm071072.htm). Although this document is primarily for medical devices, it does provide guidance on statistical aspects of the design and analysis of Bayesian clinical trials in general. It also lays out detailed guidance on the determination of sample size in a Bayesian clinical trial and provides guidance on the evaluation of the operating characteristics of a Bayesian clinical trial design. Specifically, the evaluation of a Bayesian clinical trial design should include type I error, type II error, and power.

The recent guidance document from the FDA for the evaluation of new therapies in the treatment of T2DM (www.fda.gov/downloads/Drugs/GuidanceComplianceRegulatoryInformation/Guidances/ucm071627.pdf) calls for a program-wide meta-analysis of CV outcomes. This guidance requires that before submission of a new drug application, one must show that the upper bound of the two-sided 95% confidence interval (CI) for the estimated risk ratio for comparing the incidence of important CV events occurring in the investigational agent to that of the control group is less than 1.8. This can be demonstrated either by performing a meta-analysis of the randomized phase 2 and 3 studies or by conducting an additional single, large postmarketing safety trial. If the premarketing application contains clinical data from completed studies showing that the upper bound of the 95% CI for the estimated risk ratio is between 1.3 and 1.8 and the overall risk–benefit profile supports approval, a postmarketing trial generally will be necessary to definitively show that the upper bound of the 95% CI is less than 1.3. On the other hand, if the premarketing clinical data show that the upper bound of the 95% CI is less than 1.3, then a postmarketing CV trial generally may not be necessary. This requirement will most likely necessitate the performance of a specific CV outcome study for any new therapy, but would also almost certainly include integrating data across randomized phase 2 and 3 studies. This particular guidance document establishes, for the first time, that the FDA accepts prospectively designed, formal meta-analysis in the regulatory approval path.

In 2009, the Safety Planning, Evaluation and Reporting Team (SPERT) (Crowe et al., 2009), formed in 2006 by the Pharmaceutical Research and Manufacturers of America (PhRMA) with FDA participation, gave detailed recommendations for a well-planned and systematic approach to proactively plan for a meta-analysis of the program-level safety data. In particular, SPERT recommended that when it is not feasible at the individual study level, power for a particular safety outcome should be considered at the program level using the integrated safety database across clinical trials for a particular product.

2.3 Motivating Case Study

In this motivating case study, we consider a hypothetical drug development program for treating T2DM. The program includes eight randomized controlled efficacy superiority trials (Category 1) and a large randomized controlled CV outcome noninferiority trial (Category 2). Category 1 trials target subjects with low or moderate CV risk with an assumed annualized event rate of 1.2%. Category 2 targets subjects with high or enriched CV risk with an assumed annualized event rate of 1.5% as the FDA guidance advocates the inclusion of such a population.

These trials in Category 1 include multiple dose groups for experimental drug and either active control or placebo as comparator(s). For the analysis of evaluating CV risk, the placebo and active controls are combined into one "control group," and all active dose arms are combined as the "experimental drug." Thus, the sample sizes are not the same between the control group and the experimental drug group. The enrolled subjects in those trials are generally at low or moderate CV risk. The Category 2 trial has two treatment arms (experimental drug and control group). This trial is assumed to have steady enrollment for 2 years, and the last patient enrolled has a minimum of 2 years of follow-up. Detailed specifications of these meta-trials are given in Table 2.1. In this table, for the Category 1 trials, only two of them are active placebo controlled, which are the phase 2b trial (24 weeks, 3 doses, 350 subjects) and the phase 2b trial (104 weeks, 3 doses, 3500 subjects), while all other trials are placebo controlled.

TABLE 2.1

Design of Meta-Trials with Two Categories for Evaluating the CV Risk

	Control Group (n_{1k})	Experimental Drug (n_{2k})	Total (n_k)
Category 1: Randomized efficacy superiority trials			
Individual trial			
Phase 2a—4 weeks (5 doses)	25	125	150
Phase 2b—24 weeks (3 doses)	140	210	350
Phase 2b—104 weeks (3 doses)	1400	2100	3500
Phase 3—24 weeks (3 doses)	100	300	400
Phase 3—24 weeks (4 doses)	75	300	375
Phase 3 add-on therapy—24 weeks (3 doses)	185	555	740
Phase 3 add-on therapy—24 weeks (2 doses)	250	500	750
Phase 3 add-on therapy—24 weeks (2 doses)	188	376	564
Aggregated level			
Total sample size of these eight trials	2363	4466	6829
Assumed annualized event rate (%)	1.2	1.2	1.2
Expected endpoints	38	62	100
Probability of upper 95% CI on HR < 1.3		24%	
Category 2: Randomized CV outcome trial			
Sample size	4000	4000	8000
Assumed annualized event rate (%)	1.5	1.5	1.5
Expected endpoints	181	181	362
Probability of upper 95% CI on HR < 1.3		70%	
Combined categories 1 and 2			
Expected endpoints	219	243	462
Probability of upper 95% CI on HR < 1.3		80%	

The development plan aims to use phase 2 and 3 trials to evaluate if the upper bound of a 95% CI of the hazard ratio (HR) of the experimental drug to the control group is below 1.3. A composite endpoint is considered: time to CV deaths, stroke, or myocardial infarction (MI), whichever occurs first. Under this setting, the power is the probability that the upper 95% CI of the HR < 1.3, when the true HR = 1.

Following Spiegelhalter et al. (2004), a naïve approach of power calculations is to simply pool the sample size across trials and to use the following approximation:

$$\log(\widehat{HR}) \sim N(\log(HR), 4/N), \tag{2.1}$$

where

N is the total number of endpoints observed from the pooled trials
$\widehat{HR}$ is an estimate of the HR based on the pooled data

Assuming the development plan will contain trials in two categories, the initial power calculations are summarized in Table 2.1.

Assuming the annualized event rate is 1.2%, the pooled data from the Category 1 trials are expected to have a total of 100 events with a power of 24%. Such low power implies that a large-scale CV outcome study is needed. Assuming an annualized event rate of 1.5%, the Category 2 trial is expected to have a total of 362 events with power of 70%. By combining the trials from both categories, it is expected to have a total of 462 events with 80% power to detect the upper 95% CI of the HR < 1.3. These power calculations are based on pooling sample sizes across studies using (2.1) and therefore are susceptible to issues related to the Simpson/Yule paradox. In addition, they do not account for the between-study variability in the meta-analysis framework and do not allow for different baseline event rates.

2.4 Meta-Regression Models for Survival Data

We consider K randomized trials where each trial has two treatment arms ("control" or "inv drug"). Here "inv drug" denotes investigational drug. Let $trt_{jk} = 0$ if $j = 1$ (control/placebo) and 1 if $j = 2$ ("inv drug"), and also let x_{jk} denote a trial-level binary covariate, where $x_{jk} = 1$ if the kth trial recruits subjects with low or moderate CV risk for the jth treatment and $x_{jk} = 0$ if the kth trial recruits subjects with high CV risk for the jth treatment for $j = 1, 2$. Let y_{jk} and v_{jk} denote the total subject year duration and the total number of events from n_{jk} subjects for $j = 1, 2$ and $k = 1, ..., K$. Let $D_k = \{(y_{jk}, v_{jk}, n_{jk}, trt_{jk}, x_{jk}), j = 1, 2, k = 1, ..., K\}$ denote the observed aggregate meta-survival data.

Assuming that the individual level failure time follows an exponential distribution with mean $1/\lambda_{jk}$, the likelihood function of (y_{jk}, v_{jk}) is given by

$$L(\lambda_{jk} \mid y_{jk}, v_{jk}) = \lambda_{jk}^{v_{jki}} \exp(-\lambda_{jk} y_{jk}). \tag{2.2}$$

We consider both random effects and fixed effects models for λ_{jk} in (2.2).

The log-linear random effects model for λ_{jk} assumes

$$\log \lambda_{jk} = \gamma_0 + \gamma_1 trt_{jk} + \theta x_{jk} + \xi_k, \ \xi_k \sim N(0, \tau^2), \tag{2.3}$$

where $trt_{jk} = 1$ if $j = 2$ ("inv drug") and 0 if $j = 1$ (control/placebo) for $k = 1, ..., K$. In (2.3), τ^2 captures the between-trial variability, and ξ_k also captures the trial dependence between y_{1k} and y_{2k}. Note that under the exponential model, the design parameter $\exp(\gamma_1)$ is precisely the HR of the treatment, and θ quantifies the CV risk effect (low or moderate CV risk versus high CV risk). Let $\xi = (\xi_1, \ldots, \xi_K)'$ and $\gamma = (\gamma_0, \gamma_1)'$. Then, under the random effects model, the complete data likelihood function based on the meta-survival data D_K is given by

$$L_R\left(\gamma, \theta, \xi, \tau^2 \mid D_K\right)$$

$$= \prod_{k=1}^{K} \prod_{j=1}^{2} \left(\exp\left\{ v_{jk}\left(\gamma_0 + \gamma_1 trt_{jk} + \theta x_{jk} + \xi_k\right)\right\} \right.$$

$$\left. \times \exp\left[-\exp\left\{\gamma_0 + \gamma_1 trt_{jk} + \theta x_{jk} + \xi_k\right\} y_{jk}\right] \frac{1}{\sqrt{2\pi}\tau} \exp\left\{-\frac{\xi_k^2}{2\tau^2}\right\} \right). \qquad (2.4)$$

We see from (2.4) that under the exponential model, the likelihood function is solely based on the treatment-level meta-survival data D_k. We also see from (2.4) that under the exponential model, the likelihood function does not depend on the n_{jk}'s.

The log-linear fixed effects model for λ_{jk} assumes

$$\log \lambda_{jk} = \gamma_0 + \gamma_1 trt_{jk} + \theta_k, \qquad (2.5)$$

for $k = 1, 2, \ldots, K$. To ensure identifiability in (2.5), we assume that $\sum_{k=1}^{K} \theta_k = 0$. For simplicity, we take $\theta_K = -\sum_{k=1}^{K-1} \theta_k$. Unlike the random effects model, the θ_ks in (2.5) capture the differences among the trials. Since the θ_k's are unknown, a necessary condition for existence of the maximum likelihood estimates of the θ_k's is that there is at least one event, that is, $v_{1k} + v_{2k} > 1$, for each trial. Let $\theta = (\theta_1, \ldots, \theta_K)'$. Then, similar to (2.4), under the fixed effects model, the likelihood function based on the meta-survival data D_K is given by

$$L_F\left(\gamma, \theta \mid D_K\right) = \prod_{k=1}^{K} \prod_{j=1}^{2} \left\{ \exp\left\{ v_{jk}\left(\gamma_0 + \gamma_1 trt_{jk} + \theta_k\right)\right\} \right.$$

$$\left. \times \exp\left[-\exp\left\{\gamma_0 + \gamma_1 trt_{jk} + \theta_k\right\} y_{jk}\right] \right\}. \qquad (2.6)$$

2.5 General Methodology for Bayesian Meta-Design

In this section, we present a general methodology for Bayesian meta-design only for the log-linear random effects or fixed effects regression models. We assume that the hypotheses for "noninferiority" testing can be formulated as follows:

$$H_0 : \exp(\gamma_1) \geq \delta \quad versus \quad H_1 : \exp(\gamma_1) < \delta. \tag{2.7}$$

The meta-trials are successful if H_1 is accepted. As discussed in Section 2.3, $\delta = 1.3$ as specified in the FDA guidelines.

Let $\mathbf{\Psi}_R = (\gamma, \theta, \tau^2)$ denote the collection of parameters for the random effects model and $\mathbf{\Psi}_F = (\gamma, \theta)$ denote the collection of parameters for the fixed effects model. Following Wang and Gelfand (2002), let $\pi_R^{(s)}(\mathbf{\Psi}_R)$ and $\pi_F^{(s)}(\mathbf{\Psi}_F)$ denote the sampling priors under the random and fixed effects models, respectively. Each of these sampling priors captures a certain specified portion of the parameter space in achieving a certain level of performance in the Bayesian meta-design. Also, let $\pi_R^{(f)}(\mathbf{\Psi}_R)$ and $\pi_F^{(f)}(\mathbf{\Psi}_F)$ denote the respective fitting priors under these two types of meta-regression models. As discussed in Wang and Gelfand (2002) and Chen et al. (2011), the sampling priors, $\pi_R^{(s)}(\mathbf{\Psi}_R)$ and $\pi_F^{(s)}(\mathbf{\Psi}_F)$, are used to generate the meta-survival data D_K, and the fitting priors, $\pi_R^{(f)}(\mathbf{\Psi}_R)$ and $\pi_F^{(f)}(\mathbf{\Psi}_F)$, are used to fit the random and fixed effects regression models once the meta-survival data are generated.

Given the data D_k, the fitting posterior distribution of $\mathbf{\Psi}_R$ under the random effects model takes the form

$$\pi_R^{(f)}\left(\mathbf{\Psi}_R \mid D_K\right) \propto \left\{ \prod_{k=1}^{K} \int \left[\prod_{j=1}^{2} \exp\left\{ v_{jk}\left(\gamma_0 + \gamma_1 \mathrm{trt}_{jk} + \theta x_{jk} + \xi_k\right)\right\} \right.\right.$$

$$\times \exp\left\{ -\exp\left(\gamma_0 + \gamma_1 \mathrm{trt}_{jk} + \theta x_{jk} + \xi_k\right) y_{jk}\right\}$$

$$\left.\left.\times \left[\frac{1}{\sigma} \exp\left\{ -\frac{1}{2\sigma^2}\xi_k^2\right\}\right] d\xi_k \right\} \pi_R^{(f)}\left(\mathbf{\Psi}_R\right) \right. \tag{2.8}$$

and the fitting posterior distribution of $\mathbf{\Psi}_F$ under the fixed effects model is given by

$$\pi_F^{(f)}\left(\mathbf{\Psi}_F \mid D_K\right) \propto \prod_{k=1}^{K} \prod_{j=1}^{2} \left\{ \exp\left\{ v_{jk}\left(\gamma_0 + \gamma_1 \mathrm{trt}_{jk} + \theta_k\right)\right\}\right.$$

$$\left.\times \exp\left[-\exp\left\{\gamma_0 + \gamma_1 \mathrm{trt}_{jk} + \theta_k\right\} y_{jk}\right]\right\} \pi_F^{(f)}\left(\mathbf{\Psi}_F\right). \tag{2.9}$$

We note that the fitting priors, $\pi_R^{(f)}(\Psi_R)$ and $\pi_F^{(f)}(\Psi_F)$, may be improper as long as the resulting fitting posteriors, $\pi_R^{(f)}(\Psi_R \mid D_K)$ and $\pi_F^{(f)}(\Psi_F \mid D_K)$, are proper.

To make the design using the meta-regression models more feasible, we assume $n_{jk} = \phi_{jk} n_k$ for $j = 1, \ldots, q$ and $n_k = \kappa_k n$ for $k = 1, \ldots, K$, where both ϕ_{jk} and κ_k are prespecified nonnegative constants such that $\sum_{j=1}^{2} \phi_{jk} = 1$ and $\sum_{k=1}^{K} \kappa_k = 1$. Under this setting, the total sample size based on the entire meta-analysis model is n. This sample size allocation is quite general and flexible, which allows $\phi_{jk} = 0$ for certain treatment arms and also allows an unbalanced design for certain trials. In practice, the n_{jk}'s are often determined by the data analyst based on certain constraints in the trials. An example of such a meta-design is discussed in Section 2.3. For the meta-trials shown in Table 2.1, $q = 2$ and $K = 9$, and the sample size allocation is specified as follows: (1) $n_1 = 150$, $n_2 = 350$, $n_3 = 3{,}500$, $n_4 = 400$, $n_5 = 375$, $n_6 = 740$, $n_7 = 750$, $n_8 = 564$, and $n_9 = 8{,}000$ so that the total sample size $n = 14{,}829$ and (2) $\phi_{11} = 1/6$, $\phi_{12} = 2/5$, $\phi_{13} = 2/5$, $\phi_{14} = 1/4$, $\phi_{15} = 1/5$, $\phi_{16} = 1/4$, $\phi_{17} = 1/3$, $\phi_{18} = 1/3$, and $\phi_{19} = 1/2$.

To complete the Bayesian meta-design, we need to specify two sampling priors for each of $\pi_R^{(s)}(\Psi_R)$ and $\pi_F^{(s)}(\Psi_F)$, denoted by $\pi_{R0}^{(s)}(\Psi_R)$ and $\pi_{R1}^{(s)}(\Psi_R)$, or $\pi_{F0}^{(s)}(\Psi_F)$ and $\pi_{F1}^{(s)}(\Psi_F)$, where $\pi_{R0}^{(s)}(\Psi_R)$, $\pi_{R1}^{(s)}(\Psi_R)$, $\pi_{F0}^{(s)}(\Psi_F)$, and $\pi_{F1}^{(s)}(\Psi_F)$ are proper priors, which are defined on the subsets of the parameter spaces induced by hypotheses H_0 and H_1. Following Chen et al. (2011), we define the key quantity

$$\beta_{s\ell}^{(n)} = E_{s\ell}\left[1\left\{ P\left(\exp(\gamma_1) < \delta \mid D_K, \pi^{(f)} \right) \geq \eta_0 \right\} \right], \tag{2.10}$$

where the indicator function $1\{A\}$ is 1 if A is true and 0 otherwise; $0 < \eta_0 < 1$ is a prespecified Bayesian credible level; the posterior probability $P(\exp(\gamma_1) < \delta \mid D_k, \pi^{(f)})$ is computed with respect to the posterior distribution of γ_1 given the data D_k under the fitting prior $\pi^{(f)}(\cdot)$, which is either $\pi_R^{(f)}(\Psi_R)$ or $\pi_F^{(f)}(\Psi_F)$; and the expectation $E_{s\ell}$ is taken with respect to the predictive marginal distribution of D_k under the sampling prior $\pi_{R\ell}^{(s)}(\Psi_R)$ or $\pi_{F\ell}^{(s)}(\Psi_F)$ for $\ell = 0, 1$.

For given $\alpha_0 > 0$ and $\alpha_1 > 0$, we compute

$$n_{\alpha_0} = \min\left\{ n : \beta_{s0}^{(n)} \leq \alpha_0 \right\} \quad \text{and} \quad n_{\alpha_1} = \min\left\{ n : \beta_{s1}^{(n)} \geq 1 - \alpha_1 \right\}, \tag{2.11}$$

where $\beta_{s0}^{(n)}$ and $\beta_{s1}^{(n)}$ are given in (2.10). Then, the Bayesian meta-design sample size is given by $n_B = \max\{n_{\alpha_0}, n_{\alpha_1}\}$.

We note that the quantities $\beta_{s0}^{(n)}$ and $\beta_{s1}^{(n)}$ in (2.11) correspond to the Bayesian type I error and power, respectively. Common choices of α_0 and α_1 are $\alpha_0 = 0.05$ and $\alpha_1 = 0.20$. We choose η_0 to be sufficiently large, say $\eta_0 > 0.95$, so that the Bayesian meta-design sample size n_B ensures that the type I error rate is at most $\alpha_0 = 0.05$ and the power is at least $1 - \alpha_1 = 0.80$.

2.6 Specification of Prior Distributions

2.6.1 Historical Data

Around the time that the FDA CV risk guidance document was published, the final results of the two large-scale randomized controlled CV outcome trials in T2DM patients were published. The event rates of the control groups of these two trials (see Table 2.2) are used as the historical information in the case study discussed in Section 2.3. These two trials are described as follows. The Action to Control Cardiovascular Risk in Diabetes (ACCORD) trial was designed to determine whether a therapeutic strategy targeting normal glycated hemoglobin levels (i.e., below 6.0%) would reduce the rate of CV events, as compared with a strategy targeting glycated hemoglobin levels from 7.0% to 7.9% in middle-aged and older people with T2DM and either established CV disease or additional CV risk factors. The key result based on 10,251 patients with a mean of 3.5 years of follow-up was published in 2008 (The Action to Control Cardiovascular Risk in Diabetes Study Group, 2008). The Action in Diabetes and Vascular Disease: Preterax and Diamicron Modified Release Controlled Evaluation (ADVANCE) trial was designed to assess the effects on major vascular outcomes of lowering the glycated hemoglobin value to a target of 6.5% or less in a broad cross section of patients with T2DM. The key result based on 11,140 patients with a median of 5 years of follow-up was published in 2008 (The ADVANCE Collaborative Group, 2008).

Moreover, other historical data can be obtained from the briefing documents of the FDA advisory meetings for saxagliptin and liraglutide in April 2009. The key documents (Saxagliptin, 2009; Liraglutide, 2009) of this FDA advisory meeting are listed in the following:

1. http://www.fda.gov/ohrms/dockets/ac/09/briefing/2009-4422b1-01-FDA.pdf
2. http://www.fda.gov/downloads/AdvisoryCommittees/CommitteesMeetingMaterials/Drugs/Endocrinologicand MetabolicDrugsAdvisoryCommittee/UCM148109.pdf

TABLE 2.2

Historical Meta-Data for CV Events (Control Arm)

Reference/Control Type	Number of Subjects (N)	Number of CV Events	Total Patient Years	Annualized Event Rate (%)
ACCORD (2008)/standard therapy	5123	371	16,000	2.29
ADVANCE (2008)/standard therapy	5569	590	27,845	2.10
Saxagliptin (2009)/total control	1251	17	1,289	1.31
Liraglutide (2009)/placebo	907	4	449	0.89
Liraglutide (2009)/active control	1474	13	1,038	1.24

3. http://www.fda.gov/ohrms/dockets/ac/09/briefing/2009-4422b2-01-FDA.pdf
4. http://www.fda.gov/downloads/AdvisoryCommittees/CommitteesMeetingMaterials/Drugs/Endocrinologicand MetabolicDrugsAdvisoryCommittee/UCM148659.pdf

In subsequent subsections, we discuss how to elicit the fitting priors using these five historical datasets shown in Table 2.2.

2.6.2 Meta-Regression Models for Historical Data

Suppose that the historical data are available only for the control arm from K_0 previous datasets. For the historical data shown in Table 2.2, $K_0 = 5$. Let y_{0k} denote the total subject year duration and also let $v_{0k} = \sum_{i=1}^{n_{0k}} v_{0ik}$ denote the total number of events for $k = 1, \ldots, K_0$. In addition, we let x_{0k} denote a binary covariate, where $x_{0k} = 1$ if the subjects had a low or moderate CV risk and $x_{0k} = 0$ if the subjects had a high CV risk in the kth historical dataset. Suppose that only the trial-level data $D_{0K_0} = \{(y_{0k}, v_{0k}, x_{0k}), \quad k = 1, \ldots, K_0\}$ are available from the K_0 historical trials. Assume the individual level failure time follows an exponential distribution, $\text{Exp}(\lambda_{0k})$, where $\lambda_{0k} > 0$ is the hazard rate. The random effects model assumes

$$\log \lambda_{0k} = \gamma_0 + \theta_0 x_{0k} + \xi_{0k}, \xi_{0k} \sim N(0, \tau^2), \tag{2.12}$$

for $k = 1, 2, \ldots, K_0$. Let $\xi_0 = (\xi_{01}, \xi_{02}, \ldots, \xi_{0K_0})'$. Then, the complete data likelihood function based on D_{0K_0} is given by

$$
\begin{aligned}
&L_R\left(\gamma_0, \theta_0, \tau^2, \xi_0 \mid D_{0K_0}\right) \\
&= \prod_{k=1}^{k_0}\left[\exp\left\{v_{0k}\left(\gamma_0 + \theta_0 x_{0k} + \xi_{0k}\right)\right\} \exp\left\{-\exp\left(\gamma_0 + \theta_0 x_{0k} + \xi_{0k}\right) y_{0k}\right\}\right. \\
&\quad \left. \times \left(\frac{1}{\sqrt{2\pi\tau^2}}\right) \exp\left\{-\frac{1}{2\tau^2}\xi_{0k}^2\right\}\right].
\end{aligned}
\tag{2.13}
$$

Similarly, the fixed effects model assumes

$$\log \lambda_{0k} = \gamma_0 + \theta_{0k}, \tag{2.14}$$

for $k = 1, 2, ..., K_0$, where $\sum_{k=1}^{K_0} \theta_{0k} = 0$. Again, for simplicity, we assume $\theta_{0K_0} = -\sum_{k=1}^{K_0-1} \theta_{0k}$. Let $\theta_0 = (\theta_{01}, ..., \theta_{0K_0})'$. Then, the likelihood function based on D_{0K_0} is given by

$$L_F\left(\gamma_0, \theta_0 \mid D_{0K_0}\right) = \prod_{k=1}^{K_0} \exp\{v_{0k}(\gamma_0 + \theta_{0k})\} \exp\{-\exp(\gamma_0 + \theta_{0k}) y_{0k}\}. \quad (2.15)$$

Comparing (2.12) to (2.3), the models for the historical data and the current data share the common parameters γ_0 and τ^2 under the random effects model, while the CV risk effects parameters θ and θ_0 are different in these two models. Similarly, comparing (2.14) to (2.5), the models for the historical data and the current data share the common parameter γ_0 under the fixed effects model, while the fixed effects parameters θ and θ_0 are different in these two models. Thus, the strength of the historical data is borrowed through the common parameters γ_0 and τ^2 with different parameters θ and θ_0 under the random effects model, and only a single common parameter γ_0 with different fixed effects parameters θ and θ_0 under the fixed effects model provides us with greater flexibility in accommodating different CV risk effects in the current and historical data.

2.6.3 Analysis of the Historical Data

We carry out a Bayesian analysis of the historical data in Table 2.2. Under the random effects model, the posterior distribution of $(\gamma_0, \theta_0, \tau^2, \xi_0)$ based on the historical data D_{0K_0} is given by

$$\pi_R\left(\gamma_0, \theta_0, \tau^2, \xi_0 \mid D_{0K_0}\right) \propto L_R\left(\gamma_0, \theta_0, \tau^2, \xi_0 \mid D_{0K_0}\right) \pi_{R0}\left(\gamma_0, \theta_0, \tau^2\right), \quad (2.16)$$

where $L_R(\gamma_0, \theta_0, \tau^2, \xi_0 \mid D_{0K_0})$ is given in (2.13) and $\pi_{R0}(\gamma_0, \theta_0, \tau^2)$ is an initial prior. Under the fixed effects model, the posterior distribution of $(\gamma_0, \theta_0, \tau^2, \xi_0)$ based on the historical data D_{0K_0} is given by

$$\pi_F\left(\gamma_0, \theta_0 \mid D_{0K_0}\right) \propto L_F\left(\gamma_0, \theta_0 \mid D_{0K_0}\right) \pi_{F0}\left(\gamma_0, \theta_0\right), \quad (2.17)$$

where $L_F(\gamma_0, \theta_0 \mid D_{0K_0})$ is defined in (2.15) and $\pi_{F0}(\gamma_0, \theta_0)$ is an initial prior.

In (2.16) and (2.17), we assume independent priors for all parameters, and an initial $N(0, 10)$ prior is specified for each of γ_0, θ_0, and θ_{0k} for $k = 1, ..., K_0 - 1$. In addition, an initial prior for τ^2 is taken to be $\pi_{R0}(\tau^2) \propto [1/(\tau^2)^{2.001 + 1}] \exp\{-0.1/\tau^2\}$. We generated 20,000 iterations from each of the posterior distributions in (2.16) and (2.17) using the Markov chain Monte Carlo (MCMC) sampling algorithm discussed in the next section. For the historical data

TABLE 2.3

Prior Estimates of Parameters Based on the Meta-Historical Data in Table 1.2

Model	Parameter	Posterior Mean	SD	95% HPD Interval
Random	γ_0	−3.799	0.153	(−4.106, −3.500)
effects	θ_0	−0.641	0.269	(−1.182, −0.112)
	τ^2	0.054	0.048	(0.009, 0.131)
Fixed	γ_0	−4.241	0.128	(−4.492, −3.994)
effects	θ_{01}	−0.133	0.231	(−0.596, 0.311)
	θ_{02}	−0.544	0.420	(−1.398, 0.236)
	θ_{03}	−0.186	0.250	(−0.684, 0.305)
	θ_{04}	0.477	0.134	(0.222, 0.745)

in Table 2.2, the posterior estimates, including posterior means; standard deviations (SDs) and 95% highest posterior density (HPD) intervals of γ_0, θ_0, which were computed using the Monte Carlo method of Chen and Shao (1999); and τ^2 under the random effects model as well as γ_0 and $\boldsymbol{\theta}_0 = (\theta_{01}, \ldots, \theta_{04})'$ for the fixed effects model, are given in Table 2.3. Those posterior estimates will be used as the guide values of the model parameters in specifying the sampling priors.

2.6.4 Fitting Priors

Under the random effects model, using (2.13), and extending the power prior of Ibrahim and Chen (2000), the fitting prior for $\boldsymbol{\Psi}_R$ under the random effects model is given by

$$\pi_R^{(f)}\left(\boldsymbol{\Psi}_R \mid D_{0K_0}, a_0\right)$$

$$\propto \int \left\{ \prod_{k=1}^{K_0} \left[\exp\left\{v_{0k}\left(\gamma_0 + \theta_0 x_{0k} + \xi_{0k}\right)\right\} \exp\left\{-\exp\left(\gamma_0 + \theta_0 x_{0k} + \xi_{0k}\right) y_{0k}\right\} \right]^{a_0} \right.$$

$$\times \left. \left[\prod_{k=1}^{K_0} \left(\frac{1}{\tau}\right) \exp\left(-\frac{1}{2\tau^2}\xi_{0k}^2\right) \right] d\xi_0 \right\} \pi_{R0}^{(f)}\left(\theta_0\right) d\theta_0 \pi_{R0}^{(f)}\left(\boldsymbol{\Psi}_R\right), \tag{2.18}$$

where $0 \le a_0 \le 1$ and $\pi_{R0}^{(f)}(\theta_0)$ and $\pi_{R0}^{(f)}(\boldsymbol{\Psi}_R)$ are initial priors. In (2.18), we further specify independent initial priors for $(\gamma_0, \gamma_1, \theta, \theta_0, \tau^2)$ as follows: (a) a normal prior $N(0, \tau_{0f}^2)$ is assumed for each of $\gamma_0, \gamma_1, \theta,$ and θ_0, where $\tau_{0f}^2 > 0$ is a prespecified hyperparameter, and (b) we specify an inverse gamma prior for τ^2, which is given by $\pi_0^{(f)}(\tau^2) \propto \left[1/(\tau^2)^{d_{0f1}+1}\right] \exp\{-d_{0f2}/\tau^2\}$, where $d_{0f1} > 0$ and $d_{0f2} > 0$ are prespecified hyperparameters.

Under the fixed effects model, the fitting prior for Ψ_F is given by

$$\pi_F^{(f)}\left(\Psi_F \mid D_{0K_0}, a_0\right) \propto \int\left\{\prod_{k=1}^{K_0}\left[\exp\left\{v_{0k}\left(\gamma_0 + \theta_{0k}\right)\right\}\right.\right.$$

$$\left.\left.\times \exp\left\{-\exp\left(\gamma_0 + \theta_{0k}\right)y_{0k}\right\}\right]^{a_0} \pi_{F0}^{(f)}\left(\theta_0\right)\right\} d\theta_0 \times \pi_{F0}^{(f)}\left(\Psi_F\right), \quad (2.19)$$

where $0 \le a_0 \le 1$ and $\pi_{F0}^{(f)}(\theta_0)$ and $\pi_{F0}^{(f)}(\Psi_F)$ are initial priors. In (2.19), independent initial normal priors, $N(0, \tau_{0f}^2)$, are assumed for $\gamma_0, \gamma_1, \theta_k,$ and θ_{0k}.

We note that we may choose different values for the hyperparameter τ_{0f}^2 for those location parameters including $\gamma_0, \gamma_1, \theta,$ and θ_0 under the random effects model and $\gamma_0, \gamma_1, \theta_k,$ and θ_{0k} under the fixed effects model. We also note that there is no information about γ_1 and θ (or $\boldsymbol{\theta}$) in the historical data.

In (2.18) and (2.19), the parameter a_0 controls the influence of the historical meta-survival data D_{0K_0} on the fitting prior for γ_0 and σ^2 under the random effects model or for γ_0 under the fixed effects model in $\pi_R^{(f)}(\Psi_R \mid a_0, D_{0K_0})$ and $\pi_F^{(f)}(\Psi_F \mid a_0, D_{0K_0})$. The parameter a_0 can be interpreted as a relative precision parameter for the historical meta-survival data. It is reasonable to restrict the range of a_0 to be between 0 and 1, and thus we take $0 \le a_0 \le 1$. One of the main roles of a_0 is that it controls the heaviness of the tails of the prior for (γ_0, τ^2) under the random effects model or for γ_0 under the fixed effects model. As a_0 becomes smaller, the tails of (2.18) or (2.19) become heavier. When $a_0 = 1$, (2.18) or (2.19) corresponds to the update of $\pi_{R0}^{(f)}(\gamma_0, \tau^2)$ or $\pi_F^{(f)}(\gamma_0)$ using Bayes' theorem. That is, with $a_0 = 1$, (2.18) or (2.19) corresponds to the posterior distribution of (γ_0, τ^2) or (γ_0) based on the historical meta-survival data. If $a_0 = 0$, then the prior does not depend on the historical meta-survival data. That is, $a_0 = 0$ is equivalent to a prior specification with no incorporation of historical meta-survival data. Thus, the parameter a_0 controls the influence of the historical meta-survival data on the current study. Such control is important in cases where there is heterogeneity between the historical and current meta-survival data or when the sample sizes of the historical and current meta-survival data are quite different.

In (2.18) and (2.19), we consider a fixed a_0. When a_0 is fixed, we know exactly how much historical meta-data are incorporated in the new metatrial and also how the type I error and power are related to a_0. As shown in our simulation study in Section 2.7, a fixed a_0 provides us additional flexibility in controlling the type I error. In addition, our informative prior specification only allows us to borrow historical meta-data for the control arm. Thus, the historical meta-data have the most influence on γ_0 but not on γ_1.

However, the historical meta-data does have a certain degree of influence on the new treatment through τ^2 under the random effects model, but not under the fixed effects model. Under both types of meta log-linear regression models, when analyzing the current data, the historical data can be borrowed only through the common parameters, namely, (γ_0, τ^2) or γ_0. For this reason, Ibrahim et al. (2012) called the power priors given in (2.18) and (2.19) *partial borrowing power priors*.

2.6.5 Sampling Priors

We now discuss how to specify the sampling priors. Under the random effects model, for the sampling prior, $\pi_{R\ell}^{(s)}(\Psi_R)$, $\ell = 0, 1$, we take $\pi_{R\ell}^{(s)}(\Psi_R) \propto \pi_R^{(s)}(\gamma_0)\,\pi_R^{(s)}(\theta)\,\pi_R^{(s)}(\tau^2)\,\pi_\ell^{(s)}(\gamma_1)$. In the sampling prior, we first specify a point mass prior for $\pi_\ell^{(s)}(\gamma_1)$ as

$$\pi_\ell^{(s)}(\gamma_1) = \begin{cases} \Delta_{\{\gamma_1 = \log(\delta)\}} & \text{for } \ell = 0 \\ \Delta_{\{\gamma_1 = 0\}} & \text{for } \ell = 1, \end{cases} \tag{2.20}$$

where $\Delta_{\{\gamma_1 = \gamma_{10}\}}$ denotes a degenerate distribution at $\gamma_1 = \gamma_{10}$, that is, $P(\gamma_1 = \gamma_{10}) = 1$. We note that under the random effects model, $1 - \exp\{-\exp(\gamma_0)\}$ corresponds to the annualized event rate for a subject with high CV risk and $1 - \exp\{-\exp(\gamma_0 + \theta)\}$ corresponds to the annualized event rate for a subject with low CV risk in the control group. For the meta-design given in Table 2.1, letting $1 - \exp\{-\exp(\gamma_0)\} = 0.015$ and $1 - \exp\{-\exp(\gamma_0 + \theta)\} = 0.012$, we obtain $\gamma_0 = \log[-\log(1 - 0.015)] = -4.1922$ and $\theta = \log[-\log(1 - 0.012)] - \log[-\log(1 - 0.015)] = -0.22466$. We then specify point mass sampling priors at these design values or γ_0 and θ, namely, $\pi_R^{(s)}(\gamma_0) = \Delta_{\{\gamma_0 = -4.1922\}}$ and $\pi_R^{(s)}(\theta) = \Delta_{\{\theta = -0.22466\}}$. In addition, we specify a point mass sampling prior for τ^2 as $\pi_R^{(s)}(\tau^2) = \Delta_{\{\tau^2 = \tilde{\tau}^2\}}$, where $\tilde{\tau}^2$ is an estimate of τ^2 from the historical data. Using Table 2.3, we take $\tilde{\tau}^2 = 0.054$.

Under the fixed effects model, for the sampling prior, $\pi_{F\ell}^{(s)}(\Psi_F)$, we assume $\pi_{F\ell}^{(s)}(\Psi_F) = \pi_F^{(s)}(\gamma_0)\,\pi_F^{(s)}(\theta)\,\pi_\ell^{(s)}(\gamma_1)$. We take the sampling prior (2.20) as the one for the random effects model for $\pi_\ell^{(s)}(\gamma)$. Again, we specify a point mass sampling prior for $\pi_F^{(s)}(\gamma_0)$ and $\pi_F^{(s)}(\theta)$. For the meta-design given in Table 2.1, we take $\pi^{(s)}(\gamma_0) = \Delta_{\{\gamma_0 = -4.392\}}$, $\pi^{(s)}(\theta_k) = \Delta_{\{\theta_k = -0.025\}}$ for $k = 1, \ldots, K - 1$, and $\pi^{(s)}(\theta_K) = \Delta_{\{\theta_K = 0.1997\}}$, where $K = 9$. Note that these design values are obtained by solving (γ_0, θ) from the following simultaneous equations: $1 - \exp\{-\exp(\gamma_0 + \theta_k)\} = 0.012$ for $k = 1, \ldots, K - 1$, $1 - \exp\{-\exp(\gamma_0 + \theta_K)\} = 0.015$, and $\sum_{k=1}^{K} \theta_k = 0$.

From Table 2.3, we see that the posterior means of γ_0 based on the historical meta-data were -3.799 under the random effects model and -4.241 under the fixed effects model. Thus, the posterior mean of γ_0 based on the historical meta-data is more similar to the design value of γ_0 (−4.241 versus −4.392) under the fixed effects model than the one (−3.799 versus −4.1922) under the random effects model.

2.7 Computational Algorithms

2.7.1 Predictive Data Generation

We use the sampling prior, $\pi_{R\ell}^{(s)}(\Psi_R)$ or $\pi_{F\ell}^{(s)}(\Psi_F)$, $\ell = 0, 1$, to generate the predictive data D_k. In other words, we view the distribution of D_k as the prior predictive marginal distribution of the data. For ease of exposition, we present the computational algorithm only for the log-linear random effects regression model since the algorithm for the fixed effects model is very similar. The prior predictive data generation algorithm under the random effects model is given as follows. For $i = 1, \ldots, n_{jk}$, where $j = 1, 2$ and $k = 1, \ldots, K$, (1) set n_{jk} and x_{jk}; (2) generate $\Psi_R \sim \pi_R^{(s)}(\Psi_R)$; (3) generate $\xi_k \sim N(0, \sigma^2)$ independently; (4) compute $\lambda_{jk} = \exp\{\gamma_0 + \gamma_1 \text{ trt} j_k + \theta x_{jk} + \xi_k\}$; (5) generate $y_{ijk}^* \sim \exp(\lambda_{jk})$ independently; (6) specify the censoring time C_{ijk} or generate $C_{ijk} \sim g_{jk}(c_{ijk})$ independently, where $g_{jk}(c_{ijk})$ is a prespecified distribution for the censored random variable; and (7) compute $y_{jk} = \sum_{i=1}^{n_{jk}} \min\{y_{ijk}^*, C_{ijk}\}$ and $v_{jk} = \sum_{i=1}^{n_{jk}} 1\{y_{ijk}^* \leq C_{ijk}\}$.

2.7.2 Sampling from the Fitting Posterior Distributions

Under the random effects model, using (2.4) and (2.18), the fitting posterior distribution of $(\Psi_R, \xi, \theta_0, \xi_0) = (\gamma, \theta, \tau^2, \xi, \theta_0, \xi_0)$ is given by

$$\pi_R^{(f)}\left(\Psi_R, \xi, \theta_0, \xi_0 \mid D_K, D_{0K_0}, a_0\right) \propto L_R\left(\gamma, \theta, \xi, \tau^2 \mid D_K\right)$$

$$\times \prod_{k=1}^{K_0}\left[\left(\exp\{v_{0k}(\gamma_0 + \theta_0 x_{0k} + \xi_{0k})\}\exp\{-\exp(\gamma_0 + \theta_0 x_{0k} + \xi_{0k})y_{0k}\}\right)^{a_0}\right.$$

$$\left.\times\left(\frac{1}{\tau}\right)\exp\left\{-\frac{1}{2\tau^2}\xi_{0k}^2\right\}\right] \times \pi_{R0}^{(f)}(\theta_0)\pi_{R0}^{(f)}\left(\gamma_0, \tau^2, \gamma_1, \theta\right). \quad (2.21)$$

We use the Gibbs sampling algorithm to sample from the posterior distribution in (2.21) given D_k and D_{0K_0}. The full conditional distributions required in the Gibbs sampling algorithm are (1) $[\gamma \mid \theta, \xi, \theta_0, \xi_0, D_K, D_{0K_0}, a_0]$, (2) $[\theta \mid \gamma, \xi, D_K]$, (3) $[\xi \mid \gamma, \theta, \tau^2, D_K]$, (4) $[\theta_0 \mid \gamma_0, \tau^2, D_{0K_0}, a_0]$, (5) $[\xi_0 \mid \gamma_0, \theta_0, \tau^2, D_{0K_0}, a_0]$, and (6) $[\tau^2 \mid \xi, \xi_0]$. We sample $\gamma, \theta, \xi, \theta_0, \xi_0$, and τ^2 from these full conditional distributions in turn. For (1), using the initial fitting prior in (2.18), the conditional density for $[\gamma \mid \theta, \xi, \theta_0, \xi_0, D_K, D_{0K_0}, a_0]$ is given by

$$\pi_R^{(f)}\left(\gamma \mid \theta, \xi, \theta_0, \xi_0, D_K, D_{0K_0}, a_0\right)$$

$$\propto \prod_{k=1}^{K}\left[\prod_{j=1}^{2} \exp\left\{v_{jk}\left(\gamma_0 + \gamma_1 \mathrm{trt}_{jk}\right)\right\} \exp\left\{-\exp\left(\gamma_0 + \gamma_1 \mathrm{trt}_{jk} + \theta x_{jk} + \xi_k\right) y_{jk}\right\}\right]$$

$$\times \prod_{k=1}^{K_0}\left[\exp\left\{a_0 v_{0k}\left(\gamma_0 + \theta_0 x_{0k} + \xi_{0k}\right)\right\} \exp\left\{-a_0 \exp\left(\gamma_0 + \theta_0 x_{0k} + \xi_{0k}\right) y_{0k}\right\}\right]$$

$$\times \exp\left(-\frac{1}{2\tau_{0f}^2} \gamma' \gamma\right). \tag{2.22}$$

For (2) and (4), the full conditional densities are given, respectively, as follows:

$$\pi_R^{(f)}\left(\theta \mid \gamma, \xi, D_K\right) \propto \prod_{k=1}^{K} \prod_{j=1}^{2}\left[\exp\left(v_{jk}\theta x_{jk}\right)\right.$$

$$\left. \times \exp\left\{-\exp\left(\gamma_0 + \gamma_1 \mathrm{trt}_{jk} + \theta x_{jk} + \xi_k\right) y_{jk}\right\}\right] \times \exp\left(-\frac{1}{2\tau_{0f}^2}\theta^2\right) \tag{2.23}$$

and

$$\pi_R^{(f)}\left(\theta \mid \gamma_0, \tau^2, D_{0K_0}, a_0\right) \propto \prod_{k=1}^{K_0}\left[\exp\left(a_0 v_{0k}\theta_0 x_{0k}\right)\right.$$

$$\left. \times \exp\left\{-a_0 \exp\left(\gamma_0 + \theta_0 x_{0k} + \xi_{0k}\right) y_{0k}\right\}\right] \times \exp\left(-\frac{1}{2\tau_{0f}^2}\theta_0^2\right). \tag{2.24}$$

For (3), given γ, θ, τ^2, and D_{Ki}, the ξ_k's are conditionally independent and the conditional density for ξ_k is given by

$$\pi_R^{(f)}\left(\xi_k \mid \gamma, \theta, \tau^2, D_K\right) \propto \prod_{j=1}^{2}\left[\exp\left(v_{jk}\xi_k\right)\right.$$

$$\left. \times \exp\left\{-\exp\left(\gamma_0 + \gamma_1 \mathrm{trt}_{jk} + \theta x_{jk} + \xi_k\right) y_{jk}\right\}\right] \times \exp\left(-\frac{1}{2\tau^2}\xi_k^2\right) \tag{2.25}$$

for $k = 1, \ldots, K$. Similarly, for (4), given $\gamma_0, \theta_0, \tau^2, D_{0K_0}$, and a_0, the ξ_{0k}'s are conditionally independent and the conditional density for ξ_{0k} takes the form

$$\pi_R^{(f)}\left(\xi_{0k} \mid \gamma_0, \theta_0, \tau^2, D_{0K_0}, a_0\right) \propto \exp\left(a_0 v_{0k}\xi_{0k}\right)$$

$$\times \exp\left\{-a_0 \exp\left(\gamma_0 + \theta_0 x_{0k} + \xi_{0k}\right)y_{0k}\right\}\exp\left(-\frac{1}{2\tau^2}\xi_{0k}^2\right) \tag{2.26}$$

for $k = 1, \ldots, K_0$. It is easy to show that the full conditional distributions in (2.22) through (2.26) are log-concave in each of these parameters. Thus, we can use the adaptive rejection algorithm of Gilks and Wild (1992) or the localized Metropolis algorithm discussed in Chen et al. (2000) to sample $(\gamma, \theta, \xi, \theta_0, \xi_0)$. Finally, for (6), the full conditional distribution $[\tau^2 | \xi, \xi_0]$ is an inverse gamma distribution given by

$$\tau^2 \mid \xi, \xi_0 \sim \text{IG}\left(d_{0f1} + \frac{K}{2} + \frac{K_0}{2}, d_{0f2} + \frac{1}{2}\sum_{k=1}^{K}\xi_k^2 + \frac{1}{2}\sum_{k=1}^{K_0}\xi_{0k}^2\right), \tag{2.27}$$

and, hence, sampling τ^2 is straightforward.

Under the fixed effects model, using (2.6) and (2.19), the fitting posterior distribution of $(\Psi_F, \theta_0) = (\gamma, \theta, \theta_0)$ is given by

$$\pi_F^{(f)}\left(\Psi_F, \theta_0 \mid D_K, D_{0K_0}, a_0\right)$$

$$\propto L_F\left(\gamma, \theta \mid D_K\right)\prod_{k=1}^{K_0}\left[\exp\left\{v_{0k}\left(\gamma_0 + \theta_{0k}\right)\right\}\exp\left\{-\exp\left(\gamma_0 + \theta_{0k}\right)y_{0k}\right\}\right]^{a_0}$$

$$\times \pi_{F0}^{(f)}\left(\theta_0\right)\pi_{F0}^{(f)}\left(\gamma, \theta\right). \tag{2.28}$$

We again use the Gibbs sampling algorithm to sample from the posterior distribution in (2.28) given D_k and D_{0K_0}. The full conditional distributions required in the Gibbs sampling algorithm are (a) $[\gamma | \theta, \theta_0, D_K, D_{0K_0}, a_0]$, (b) $[\theta | \gamma, D_K]$, and (c) $[\theta_0 | \gamma_0, D_{0K_0}, a_0]$. We sample γ, θ, and θ_0 from these full conditional distributions in turn. For (a), using the initial fitting prior in (2.19), the conditional density for $[\gamma | \theta, \theta_0, D_K, D_{0K_0}, a_0]$ is given by

$$\pi_F^{(f)}\left(\gamma \mid \theta, \theta_0, D_K, D_{0K_0}, a_0\right)$$

$$\propto \prod_{k=1}^{K}\prod_{j=1}^{2}\left[\exp\left\{v_{jk}\left(\gamma_0 + \gamma_1 \mathrm{trt}_{jk} + \theta_k\right)\right\}\exp\left\{-\exp\left(\gamma_0 + \gamma_1 \mathrm{trt}_{jk} + \theta_k\right)y_{jk}\right\}\right]$$

$$\times \prod_{k=1}^{K_0}\exp\left\{a_0 v_{0k}\left(\gamma_0 + \theta_{0k}\right)\right\}\exp\left\{-a_0\exp\left(-\gamma_0 + \theta_{0k}\right)y_{0k}\right\}\times\exp\left(-\frac{1}{2\tau_{0f}^2}\gamma'\gamma\right).$$

$$(2.29)$$

For (b) and (c), the full conditional densities are given, respectively, as follows:

$$\pi_F^{(f)}\left(\theta \mid \gamma, D_K\right)\propto \prod_{k=1}^{K}\prod_{j=1}^{2}\left[\exp\left\{v_{jk}\left(\gamma_0 + \gamma_1 \mathrm{trt}_{jk} + \theta_k\right)\right\}\right.$$

$$\left.\times\exp\left\{-\exp\left(\gamma_0 + \gamma_1 \mathrm{trt}_{jk} + \theta_k\right)y_{jk}\right\}\right]\times\exp\left(-\frac{1}{2\tau_{0f}^2}\theta'\theta\right) \quad (2.30)$$

and

$$\pi_F^{(f)}\left(\theta_0 \mid \gamma_0, D_{0K_0}, a_0\right)\propto \left[\prod_{k=1}^{K_0}\exp\left\{a_0 v_{0k}\left(\gamma_0 + \theta_{0k}\right)\right\}\right.$$

$$\left.\times\exp\left\{-a_0\exp\left(\gamma_0 + \theta_{0k}\right)y_{0k}\right\}\right]\times\exp\left(-\frac{1}{2\tau_{0f}^2}\theta_0'\theta_0\right). \quad (2.31)$$

We can show that the full conditional distributions in (2.29) through (2.31) are log-concave in each of these parameters. Therefore, we can use the adaptive rejection algorithm of Gilks and Wild (1992) or the localized Metropolis algorithm discussed in Chen et al. (2000) to sample $(\gamma, \theta, \theta_0)$.

2.7.3 Power Calculation

Let $\left\{\gamma_1^{(m)}, m = 1,\ldots,M\right\}$ denote a Gibbs sample of γ_1 from the fitting posterior distribution $\pi_R^{(f)}(\Psi_R, \xi, \theta_0, \xi_0 \mid D_K, D_{0K_0}, a_0)$ or $\pi_F^{(f)}(\Psi_F, \theta_0 \mid D_K, D_{0K_0}, a_0)$. Using this Gibbs sample, a Monte Carlo estimate of $P(\exp(\gamma_1) < \delta \mid D_K, \pi^{(f)})$ is given by $\hat{P}_j = \frac{1}{M}\sum_{m=1}^{M}1\left\{\exp\left(\gamma_1^{(m)}\right) < \delta\right\}$, where $\pi^{(f)}$ is either $\pi_R^{(f)}(\Psi_R \mid D_{0K_0}, a_0)$ in (2.18) or $\pi_F^{(f)}(\Psi_F \mid D_{0K_0}, a_0)$ in (2.19). To compute $\beta_{s\ell}^{(n)}$ in (2.10), we use the following computational algorithm: (1) set $\ell = 0$ or 1, n_{jk}, x_{jk}, τ_0, M (the Gibbs sample size), and N (the number of simulation runs); (2) generate $\Psi_R \sim \pi_{R\ell}^{(s)}(\Psi_R)$ or $\Psi_F \sim \pi_{F\ell}^{(s)}(\Psi_F)$; (3) generate D_K via the predictive data generation algorithm in Section 2.7.1; (4) run the Gibbs sampler to generate a Gibbs sample $\{\gamma_1^{(m)}, m = 1, \ldots, M\}$ of size M from the fitting posterior distribution $\pi_R^{(f)}$ or $\pi_F^{(f)}$; (5) compute $\hat{P}_f$; (6) repeat (2) through (5) N times; and (7) compute the proportion of $\{\hat{P}_f \geq \eta_0\}$ in these N runs, which gives an estimate of $\beta_{s\ell}^{(n)}$.

2.8 Meta-Design for Evaluating CV Risk

We consider a meta-design discussed in Section 2.3. From Table 2.1, we have $K = 9$. Using the historical data shown in Table 2.2, we have $K_0 = 5$. The non-inferiority margin was set to $\delta = 1.3$. To ensure that the type I error is controlled under 5%, we chose $\eta_0 = 0.96$. The choice of $\tau_0 = 0.96$ was discussed in Chen et al. (2011) and recommended in the FDA Guidance discussed in Section 2.2. In the fitting prior (2.18), we chose an initial $N(0, 10)$ prior for each of γ_0, γ_1, θ, and θ_0 and $d_{0f1} = 2.001$ and $d_{0f2} = 0.1$ for the initial prior for τ^2 under the random effects model. Note that this choice of (d_{0f1}, d_{0f2}) leads to a prior variance of 9.98 (about 10) for τ^2. Similarly, in the fitting prior (2.19), we chose an initial $N(0, 10)$ prior for each of γ_0, γ_1, θ_k, and θ_{0k} under the fixed effects model. Thus, we specify relatively vague initial priors for all the parameters.

In all the simulations, we generated the data under the random effects model using the sampling prior specified in Section 2.6.5 and the predictive data generation algorithm given in Section 2.7.1. In Table 2.1, $n_{19} = n_{29} = 4000$. However, we consider several values of $n_{19} = n_{29}$, which may be smaller than 4000. The design strategy is to find a minimum sample size of $n_{19} = n_{29}$ and a value of a_0 so that the power is at least 80% and the type I error is controlled at 5% when the n_{jk} for $j = 1, 2$ and $k = 1, 2, \ldots, 8$ are fixed in Table 2.1. The powers and type I errors for various values of a_0 and $n_{19} = n_{29}$ are shown in Table 2.4 and plotted in Figure 2.1. In all computations of the type I errors and powers, $N = 10,000$ simulations and $M = 10,000$ with 1,000 burn-in iterations within each simulation were used.

From Table 2.4, we see that the powers and type I errors with no incorporation of historical meta-survival data are 0.7396 and 0.0411 when $n_{19} = n_{29} = 2750$, 0.7656 and 0.0389 when $n_{19} = n_{29} = 3000$, and 0.7868 and 0.0394 when $n_{19} = n_{29} = 3250$, respectively, under the random effects model. Similarly, without incorporation of historical meta-survival data, the powers and type I errors are 0.7443 and 0.0425 when $n_{19} = n_{29} = 2750$, 0.7677 and 0.0392 when $n_{19} = n_{29} = 3000$, and 0.7914 and 0.0402 when $n_{19} = n_{29} = 3250$ under the fixed

TABLE 2.4

Powers and Type I Errors for the Meta-Design in Table 1.1

Model	a_0	$n_{19} = n_{29} = 2750$		$n_{19} = n_{29} = 3000$		$n_{19} = n_{29} = 3250$	
		Power	Type I Error	Power	Type I Error	Power	Type I Error
Random	0	0.7396	0.0411	0.7656	0.0389	0.7868	0.0394
effects	0.005	0.7592	0.0444	0.7804	0.0410	0.8037	0.0418
	0.015	0.7807	0.0489	0.8019	0.0450	0.8225	0.0469
	0.0215	0.7868	0.0519	0.8082	0.0470	0.8289	0.0497
	0.025	0.7911	0.0526	0.8111	0.0473	0.8326	0.0503
	0.03	0.7931	0.0529	0.8138	0.0480	0.8341	0.0517
	0.05	0.8017	0.0557	0.8209	0.0508	0.8385	0.0540
Fixed	0	0.7443	0.0425	0.7677	0.0390	0.7914	0.0402
effects	0.10	0.7651	0.0440	0.7846	0.0402	0.8050	0.0424
	0.15	0.7782	0.0483	0.7973	0.0437	0.8179	0.0466
	0.20	0.7906	0.0515	0.8093	0.0474	0.8296	0.0494
	0.25	0.8030	0.0554	0.8198	0.0498	0.8381	0.0517
	0.30	0.8136	0.0583	0.8307	0.0539	0.8456	0.0559

effects model. Thus, without incorporation of historical meta-survival data, the sample size $n_{19} = n_{29} = 3250$ is not large enough to yield 80% power, although the type 1 errors are controlled at 5% under both the random and fixed effects models.

With incorporation of historical meta-survival data and controlling the type I error rate at 5%, the powers are 0.8138 when $a_0 = 0.03$ and $n_{19} = n_{29} = 3000$ and 0.8289 when $a_0 = 0.0215$ and $n_{19} = n_{29} = 3250$, respectively, under the random effects model, and these powers become 0.8198 when $a_0 = 0.25$ and $n_{19} = n_{29} = 3000$ and 0.8296 when $a_0 = 0.20$ and $n_{19} = n_{29} = 3250$, respectively, under the fixed effects model. These results imply that the gains in power are about 4.8% and 4.2% with incorporation of 3.0% and 2.15% of the historical data for $n_{19} = n_{29} = 3000$ and 3250 under the random effects model and the gains in power are about 5.2%–3.8% with incorporation of 25% and 20% of the historical data for $n_{19} = n_{29} = 3000$ and 3250 under the fixed effects model. We note that these powers under the random effects model are very similar to those under the fixed effects model. From Table 2.4, we also see that when $n_{19} = n_{29} = 2750$, and controlling the type I error rate at 5%, the maximum powers that can be achieved are 0.7807 under the random effects model and 0.7782 under the fixed effects model. Thus, under both the random and fixed effects models, the sample size $n_{19} = n_{29} = 2750$ is not large enough to yield 80% power, while the sample size $n_{19} = n_{29} = 3000$ is sufficient to yield 80% power. We note that using the naïve approach of power calculation discussed in Section 2.3, $n_{19} = n_{29} = 4000$ is required to achieve 80% power. Therefore, the Bayesian meta-design yields a reduction of the total sample size by 2000 patients.

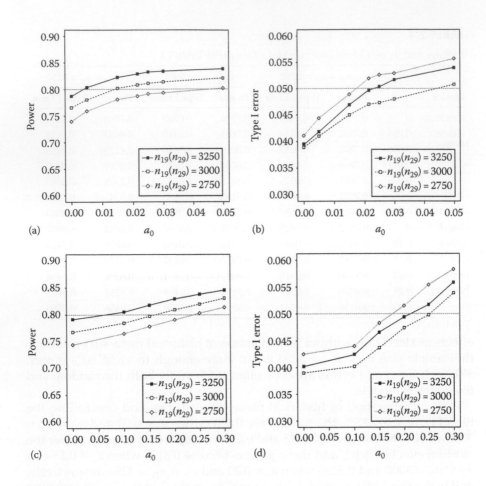

FIGURE 2.1
Plots of powers under random effects model (a) and fixed effects model (c) and their type I errors (b and d) versus a_0 for $n_{19} = n_{29} = 2750, 3000, 3250$.

As discussed earlier, the historical meta-survival data can be partially borrowed through the two common parameters γ_0 and τ^2 under the random effects model but through only one common parameter γ_0 under the fixed effects model. This implies that under the same value of a_0, that is, the same amount of incorporation of the historical meta-survival data, the power is higher under the random effects model than under the fixed effects model. On the other hand, the design value of γ_0 in the current metatrial is more comparable to the value of γ_0 in the historical meta-data under the fixed effects model than under the random effects model, which explains why more historical meta-survival data can be allowed to be borrowed under the fixed effects model than the random effects model. From Figure 2.1, it is interesting to see that (1) both the power and type I error increase in a_0; (2) the power is roughly quadratic in a_0 under the random effects model and roughly linear in a_0 under the fixed effects model; and (3) the powers increase in $n_{19} = n_{29}$ as expected. These phenomena are also observed in Ibrahim et al. (2012) under different design settings.

2.9 Discussion

In this chapter, we have presented a general Bayesian methodology for the design of meta noninferiority clinical trials for survival data by extending the Bayesian sample size determination methods of Wang and Gelfand (2002) and Chen et al. (2011). The Bayesian method discussed in this chapter not only allows for planning the sample size for a phase 2/3 development program in the meta-analysis framework by accounting for between-study heterogeneity, but it also allows for incorporation of prior information for the underlying risk in the control population. We have also discussed the partial borrowing power prior with a fixed a_0 to incorporate the historical meta-survival data. The fixed a_0 approach greatly eases the computational implementation and also provides greater flexibility in controlling the type I error rates as empirically shown in Section 2.8.

Several points require further discussion. First, in the types of trials considered in this chapter, since the CV event rates are so low, it appears that metatrial design is the only feasible way to proceed as mandated by the FDA. A common clinical development plan should have multiple phase 2 and 3 trials. In the area of developing T2DM drugs, the common practice is to evaluate efficacy from multiple studies enrolling various patients in different disease status and/or background therapies. With the need of having multiple studies, the concept of meta-design is a reasonable approach to enhance the evaluations of the integrated safety database, which is important for both the sponsor and regulatory agencies in order to assess the risks to the population. Second, throughout this chapter, we only consider the exponential regression model, which is one of the limitations of the current approach. However, the exponential model is attractive in the sense that the individual patient level (IPD) survival data are not required. Another major motivation for using the exponential model in this chapter is that the historical data were only available in aggregate form; thus, the exponential model was most natural in this setting, even if we had IPD data for the trials themselves. If we use a different survival model such as the Weibull regression model for the design, IPD data would be needed in the design since the likelihood function of the Weibull regression model does depend on subject-level failure times and censoring indicators. If IPD meta-survival data are available, then the Bayesian methodology discussed in this chapter can be easily extended to other survival regression models discussed in Ibrahim et al. (2001). Third, under both the random and fixed effects models, the treatment effect is assumed to be constant across all trials. This assumption may be reasonable since (1) it is not known whether the treatment effects are different across trials, and (2) as shown in Table 2.1, Category 2 trial has a large sample size, which may play a dominant role in determining the treatment effect. We also note that the historical data are available only for the control arm. Thus, there is no *a priori* information available regarding the magnitude of among-trial variability for the treatment effects from previous trials. This poses a major challenge in meta-design clinical trials. However, the Bayesian methodology can be extended to allow for nonconstant random treatment effects. Fourth, to account for heterogeneity among meta-studies, we have presented both the log-linear random effects and fixed effects regression models. As empirically shown in the simulation study in Section 2.8, both types of regression models yield similar sample sizes of meta-trials to achieve a prespecified power (80%) under a prespecified type I error (5%). The advantages of the random effects model are that it provides much greater flexibility for borrowing historical data than the fixed effects model and it has more parameters that control the borrowing compared to the fixed effects model. As a result, the random effects model requires a smaller a_0 than the fixed effects model for achieving a desired power and type I error. From a computational point of view, the fixed effects model is more convenient to implement, while the random effects model is more computationally intensive. Fifth, in our simulation studies, we observed zero events for several small trials. Thus, under the fixed effects regression model, a frequentist approach for the sample size calculation does not work when there are zero events for small trials. Under the Bayesian paradigm, we use a proper prior for θ_k or θ_{0k} to circumvent this zero event issue, thereby yielding reasonable sample sizes.

Finally, we mention that the simulation-based algorithms discussed in Section 2.7 have been implemented using the FORTRAN 95 software with double precision and IMSL subroutines. We have also written very user-friendly SAS code using PROC MCMC for the fixed effects model discussed in this chapter. We are now in the process of generalizing this SAS code to the random effects model as well as developing a user-friendly R interface of our available FORTRAN code. Both the FORTRAN 95 code and the SAS macro are available from the authors upon request.

Acknowledgment

Drs. Ibrahim and Chen's research was partially supported by U.S. National Institutes of Health (NIH) grants #GM 70335 and #CA 74015.

References

Chen, M.-H., Ibrahim, J. G., Lam, P., Yu, A., and Zhang, Y. (2011) Bayesian design of non-inferiority trials for medical devices using historical data, *Biometrics* **67**, 1163–1170.

Chen, M.-H. and Shao, Q.-M. (1999) Monte Carlo estimation of Bayesian credible and HPD intervals, *Journal of Computational and Graphical Statistics* **8**, 69–92.

Chen, M.-H., Shao, Q.-M., and Ibrahim, J. G. (2000) *Monte Carlo Methods in Bayesian Computation*. New York: Springer-Verlag.

Crowe, B. J., Xia, H. A., Berlin, J. A., Watson, D.J., Shi, H., Lin, S. L., Kuebler, J., Schriver, R. C., Santanello, N. C., Rochester, G., Porter, J. B., Oster, M., Mehrotra, D. V., Li, Z., King, E. C., Harpur, E. S., and Hall, D. B. (2009) Recommendations for safety planning, data collection, evaluation and reporting during drug, biologic and vaccine development: A report of the safety planning, evaluation and reporting Team, *Clinical Trials* **6**, 430–440.

De Santis, F. (2007) Using historical data for Bayesian sample size determination, *Journal of the Royal Statistical Society, Series A* **170**, 95–113.

Gilks, W. R. and Wild, P. (1992) Adaptive rejection sampling for Gibbs sampling, *Applied Statistics* **41**, 337–348.

Ibrahim, J. G. and Chen, M.-H. (2000) Power prior distributions for regression models, *Statistical Sciences* **15**, 46–60.

Ibrahim, J. G., Chen, M.-H., and Sinha, D. (2001) *Bayesian Survival Analysis*. New York: Springer-Verlag.

Ibrahim, J. G., Chen, M.-H., Xia, H. A., and Liu, T. (2012) Bayesian meta-experimental design: Evaluating cardiovascular risk in new antidiabetic therapies to treat type 2 diabetes, *Biometrics* **68**, 578–586.

Inoue, L. Y. T., Berry, D. A., and Parmigiani, G. (2005) Relationship between Bayesian and frequentist sample size determination, *The American Statistician* **59**, 79–87.

Liraglutide (2009) http://www.fda.gov/downloads/AdvisoryCommittees/Committees MeetingMaterials/Drugs/EndocrinologicandMetabolicDrugsAdvisoryCommittee/ UCM148109.pdf.

M'Lan, C. E., Joseph, L., and Wolfson, D. B. (2008) Bayesian sample size determination for binomial proportions, *Bayesian Analysis* **3**, 269–296.

Saxagliptin (2009) http://www.fda.gov/ohrms/dockets/ac/09/brie'ng/2009-4422b1-01-FDA.pdf.

Spiegelhalter, D. J., Abrams, K. R., and Myles, J. P. (2004) *Bayesian Approaches to Clinical Trials and Health-Care Evaluation*. New York: Wiley.

Sutton, A. J., Cooper, N. J., Jones, D. R., Lambert, P. C., Thompson, J. R., and Abrams, K. R. (2007) Evidence-based sample size calculations based upon updated meta-analysis, *Statistics in Medicine* **26**, 2479–2500.

The Action to Control Cardiovascular Risk in Diabetes Study Group. (2008) Effects of intensive glucose lowering in type 2 diabetes, *The New England Journal of Medicine* **358**, 2545–2559.

The ADVANCE Collaborative Group. (2008) Intensive blood glucose control and vascular outcomes in patients with type 2 diabetes, *The New England Journal of Medicine* **358**, 2560–2572.

Wang, F. and Gelfand, A. E. (2002) A simulation-based approach to Bayesian sample size determination for performance under a given model and for separating models, *Statistical Sciences* **17**, 193–208.

3

Non-Inferiority Study Design and Analysis for Safety Endpoints

Steven Snapinn and Qi Jiang

CONTENTS

3.1 Introduction to Non-Inferiority Trials for Efficacy and Safety

A non-inferiority trial attempts to demonstrate that the effect of a treatment is either superior to that of a control or similar to it (Snapinn 2000). The concepts of equivalence and non-inferiority are closely related: in the former case, an important difference between treatments in either direction is of concern, while in the latter case an important difference in one direction is of concern, while any difference in the other direction (toward superiority) is not. Trials evaluating clinical outcomes are more typically classified as non-inferiority trials than equivalence trials, since superiority would not typically be of concern. In some contexts, however, equivalence trials are more common. One such context is bioequivalence, where the objective is to demonstrate that the pharmacokinetic properties of one drug formulation are similar to those of another formulation of the same drug. In addition, regulatory guidance for the development of a biosimilar (a drug that is designed to be similar to, but not identical to, an innovative biologic therapy) requires demonstration of equivalence with respect to clinical endpoints (U.S. Food and Drug Administration 2012); the rationale is that superiority with respect to the clinical endpoint is of concern in this context, since it could be accompanied by increased safety risks.

Note the use of the word "important" as a modifier of "difference" in the previous paragraph. It is generally not feasible, or even possible, to demonstrate conclusively that there is no difference at all between two treatments. Therefore, non-inferiority and equivalence are often defined through the use of a margin, typically designated by δ, to indicate the smallest difference between treatments that would be considered "important." (In the cases of equivalence, different margins for superiority and inferiority are sometimes used.)

This concept can be defined through notation. Let the letter T represent the experimental treatment and C represent the control, which may or may not be a placebo. Suppose that the variable of interest is continuous and its mean values with the two treatments are μ_T and μ_C, respectively, and assume that larger values of the variable are favorable. (This concept can be easily extended to other types of endpoints.) Based on the prespecified margin, δ, the null and alternative hypotheses for a non-inferiority trial are

$$H_0 : \mu_T - \mu_C \leq -\delta$$

$$H_A : \mu_T - \mu_C > -\delta$$

Note that the alternative hypothesis allows for some degree of inferiority of treatment T relative to treatment C, as long as the degree of inferiority does not exceed the margin, δ. A non-inferiority trial involves estimating the mean values for the two treatments, calculating the confidence interval for the difference between them, and demonstrating that the confidence interval rules out inferiority of a magnitude larger than the margin. Let $\hat{\mu}_{TC}$ represent the estimate of the mean difference between the treatments, and let $\hat{\sigma}_{TC}$ represent its estimated standard error. Therefore, using a 97.5% confidence bound (corresponding to the lower limit of a two-sided 95% confidence interval), non-inferiority is demonstrated if

$$\hat{\mu}_{TC} - 1.96\,\hat{\sigma}_{TC} > -\delta$$

This concept is illustrated in Figure 3.1 by presenting a set of hypothetical point estimates and two-sided 95% confidence intervals for μ_{TC}. In Cases 1–3, non-inferiority has not been demonstrated because the confidence intervals extend below $-\delta$. This is despite the fact that the point estimate is in favor

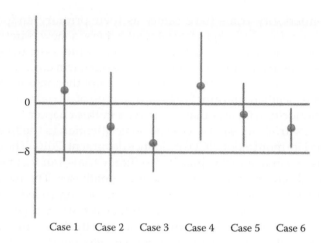

FIGURE 3.1
Six hypothetical point estimates and confidence intervals for the difference between treatment *T* and treatment *C*.

of the experimental treatment in Case 1. In addition, Case 3 demonstrates significant inferiority, because the confidence interval is entirely below 0. In Cases 4 and 5, non-inferiority is demonstrated because the confidence interval is entirely above $-\delta$. This is despite the fact that the point estimate is in favor of the control in Case 5. Case 6 is a paradoxical example in which both non-inferiority and inferiority have been demonstrated, since the experimental treatment appears to be inferior to the control by a magnitude greater than 0 but less than δ.

Most non-inferiority trials have efficacy as their primary endpoint, and most papers and regulatory guidance documents focus on this primary efficacy endpoint. Safety is often discussed in the context of an ancillary benefit; that is, while superior efficacy is desirable, similar efficacy, or even some loss of efficacy, can be acceptable if the new treatment has other benefits, such as superior safety, convenience, or cost (CHMP 2005; Kaul and Diamond 2006; U.S. Food and Drug Administration 2010). For example, angiotensin II–converting enzyme inhibitors are effective treatments for hypertension, but cause a distinctive cough in some patients. Angiotensin receptor blockers have a similar mechanism of action, but are less likely to cause cough. Therefore, demonstrating similar antihypertensive efficacy with an angiotensin receptor blocker as with an angiotensin II–converting enzyme inhibitor could represent an important advantage for patients. While in most cases the ancillary benefit either has been demonstrated in previous trials or is assumed, D'Agostino et al. (2003) proposed a design to jointly test for non-inferiority with respect to the efficacy endpoint and superiority with respect to the safety endpoint.

Some non-inferiority trials have safety as their primary endpoint. This is often the case for a marketed treatment where a new safety concern has arisen. As in the case of an efficacy endpoint, unless the drug actually improves safety, no trial will ever be able to rule out harm entirely; therefore, the goal of these trials is to determine whether or not the magnitude of harm exceeds the prespecified margin, δ. This is the type of a trial we refer to as a safety non-inferiority trial and that we discuss in this chapter.

Non-inferiority trials for safety received some attention in the International Conference on Harmonisation (ICH) guidance document (2010). The document states that it is meant to apply primarily to efficacy trials, but that many of the considerations discussed also apply to safety hypotheses. The most relevant comment in that document is the following: "Placebo-controlled trials seek to show a difference between treatments when they are studying effectiveness, but may also seek to show lack of difference (of specified size) in evaluating a safety measurement. In that case, the question of whether the trial could have shown such a difference if there had been one is critical (see Section 1.5)." Note that Section 1.5 refers to the issue of assay sensitivity, which we discuss later.

In a paper commemorating the 10-year anniversary of ICH E10, Rockhold and Enas (2011) discussed the relevance of E10 for safety non-inferiority trials and noted that, despite the inherent complexities in their design, it is likely that the use of non-inferiority designs for safety will continue. Safety non-inferiority trials were also discussed in a review paper by Fleming (2008).

Of course, almost every clinical trial, including most non-inferiority trials with efficacy primary endpoints, will collect and analyze safety data. The safety analyses done in an efficacy non-inferiority trial are essentially no different than those in an efficacy superiority trial. However, one possible distinction relates to the interpretation of the safety information in assessing the experimental treatment's relative benefits and harms. As stated previously, in some cases, improvement in safety might be an expected ancillary benefit of a treatment, and therefore, the lack of improvement might be seen as problematic. In addition, any unexpected harm caused by the experimental treatment might be less acceptable if the experimental treatment has simply demonstrated similar efficacy to a control than if it had demonstrated superior efficacy.

In a sense, virtually, all clinical trials can be considered to be safety non-inferiority trials, since virtually, all trials collect safety information and therefore provide evidence as to whether or not there is an important difference in safety between treatments. However, there are two key distinguishing characteristics of the trials that we refer to as safety non-inferiority trials in this chapter: a safety endpoint is designated as primary and a non-inferiority margin is prespecified. In the next section, we will discuss some of the issues associated with non-inferiority trials in general, focusing on the factors that are unique to safety non-inferiority trials.

3.2 Issues in Non-Inferiority Trials

3.2.1 Selection of the Non-Inferiority Margin and Indirect Comparisons

The concept of margin selection has mostly been discussed in the context of efficacy non-inferiority trials, and this is perhaps the most important distinction between efficacy and safety trials. The earliest examples of non-inferiority trials typically used margins based on clinical considerations. The weakness of this approach becomes apparent when one considers the possibility that the margin chosen in this way could be greater than the effect of the active control. Let μ_P represent the mean value of the variable of interest when patients are treated with a placebo. When $\delta > (\mu_C - \mu_P)$, the condition $\mu_T - \mu_C > -\delta$ can be met when $(\mu_T - \mu_P) < 0$. In other words, unless one can guarantee that the margin is no larger than the effect size of the active control (as defined by the difference between that treatment and placebo), an experimental drug that is actually harmful (i.e., worse than placebo) can be shown to be non-inferior. For that reason, ICH E10 contains the following statement: "The margin chosen for a non-inferiority trial cannot be greater than the *smallest effect size that the active drug would be reliably expected to have* compared with placebo in the setting of the planned trial."

Based on this ICH guidance, margins for efficacy non-inferiority trials are now generally based on data from historical placebo-controlled trials involving the active control. This approach is also discussed in guidance documents from the Food and Drug Administration (FDA) (U.S. Food and Drug Administration 2010) and the European Medicines Agency (EMA) (CHMP 2005). Specifically, a meta-analysis of such trials is performed, and the margin is based on the lower bound of the 95% confidence interval for the effect of the active control. Note that while the point estimate is the best estimate of the effect of the active control, the lower end of the confidence interval is chosen in order to be more conservative, which is consistent with the word "reliably" in the ICH guidance document. In some cases, the margin is set equal to the lower bound, but in other cases, the margin is reduced further, due to a requirement for preservation of effect or to an account for uncertainty in the key assumptions supporting the validity of non-inferiority trials (see Sections 3.2.2 and 3.2.3 for a discussion of these issues).

The approach described previously is clearly meant to ensure that a conclusion of non-inferiority can also be interpreted as a conclusion that $(\mu_T - \mu_P) > 0$, that is, that the experimental treatment has at least some efficacy. In a sense, therefore, an efficacy non-inferiority trial can be thought of as part of an indirect comparison of the experimental treatment to placebo, with a null hypothesis $H_0: (\mu_T - \mu_P) = 0$ versus the alternative $H_A: (\mu_T - \mu_P) > 0$. This concept is discussed in detail by Snapinn and Jiang (2011).

When considering a safety non-inferiority trial, on the other hand, this approach to defining a margin is clearly not relevant. Typically, these trials are themselves placebo controlled, but even when an active control is used, the goal of the trial is to rule out excess harm specifically relative to the active control; the magnitude of any benefit or harm due to the active control is irrelevant. Therefore, there is no indirect comparison in a safety non-inferiority trial, and the margin must be chosen based solely on clinical grounds. This is such an important distinction between efficacy and safety non-inferiority trials that it seems misleading and potentially confusing to use the same term ("non-inferiority") to describe both. In fact, the term "non-inferiority" seems to apply better to the case of a safety endpoint than an efficacy endpoint due to the indirect comparison aspect in the latter case. Perhaps, the term "active-controlled efficacy trial" would make more sense for an efficacy non-inferiority trial.

The fact that an efficacy non-inferiority trial involves an indirect comparison between the experimental treatment and placebo, while a safety non-inferiority trial does not, underlies nearly all of the key differences between these types of trials, as discussed in the following.

3.2.2 Preservation of Effect

For many efficacy non-inferiority trials, it is considered to be insufficient to simply demonstrate indirectly that the experimental treatment is superior to placebo; rather, it is sometimes required to demonstrate that the experiment treatment preserves some fraction of the active control's effect (Everson-Stewart and Emerson 2010; Snapinn and Jiang 2008b). The null hypothesis for this situation can be framed in terms of a fixed margin or in terms of an indirect comparison, as follows:

$$H_0 \colon \mu_T - \mu_C \leq -(1-f)^* \delta$$

or

$$H_0 \colon (\mu_T - \mu_P) \leq f^* (\mu_C - \mu_P)$$

where f is the preservation factor and lies between 0 and 1. When $f = 0$, this reduces to the situation without preservation of effect described previously, while when $f = 1$, this corresponds to an evaluation of superiority of the experimental treatment to the active control. Preservation of effect is typically implemented by reducing the margin, δ, by the factor $(1 - f)$ and proceeding to demonstrate that $\hat{\mu}_{TC} - 1.96 \hat{\sigma}_{TC} > -(1-f)\delta$.

While preservation of effect is an important consideration for efficacy non-inferiority trials and is discussed in both the FDA and EMA guidance documents, it is not relevant to the design of a safety non-inferiority trial.

3.2.3 Assay Sensitivity and Constancy

Much of the controversy around NI trials relates to their reliance on untestable assumptions for their validity. In this section, we will discuss the two key assumptions.

Assay sensitivity is the ability of a trial to detect differences between treatments if they exist. There are many potential reasons to be concerned about assuming assay sensitivity, such as lack of compliance with the study medications, inclusion of inappropriate patients in the trial, and poor assessment of the study endpoints. For example, if a large fraction of subjects in each treatment arm failed to take their study medication, then the results for the two treatments might appear similar, resulting in a conclusion of non-inferiority, even if the experimental treatment were in fact inferior to the control. This is just as much a concern in safety trials as in efficacy trials.

There is in general no direct approach to assess assay sensitivity, although there are a variety of indirect approaches. One such approach is to include a third treatment arm in the trial; if this third arm can be distinguished from the control, then it would seem reasonable to infer that the trial would have been able to distinguish the experimental treatment from the control if such a difference existed. Other study design and conduct features that can reduce concern with the assumption of assay sensitivity include blinding, objective study endpoints or independent adjudication of study endpoints, careful assessment of compliance with study medication, and emphasis on careful study monitoring and data cleaning.

Concern with the assumption of assay sensitivity can also be addressed through performing both intention-to-treat and per-protocol analyses (Sanchez and Chen 2006). In a superiority trial, the intention-to-treat analysis is generally preferred, in large part because it is conservative; that is, factors that reduce confidence in assay sensitivity also make it more difficult to demonstrate superiority using this type of analysis. For example, consider the possibility that patients in both treatment arms have poor compliance with their study medications. In an intention-to-treat approach, all subjects are included in the analysis, regardless of compliance. This tends to diminish the estimated difference between treatment arms, thus making it more difficult to demonstrate superiority. In a non-inferiority trial, on the other hand, inclusion of noncompliant patients in the analysis can make an inferior experimental treatment appear similar to the control, thus inflating the probability of falsely concluding non-inferiority. For this reason, non-inferiority trials often include a per-protocol analysis. Per-protocol analyses have various definitions, but in general, they exclude subjects with major protocol violations and subjects with poor compliance to study medications. When prespecified, per-protocol analyses do address the concern with the assay sensitivity assumption to some extent, and for this reason, some authors have proposed that they be the primary analysis approach for non-inferiority trials. However, by excluding subjects based on postrandomization information, these analyses are subject to other biases that can be more important than those caused by lack of assay sensitivity. Therefore, there is no universally accepted best analysis approach for non-inferiority trials, and the most common recommendation is to do both an intention-to-treat and per-protocol analysis and requires consistent demonstration of non-inferiority with both analyses.

The constancy assumption is that the effect of the active control is the same in the non-inferiority trial as it was in the historical trials used to derive the margin. There are many reasons to have concerns with the constancy assumption, such as differences in characteristics of the patients enrolled in the trials, dosages of the control treatment, endpoint definitions, study duration, and background therapies used by patients. For example, suppose that in the time between when the historical trials were performed and the time the non-inferiority trial is performed, a particular background therapy has become the standard of care. Further, suppose that the effect of the control treatment is of a smaller magnitude in patients taking this background therapy than in patients not taking it. Therefore, a margin derived from historical trials could be larger than the effect size of the control treatment in the context of the non-inferiority trial, thus inflating the possibility that an ineffective experimental treatment could be found to be non-inferior to the control.

While the constancy assumption has important implications for efficacy non-inferiority trials, this is an example of an issue that is related to the indirect comparison aspect of those trials and therefore has no relevance for safety non-inferiority trials.

Because of concern with the assumption of assay sensitivity and constancy, the non-inferiority margin used in efficacy non-inferiority trials is sometimes reduced or discounted (Snapinn 2004; Snapinn and Jiang 2008a; U.S. Food and Drug Administration 2010). Since safety non-inferiority trials also rely on the assumption of assay sensitivity, the concept of discounting could apply in this setting as well. However, to our knowledge, this has not been done.

3.2.4 Analysis Method

The analysis approach described earlier for assessing non-inferiority is to define a margin and determine whether or not the 95% confidence interval for the difference between the experimental treatment and control excludes that margin. That is, non-inferiority is concluded if $\hat{\mu}_{TC} - 1.96\,\hat{\sigma}_{TC} > -\delta$. This approach is known as the fixed margin method.

However, there is a second approach in common use for efficacy non-inferiority trials, and there is no consensus on which approach is more appropriate (Peterson et al. 2010; U.S. Food and Drug Administration 2010). This approach, known as the synthesis method (Hasselblad and Kong 2001; Holmgren 1999; Rothmann et al. 2003), specifically addresses the indirect comparison of the experimental treatment with placebo; that is, it is meant to test the hypothesis H_0: $(\mu_T - \mu_P) = 0$ versus the alternative H_A: $(\mu_T - \mu_P) > 0$. Using this approach, one sums the estimate of the effect of the experimental treatment relative to the control from the non-inferiority trial with the estimate of the effect of the control relative to placebo from the historical trials to obtain an indirect estimate of the effect of the experimental treatment relative to placebo. Let $\hat{\mu}_{CP}$ represent the estimate of the mean difference between the control and placebo based on the historical trials, and let $\hat{\sigma}_{CP}$ represent its estimated standard error. Therefore, one concludes non-inferiority based on the synthesis method if $(\mu_{TC} + \mu_{CP}) - 1.96\sqrt[2]{\sigma_{TC}^2 + \sigma_{CP}^2} > 0$. The synthesis method can also be extended to handle preservation of effect.

Because the synthesis method is specifically focused on the indirect comparison aspect of efficacy non-inferiority trials, it does not apply in the context of safety non-inferiority trials; therefore, all safety non-inferiority trials use the fixed margin method.

3.2.5 Demonstrating Non-Inferiority and Superiority

Often, there is interest in determining not only whether or not an experimental treatment is non-inferior to a control but also whether or not it is superior. This can be accomplished through a hierarchical testing approach where one first tests for non-inferiority and, if non-inferiority is concluded, one next tests for superiority. There is no multiplicity correction required using this approach. (Note, however, that if there are other hypotheses tested simultaneously with the superiority hypothesis, then a multiplicity correction may be required [Ke et al. 2012].)

This hierarchical approach can be prespecified and applied to either efficacy or safety non-inferiority trials. However, the superiority hypothesis is less likely to be of interest in the safety setting than in the efficacy setting, since it is less likely that an experimental treatment could be hypothesized to be superior to placebo with respect to a safety endpoint than it is that an experimental treatment could be hypothesized to be superior to an active control with respect to an efficacy endpoint.

3.2.6 Choice of the Metric

In the previous discussion, we have focused on an endpoint measured on a continuous scale. In such a case, there is generally little controversy over the choice of the mean difference between treatments as the measure of the treatment effect. However, for other endpoint types, the choice of the metric for the treatment effect can be critical and far more controversial.

Consider a binary or dichotomous endpoint such as the response rate in an oncology trial. Let the true response rates with the experimental and control treatments be denoted r_T and r_C, respectively. Suppose that the response rate with the control treatment is known to be $r_C = 0.20$ and the lowest acceptable response rate with the experimental treatment is $r_T = 0.15$. In this case, one could define the treatment effect using a variety of metrics, including the risk difference, the risk ratio, and the odds ratio. The null hypotheses and margins corresponding to these three metrics are

Risk difference

$$H_0: r_T - r_C \leq -\delta_{RD}, \quad \text{where } \delta_{RD} = 0.20 - 0.15 = 0.05$$

Risk ratio

$$H_0: r_T / r_C \leq \delta_{RR}, \quad \text{where } \delta_{RR} = 0.15 / 0.20 = 0.75$$

Odds ratio

$$H_0: r_T * (1 - r_C) / \{r_C * (1 - r_T)\} \leq \delta_{OR}, \quad \text{where } \delta_{OR} = 0.15 * 0.80 / (0.20 * 0.85) = 0.71$$

When the control rate is known, these three approaches will give consistent conclusions regarding non-inferiority. The issue arises when the control rate differs from the hypothesized rate. Suppose now that the trial was designed under the assumption that the response rate in the control group is 0.2 but in reality, in the context of the non-inferiority trial, the true response rate is 0.1. In this case, the three metrics can result in inconsistent conclusions. Consider the response rate in the experimental group that would be at the boundary of the non-inferiority definition using the three margins described previously.

Risk difference

$$r_T = r_C - \delta_{RD} = 0.050$$

Risk ratio

$$r_T = r_C * \delta_{RR} = 0.075$$

Odds ratio

$$r_T = r_C * \delta_{OR} / (1 - r_C + r_C * \delta_{OR}) = 0.073$$

In this example, the one absolute measure of effect, the risk difference, allows for a considerably lower response rate for the experimental treatment to be determined to be non-inferior to the control than the two relative measures of effect. Such a difference could have enormous implications for sample size and power and, more importantly, for the response rates with the two treatments that would represent a clinically unimportant difference.

The same issue exists for time-to-event endpoints. While the most common metric for trials with these endpoints is the hazard ratio evaluated using survival analysis (usually a Cox proportional hazards model), absolute measures, such as the difference in event rates at a fixed follow-up time, are sometimes used.

To our knowledge, there is no consensus on the most appropriate metric for binary or time-to-event endpoints, and, therefore, this is a factor that must be carefully considered by an investigator designing a non-inferiority trial with such endpoints.

3.2.7 Ethics

Efficacy non-inferiority trials are often conducted in situations where a placebo-controlled trial would not be ethical, for example, when there exists effective therapy for a serious or life-threatening condition. Despite this, the ethical basis for non-inferiority trial has been called into question. For example, Garattini and Bertele (2007) argue that "... it is unethical to leave to chance whether patients receive a treatment that is anticipated to provide no extra benefit, but could be less safe and less effective than existing treatment options." While their comments were in the context of efficacy non-inferiority trials, the issue could be even greater in the context of a safety non-inferiority trial. The primary objective of a safety non-inferiority trial is to determine whether or not an experimental treatment causes an important harm; on this basis alone, there would be little incentive for a prospective patient to participate. Patients must be fully informed of the purpose of such a trial, including the potential benefits and harms, before they can consent to participate.

3.3 Examples

Prospective Randomized Evaluation of Celecoxib Integrated Safety versus Ibuprofen or Naproxen (PRECISION) (Becker et al. 2009) is a safety non-inferiority trial evaluating the cardiovascular safety of one nonsteroidal anti-inflammatory drug, celecoxib, relative to two other such drugs, ibuprofen and naproxen. Nonsteroidal anti-inflammatory drugs are used for chronic pain management by patients with osteoarthritis or rheumatoid arthritis. Older drugs in this class, including ibuprofen and naproxen, are effective at pain management, but can cause intestinal bleeding with long-term use. Celecoxib is in a new class of nonsteroidal anti-inflammatory drugs known as COX-2 inhibitors; these drugs are believed to be safer with respect to intestinal bleeding, but have been shown to increase cardiovascular risk in placebo-controlled trials. Therefore, the purpose of PRECISION is to determine whether or not celecoxib increases cardiovascular risk relative to older nonsteroidal anti-inflammatory drugs.

Patients with osteoarthritis or rheumatoid arthritis are randomized to one of three treatments, celecoxib, ibuprofen, or naproxen, and the primary endpoint is the occurrence of a cardiovascular endpoint: a nonfatal myocardial infarction, a nonfatal stroke, or any cardiovascular death. Non-inferiority will be assessed for three different pairwise comparisons: celecoxib versus ibuprofen, celecoxib versus naproxen, and ibuprofen versus naproxen. The definition of non-inferiority differs somewhat from the fixed margin approach describe earlier in that there are separate criteria for the confidence interval and the point estimate. The hazard ratio for each comparison will be calculated, and non-inferiority will be concluded if the upper end of the one-sided 97.5% confidence interval is below the margin of 1.33 (this corresponds to the standard fixed margin approach described previously) and the point estimate is below 1.12. Non-inferiority must be demonstrated both based on an intention-to-treat approach and based on an approach in which patients are censored 30 days after permanent discontinuation of study drug; the latter is referred to as a modified intention-to-treat approach, but is consistent with the concept of a per-protocol approach as described earlier.

Thiazolidinedione Intervention with Vitamin D Evaluation (TIDE) (The TIDE Trial Investigators 2012) was a trial comparing the cardiovascular effects of two thiazolidinediones (TZDs), rosiglitazone and pioglitazone, in patients with diabetes. This was actually a 3 × 2 factorial trial that was also designed to evaluate the effects of vitamin D on cancers and mortality, but we will restrict our attention to the evaluation of TZDs. The trial was mandated by the FDA in response to concerns that TZDs adversely affect cardiovascular risk. Patients with diabetes were randomized to rosiglitazone, pioglitazone, or placebo, and the primary endpoint was the occurrence of myocardial infarction, stroke, or cardiovascular death (as in PRECISION). The trial had 90% power to detect a 20% or greater reduction in risk of the primary outcome for the combined TZD groups relative to placebo (i.e., superiority) and was also designed to exclude a 30% increase in cardiovascular risk for rosiglitazone relative to placebo (i.e., non-inferiority). However, TIDE was terminated by the FDA prior to its completion based on an assessment of all available data on the cardiovascular risk of rosiglitazone (Woodcock et al. 2010).

The number of safety non-inferiority trials to evaluate cardiovascular risk for new treatments for diabetes will likely increase due to a recent guidance from the FDA (U.S. Food and Drug Administration 2008). This document requires that the hazard ratio for a new treatment with respect to cardiovascular endpoints be below a prespecified margin. A margin of 1.8 (i.e., excluding an 80% increase in risk) is required to obtain approval, but if the upper bound of the 95% confidence interval is between 1.3 and 1.8, then it will be necessary to provide additional postmarketing information to definitively show that the upper bound of the 95% confidence interval is below 1.3. This can be done by conducting a new stand-alone safety non-inferiority trial or by combining the results of the premarketing data with those from a new postmarketing safety non-inferiority trial.

3.4 Summary

Non-inferiority trials are designed to demonstrate that one treatment has an effect that is either superior to that of another treatment, or similar to it. Safety non-inferiority trials are non-inferiority trials with a primary safety endpoint and with a prespecified margin. Non-inferiority trials for efficacy and safety endpoints have some similarities, but also some fundamental differences. The most important distinction is that efficacy non-inferiority trials inherently involve an indirect comparison between the experimental treatment and placebo, while safety non-inferiority trials are simple direct comparisons between the experimental treatment and the control, which is often a placebo. For this reason, some issues that are important in the design and analysis of efficacy non-inferiority trials do not apply to safety non-inferiority trials. These include the need to define the non-inferiority margin based on historical placebo-controlled trials, the requirement to preserve a fraction of the control treatment's effect, and reliance on the constancy assumption for the trial's validity. Like efficacy non-inferiority trials, safety non-inferiority trials rely on the assumption of assay sensitivity, and, for binary and time-to-event endpoints, the choice of the metric for the treatment effect is critical.

References

Becker, M.C., Wang, T.H., Wisniewski, L., Wolski, K., Libby, P., Lüscher, T.F., Borer, J.S. et al. Rationale, design, and governance of prospective randomized evaluation of celecoxib integrated safety versus ibuprofen or naproxen (PRECISION), a cardiovascular end point trial of nonsteroidal antiinflammatory agents in patients with arthritis. *American Heart Journal* **157**:606–612, 2009.

Committee for Medicinal Products for Human Use (CHMP). Points to consider on the choice of non-inferiority margin, 2005. http://www.ema.europa.eu/docs/en_GB/document_library/Scientific_guideline/2009/09/WC500003636.pdf

D'Agostino Sr., R.B., Massaro, J.M., Sullivan, L.M. Non-inferiority trials: Design concepts and issues—The encounters of academic consultants in statistics. *Statistics in Medicine* **22**:169–186, 2003.

Everson-Stewart, S., Emerson, S.S. Bio-creep in non-inferiority clinical trials. *Statistics in Medicine* **29**:2769–2780, 2010.

Fleming, T.R. Current issues in non-inferiority trials. *Statistics in Medicine* **27**:317–332, 2008.

Garattini, S., Bertele, V. Non-inferiority trials are unethical because they disregard patients' interests. *The Lancet* **370**:1875–1877, 2007.

Hasselblad, V., Kong, D.F. Statistical methods for comparison to placebo in active-control trials. *Drug Information Journal* **35**:435–489, 2001.

Holmgren, E.B. Establishing equivalence by showing that a specified percentage of the effect of the active control over placebo is maintained. *Journal of Biopharmaceutical Statistics* **9**:651–659, 1999.

International Conference on Harmonisation (ICH). Topic E10: Choice of control group in clinical trials, 2000. London: European Agency for the Evaluation of Medicinal Products. http://www.ich.org/fileadmin/Public_Web_Site/ICH_Products/Guidelines/Efficacy/E10/Step4/E10 Guideline.pdf

Kaul, S., Diamond, G.A. Good enough: A primer on the analysis and interpretation of noninferiority trials. *Annals of Internal Medicine* **145**:62–69, 2006.

Ke, C., Ding, B., Jiang, Q., Snapinn, S.M. The issue of multiplicity in noninferiority studies. *Clinical Trials* **9**:730–735, 2012.

Peterson, P., Carroll, K., Chuang-Stein, C., Ho, Y.-Y., Jiang, Q., Li, G., Sanchez, M., Sax, R., Wang, Y.-C., Snapinn, S. PISC expert team white paper: Toward a consistent standard of evidence when evaluating the efficacy of an experimental treatment from a randomized, active-controlled trial. *Statistics in Biopharmaceutical Research* **2**:522–531, 2010.

Rockhold, F.W., Enas, G.G. 10 years with ICH E10: Choice of control groups. *Pharmaceutical Statistics* **10**:407–409, 2011.

Rothmann, M., Li, N., Chen, G., Chi, G. Y.H., Temple, R., Tsou, H.-H. Design and analysis of non-inferiority mortality trials in oncology. *Statistics in Medicine* **22**:239–264, 2003.

Sanchez, M.M., Chen, X. Choosing the analysis population in non-inferiority studies: Per protocol or intent-to-treat. *Statistics in Medicine* **25**:1169–1181, 2006.

Snapinn, S.M. Noninferiority trials. *Current Controlled Trials in Cardiovascular Medicine* **1**:19–21, 2000.

Snapinn, S.M. Alternatives for discounting in the analysis of noninferiority trials. *Journal of Biopharmaceutical Statistics* **14**:263–273, 2004.

Snapinn, S., Jiang, Q. Controlling the type 1 error rate in non-inferiority trials. *Statistics in Medicine* **27**:371–381, 2008a.

Snapinn, S., Jiang, Q. Preservation of effect and the regulatory approval of new treatments on the basis of non-inferiority trials. *Statistics in Medicine* **27**:382–391, 2008b.

Snapinn, S., Jiang, Q. Indirect comparisons in the comparative efficacy and non-inferiority settings. *Pharmaceutical Statistics* **10**:420–426, 2011.

The TIDE Trial Investigators. Design, history and results of the thiazolidinedione intervention with vitamin D evaluation (TIDE) randomised controlled trial. *Diabetologia* **55**:36–45, 2012.

U.S. Food and Drug Administration. Guidance for industry: Diabetes mellitus—Evaluating cardiovascular risk in new antidiabetic therapies to treat type 2 diabetes, 2008. http://www.fda.gov/downloads/Drugs/GuidanceComplianceRegulatoryInformation/Guidances/ucm071627.pdf.

U.S. Food and Drug Administration. Guidance for industry: Non-inferiority clinical trials: Draft guidance, 2010. http://www.fda.gov/downloads/Drugs/Guidances/UCM202140.pdf.

U.S. Food and Drug Administration. Guidance for industry: Scientific considerations in demonstrating biosimilarity to a reference product, 2012.

Woodcock, J., Sharfstein, J.M., Hamburg, M. Regulatory action on rosiglitazone by the US Food and Drug Administration. *New England Journal of Medicine* **363**:1489–1491, 2010. http://www.fda.gov/downloads/Drugs/GuidanceComplianceRegulatoryInformation/Guidances/UCM291128.pdf.

Section II

Safety Monitoring

Section II

Safety Monitoring

4

Program Safety Analysis Plan:
An Implementation Guide

**Brenda Crowe, H. Amy Xia, Mary Nilsson,
Seta Shahin, Wei Wang, and Qi Jiang**

CONTENTS

4.1 Introduction

In 2009, the Safety Planning, Evaluation, and Reporting Team (SPERT) (a team formed in 2006 by the Pharmaceutical Research and Manufacturers of America to recommend a pharmaceutical industry standard for safety planning, data collection, evaluation, and reporting) recommended that sponsors create a program safety analysis plan (PSAP) early in product development (Crowe et al., 2009). The PSAP is described as a living document (updated periodically and amended as needed in response to the emerging safety profile) that eventually forms the basis for the statistical analysis plan (SAP) for the Summary of Clinical Safety (SCS). The PSAP generally has two main sections: a standard data collection plan and an analytical section. It can be a stand-alone document or can be embedded in a different document.

The main purpose of the PSAP is to facilitate a well-planned and systematic approach for safety data collection and analysis. This also facilitates regular reviews of aggregate data by a multidisciplinary safety management team, as recommended by the Council for International Organisations of Medical Sciences Working Group VI (CIOMS Working Group VI, 2005). By following this approach, potential harms may be identified earlier in the drug development process, allowing data collection strategies to be modified in time to collect additional data to further understand the potential harm. The appropriate variety of postapproval activities will permit ongoing refinement of the understanding of the benefit–risk profile of a new product. At the present time, there is no firm regulatory requirement for a PSAP; however, we have had some experience with a review division of the Food and Drug Administration (FDA) requesting a quantitative safety analysis plan (QSAP), which is the same concept as a PSAP.

This chapter covers several important aspects of a PSAP. Section 4.2 includes an explanation of the key components of the PSAP. Section 4.3 discusses how the PSAP was implemented in two companies (Lilly and Amgen) and provides our recommendation for PSAP implementation. These sections are intended to help sponsors understand the most important content of a PSAP and share some of the implementation issues they may encounter. Section 4.4 contains concluding remarks. Throughout this chapter, we use the term drug or product to refer to investigational or marketed drugs, biologics, or vaccines.

4.2 Key Components of a Program Safety Analysis Plan

In this section, we delineate potential components of a PSAP. These components are primarily based on information provided in the SPERT publication (Crowe et al., 2009) and a presentation by Rochester (2009). The main components of the PSAP are outlined in Table 4.1 and discussed in Sections 4.2.1 through 4.2.5.

TABLE 4.1

Key Components of a PSAP

1. Introduction
2. Overview of studies, exposure, and power
3. AEs of special interest and potential toxicities that should be considered for all products
4. Data collection plan and standardization approaches
5. Methods for analysis and reporting

4.2.1 Introduction

The introduction describes the scope of the PSAP in the context of the purpose of the product development program (e.g., planned indications). This section may also include a high-level overview of important regulatory interactions and/or previous agreements with regulatory agencies such as the FDA with respect to safety issues. It may also include highlights from safety documents such as the Investigator's Brochure.

4.2.2 Overview of Studies, Exposure, and Power

This section provides an overview of the studies in the product development program. It also summarizes the anticipated exposure and may include a description of power. Specifically, the following information may be considered for inclusion:

- A description (e.g., in a table) of studies. Xia et al. (2011) provide an example of such a table. At a minimum, the table should include the planned Phase 2 and 3 studies. The description of studies includes study design (e.g., fixed versus flexible dose, parallel versus crossover), dosing schedule, study location, treatment groups and doses, number of subjects planned (by treatment group and doses), patient population (e.g., based on planned indication, special age groups), and duration of study.

- The approximate total number of subjects expected to be exposed to study treatment with the currently completed, ongoing, or planned studies. Additionally, it would typically include an estimate of the number of subjects expected to be exposed for 6 months and at least 12 months, respectively. It should also discuss why these exposure numbers and durations are adequate. The rationale may include factors such as product novelty for the proposed indication, the intended population (e.g., pediatrics, geriatrics), and intended duration of use (e.g., lifelong therapies, chronic versus episodic).

- A description of the power (which could be via power plots and/or confidence interval plots) for the clinical trial database to detect differences in event rates (based on, e.g., risk difference or odds ratio) for various control event rates (Cooper et al., 2008).

4.2.3 Adverse Events of Special Interest and Potential Toxicities That Should Be Considered for All Products

This section specifies the list of adverse events (AEs) of special interest for the product as well as potential toxicities that should be considered for all products. The potential toxicities that should be explicitly considered for all products are AEs that are serious and have been linked to drugs often enough that they are considered universally of interest to evaluate in depth. They are toxicities considered important based on historical precedent across drugs. Examples are alterations in cardiac electrophysiology, hepatotoxicity, and (for large molecules) immunogenicity. Typically, AEs of special interest are those events for a particular product that are identified via evidence summarized from basic research, information from pre and nonclinical investigations, similar drugs in class, epidemiologic data on diseases, and clinical trials. The following are some specific items to consider:

- The rationale for inclusion of each AE of special interest.
- Information regarding at what time point in the product's development the AE of special interest was identified, as the timing of when an emerging issue was specified as an AE of special interest may be important for analysis and interpretation of results.
- How to identify subjects with each AE of special interest. This includes specifying the medical definition of the AE of special interest (preferably, an established definition in the medical literature). Identification of subjects may include any of the following: utilization of a search strategy in the Medical Dictionary for Regulatory Activities (MedDRA) or other dictionary utilization of a special case report form and/or adjudication. For example, it may be possible to utilize a standardized MedDRA query (SMQ) (see the MedDRA site). If a special case report form is developed, details of special collection may be included in the PSAP section on "Data Collection Plan and Standardization Approaches." If the AE of special interest definition will include event adjudication, detailed information on the adjudication process or a link to more detailed information should be included. It is recommended to have a common definition for an AE of special interest for the entire development program, in order to facilitate data integration, analysis, and interpretation.
- Specification of "potential toxicities that should be explicitly considered for all products." These are potential toxicities that should be explicitly considered for all products regardless of therapeutic areas or diseases. These toxicities typically include, but are not limited to, QT prolongation, liver toxicity, nephrotoxicity, immunogenicity, and bone marrow toxicity (CIOMS Working Group VI, 2005). It may include information regarding (e.g., via a link) any special monitoring that will be utilized to ensure that the minimum data required to adequately understand the toxicity are collected, for example, a comprehensive monitoring plan for managing potential events of liver toxicity (U.S. Food and Drug Administration, 2009).

4.2.4 Data Collection Plan and Standardization Approaches

This section describes how safety data will be collected in a standardized approach for the program. It covers the principles or rules for data collection that applies to all, or nearly all, studies. The following are specific items recommended for inclusion:

- Design aspects of the clinical studies (i.e., a common, or at least compatible, visit structure). For example, if it is important to investigate the value of a laboratory parameter at 3 months, one could specify that all studies must include a 3-month visit.
- Study entry (inclusion) and exclusion criteria that apply to all studies.
- Coding procedures for AEs, medical history, concomitant medications, and any other text that needs to be coded.
- Any aspects of case report form design and instructions, operational definitions of potential findings that can be coded on the standardized forms, and standardized edits that will be applied to all studies, including processes and procedures for eliciting safety outcomes (CIOMS Working Group VI, 2005, p. 79). This is particularly useful if the sponsor is in a partnership with another company or is contracting with third-party organizations (TPOs) to run part or all of some of the studies. Each company may have company-specific special data collection forms to collect certain kinds of information (such as qualitative electrocardiogram [ECG] data together with the operational definitions of potential findings that can be coded on these standardized forms and standardized edits). In order to facilitate data integration, the forms and definitions should be consistent across all studies.
- Definitions of subgroups of subjects (i.e., risk groups of special interest) and how they will be defined in a standard fashion among protocols. For example, to investigate the safety of the product for renally impaired patients, the PSAP should indicate how such patients will be identified in a consistent fashion.
- Data monitoring committee (DMC) utilization. Consider the need for a program-level DMC or whether individual study DMCs will suffice (and which studies need a DMC).
- Data quality and assurance that go above and beyond standard processes. For example, this could include checking concordance of AE and laboratory-defined abnormalities or what will be considered concordance between solicited AEs and non-prespecified AEs. As an example of the latter, suppose all studies in the development program will utilize the Columbia Suicide Severity Rating Scale (C-SSRS). The PSAP might indicate that spontaneously collected AEs will be reviewed to make sure every suicide-related event is reflected appropriately in the C-SSRS.

- Any instructions for reporting AEs that should be included in all protocols. For example, the PSAP may state that for all protocols, abnormal lab findings without clinical significance are usually not considered AEs.

- The type of data standards that will be implemented should be stated. For example, datasets will be created using Study Data Tabulation Model (SDTM) and Analysis Data Model (ADaM) standards (as specified by the Clinical Data Interchange Standards Consortium at http://www.cdisc.org/).

- Strategies to handle different MedDRA and WHO drug versions over time and the mapping of these to a single version for the integrated safety database. For example, for MedDRA coding, will verbatim AE text be completely recoded to a current dictionary prior to submission, or will analysis in a single version map from the lowest level terms that were coded in many versions over the course of the product's development?

- How laboratory results will be standardized. This includes consideration of reference ranges for analysis. Also, if some studies use a central laboratory and others use local laboratories, the method for handling the integrated data should be specified.

4.2.5 Methods for Analysis, Presentation, and Reporting

This section describes the methods for analysis, presentation, and reporting of program-level safety data including the SCS. The PSAP may also contain the analysis strategy and principles that individual studies should follow. The following are more specific items to consider:

- Criteria that will be used to determine if a subject (or their data) is to be included in the safety analysis set. For example, this might include all subjects who received at least one dose of investigational product. This may also include decisions regarding how to handle data subsequent to discontinuation of study drug.

- Subsets of studies or study phases that will be analyzed together ("pools") and the rationale for the different pools. For example, all placebo-controlled studies could be one such pool.

- The definition of a treatment emergent adverse event (TEAE) that will be used for analysis in all studies. TEAEs may be defined in myriad ways, and it will generally be beneficial to utilize a single definition of TEAE for analysis in all studies. See, for example, Nilsson and Koke (2001) and Crowe et al. (2013) for a discussion of definitions of TEAEs.

- Analytical details (including the proper meta-analytical methods that stratify by study) for all standard analyses of safety data (e.g., AEs, SAEs, laboratory, ECGs, vital signs, reasons for discontinuation from study drug and/or study). See Xia et al. (2011) for a description of core analyses to be considered. See also the Pharmaceutical Users Software Exchange (PhUSE) Standard Scripts Working Group site where white papers defining recommended core analyses are in progress (PhUSE, 2013).

- Details of statistical analyses for potential toxicities that should be explicitly considered for all products and AEs of special interest. Analyses for these events will in general be more comprehensive than for standard safety parameters. These analyses may include subject-year adjusted rates, Cox proportional hazards analysis of time to first event, and Kaplan–Meier curves. Detailed descriptions of the models would typically be provided. For example, if Cox proportional hazards analysis is specified, a detailed description of the model(s) that will be used should be provided. This would generally include study as a stratification factor, covariates, and model selection techniques. More advanced methods, such as multiple events models or competing risk analyses, should be described if used (as appropriate). It is recommended that graphical methods also be employed, for example, forest plot and risk-over-time plot (Xia et al., 2011).
- A list of potential covariates to examine safety risk factors.
- Important details of any analysis of concomitant medication.
- Subgroup analyses that will be performed. Typically, age, sex, and race will always be included as subgroup factors, but there may be others that are particular to a product or patient population.
- Description of supportive analyses such as study drug administration, demographics, and exposure.
- Biomarkers: Some products may have identified biomarkers that impact the response to treatment. If biomarkers are being considered for safety assessment, methods to investigate the relationship among treatments, biomarkers, and safety outcomes could be described.
- The potential effects of extrinsic factors such as use of tobacco and alcohol. If these are needed for the product, a description of the analyses planned to evaluate the impact of these factors on safety could be provided.

4.3 Implementation

In this section, we discuss how two companies implemented the PSAP. Anything new requires resources to implement and to maintain. However, both companies proceeded, as they felt that the benefits of early planning were greater than the "costs" of implementation and maintenance. The product teams who implemented the PSAP found the documentation of safety planning valuable since it helped ensure strategic Phase 3 safety planning.

4.3.1 Lilly Implementation

After reviewing the recommendations from SPERT (Crowe et al., 2009), Lilly recognized that creating a PSAP would improve the ability to have a more solid product-level data collection plan, as knowing how safety data will be analyzed facilitates decisions around how the data should be collected. Furthermore, the company was looking for a way to improve communication between the Lilly statisticians and TPOs regarding the expectations for safety analyses for protocols and SAPs. The PSAP seemed to be a reasonable way to help ensure that planned analyses met regulatory and company expectations.

Lilly used the PSAP concept for analytical planning only, without details around data collection, as Lilly already had a standard operating procedure (SOP) in place for a product-level data collection plan. Lilly's format of the PSAP is very similar to an SCS SAP and contains the following sections:

- Objective and general considerations
  - Includes how studies will conceptually be combined and statistical methods that apply broadly
- A priori planned statistical methods
  - Exposure
  - Demographics
  - AEs
  - Vital signs and physical characteristics
  - Laboratory measurements
  - ECGs
  - Safety in special groups and situations
  - Special topics

The objective is generally written as follows:

> The objective of this PSAP is to provide an outline of the planned analyses for the safety sections of relevant Phase 3 protocols and the Summaries of Clinical Safety (SCSs) and/or Integrated Summaries of Safety (ISSs) for product X, indications A, B, and C.

After each section heading of the PSAP, the anticipated section placement in the SCS is included. Within each section of the planned statistical methods, language is included to specify whether the analyses will be done in the individual Phase 3 studies, the SCS, or both. Lilly does not include table shells in the PSAP.

Since the PSAP includes planned analyses for individual studies, consideration was given whether the safety section of the SAP for an individual study should simply refer to the PSAP. However, Lilly decided to maintain a full analytical description in the SAP for individual studies. The safety sections in the individual study SAPs generally need more detail to describe the analyses needed for all the phases and more study-specific details (e.g., reference to specific phases) for implementing the analyses. In contrast, a separate SAP for the SCS/integrated summaries of safety (ISS) is not required as the PSAP can serve this need. However, creating a separate SAP for the SCS/ISS is optional.

Although the focus of the PSAP is planning for the SCS and the primary controlled phases of individual Phase 3 trials, it may inform other safety-related reviews for a product (e.g., ongoing safety reviews at the trial or program level). Lilly has not yet had experience utilizing the PSAP to inform Phase 4 trials (i.e., after marketing authorization), but maintaining the PSAP for this purpose may be valuable for teams.

When the SOP related to SAPs was being updated, much consideration was given on whether to require PSAPs. Although not required by a regulatory agency, Lilly decided requiring PSAPs was justified as a means to improve communication around expectations for safety assessment. When considering the timing of the PSAP, Lilly decided to require an approved PSAP prior to protocol approval of the first Phase 3 study. Consideration was given as to whether to require the PSAP earlier in a product's life cycle; however, Lilly decided it was appropriate to start this more formal planning process after the product reached the hurdle of Phase 3 planning.

Lilly identifies statistics as the responsible function for creating the PSAP, but includes the physician responsible for safety of the product as a primary contributor. The SOP includes a step for a review by the broader study team (including regulatory, medical writing, other physicians involved with the product). Consistent with Lilly's SOP on SAPs for an individual study, the supervisor of the authoring statistician is responsible for approving the document. Teams use an example/template PSAP as a resource to make it easier to create the PSAP. The example/template PSAP includes language for the recommended standard analyses for safety, which teams can generally use directly with little modification.

Although the timing requirement for an approved PSAP is not until protocol approval of the first Phase 3 study, teams are encouraged to include a very good draft of the PSAP in the end of Phase 2 briefing documents for the FDA (consistent with SPERT's recommendation). If the PSAP is not submitted in the end of Phase 2 briefing document, teams are encouraged to send it to the FDA (and potentially to other regulatory agencies) for feedback, but this is not required in the SOP.

Lilly has encountered hurdles implementing the PSAP. The initial hurdle was to make sure teams were aware of the new requirement early enough so that time was allotted for the effort. This included proactive education to teams who were beginning protocol development for their Phase 3 studies.

The next hurdle was to make sure the PSAP authors were providing the approved PSAP to the TPOs or internal authors as their starting points for protocol and SAP development of the individual trials.

One of the solutions to this was to include an expectation in TPO work orders that the TPO should start with the PSAP for protocol and SAP development. Additionally, in the SOP for analysis development, a step was added to start with the PSAP when developing a SAP for an individual study. Also, some teams have struggled to keep the PSAP up to date. For example, some teams were making changes in the individual study protocol or SAP, but not updating the PSAP. As one of the goals of the PSAP is to keep safety-related analysis decisions documented in a single place, the value of the PSAP will diminish if teams aren't diligent in updating it. Continued proactive education is required, until the PSAP becomes more fully utilized across the company.

4.3.2 Amgen Implementation

In light of the SPERT recommendations and increasing emphasis on proactive, program-wide safety assessment throughout the product development life cycle among industry and regulatory authorities, we believe that the benefits of developing a PSAP are multifold. First, it is crucial to be proactive and to plan early. Proper planning enhances our ability to appropriately identify, characterize, and manage safety risks in a timely fashion. It also allows data collection strategies to be established early and potentially modified later in time to collect additional data that are helpful for further understanding of a safety issue. Second, a PSAP can serve as a vehicle to facilitate communications with regulatory agencies on key elements of safety data collection and evaluation. For example, the definition of an AE of special interest can be agreed upon early to ensure appropriate collection of data for the AE of special interest in Phase 3 clinical trials and the analysis of data pertinent to AEs of special interest derived from all clinical trials. Third, it enables the product team to implement a systematic and consistent way for planning, analysis, and reporting of safety data across all clinical trials in a product development program. Individual study protocols and SAPs can reference the PSAP for key elements of safety data collection and analysis. This could greatly facilitate future integrated safety data analyses at the time of the New Drug Application (NDA) or Biologic License Application (BLA) submission and thereafter throughout the product's life cycle.

In 2009, Amgen started an initiative to implement the PSAP for their late phase (i.e., starting from Phase 2) development programs. It included two steps: (1) establish an SOP for the PSAP and (2) build a PSAP template linked to the PSAP SOP.

The purpose of the PSAP SOP is to describe a process for creating and maintaining a PSAP and, therefore, to facilitate a systematic and consistent approach across programs. The SOP defines cross-functional roles and responsibilities in authoring, reviewing, approving, amending, storing, and distributing a PSAP.

The development of the PSAP is a multidisciplinary collaboration with authors from both biostatistics and safety and contributors from clinical development, regulatory, early development, and other functions as needed. Considering that many marketed products typically have well-established safety profiles and the PSAP is a planning document in nature (especially in premarketing stage), it was decided that a PSAP may not be needed for all existing products and is required for newly developed products. It is recommended in the SOP that a PSAP may be initially generated to support the end of Phase 2 milestone within the drug development process and, thereafter, it may be maintained throughout the remainder of the product's life cycle. If developed, the PSAP may be amended internally and as needed in response to the product's emerging safety profile or annually upon decision of the Global Safety Team, a cross-functional, product-specific safety management team. It is encouraged to discuss the key components of the PSAP with regulatory agencies in conjunction with a development milestone meeting such as end of Phase 2 meeting.

The content of the PSAP template at Amgen has been developed as described in Section 4.2. In addition, please see Chuang-Stein and Xia (2013) for a useful discussion of the PSAP. Having a standard PSAP template ensures that all the important elements of safety planning, evaluation, and reporting are considered and addressed and resources and time can be saved by avoiding reinventing the wheel for each team. The proposed PSAP template content was reviewed and approved by internal relevant cross-functional teams and took effect along with the PSAP SOP in 2011.

During the implementation of PSAP at Amgen, a few questions were raised by study teams. First, what is the relationship between PSAP and the integrated SAP (iSAP) (the SAP for ISS)? Although the PSAP serves as a basis for development of the iSAP in an NDA/BLA submission, it has some differences. At Amgen, the PSAP is a planning, strategic document and has a "longitudinal" feature, because it is a living document and is being maintained throughout a product life cycle. It defines the data collection and analytic principles across multiple trials in a program. On the other hand, the iSAP is a "cross-sectional" document that is constructed around a time prior to the NDA/BLA filing and will not be maintained thereafter. While the analytical principles in an iSAP generally will follow the PSAP, like any other study-specific SAPs, it defines specific analyses, datasets, and associated deliverables that will be included in the filing package. Second, should the PSAP be submitted to agencies other than the FDA, and if so, when? Amgen recommends discussing the key components of a PSAP at the end of Phase 2 and other milestone meetings, which would provide the advantage of allowing the FDA to offer suggestions at the meeting. The PSAP should be finalized shortly after end-of-Phase 2 meeting and before the first Phase 3 study begins. The decision on whether to submit to agencies other than FDA may be based on factors such as availability of regional regulatory authority guidance/guidelines requiring a PSAP or whether or not sponsors have specific safety questions to discuss with a regional regulatory authority. Third, should a single PSAP be developed for a product with multiple indications? Amgen supports a single PSAP for multiple indications based on the idea that safety assessment should be based on all available information.

So far, Amgen has successfully developed PSAPs for various programs and submitted them to the FDA (a few upon request). As with implementation of any new process, teams may initially have a lot of questions. Education, devoting resources, and having a process in place are key. From our experience, once the PSAP is developed, teams see and appreciate the benefits of this systematic, program-wide approach. Proactive planning will lead to identifying and resolving safety issues in a timely fashion. As a result, it will improve efficiency and effectiveness in the drug development process and ultimately save time and resources.

4.3.3 Implementation Recommendations

As noted in previous sections, there are many benefits to creating PSAPs. As with any new process, implementation can be challenging. Based on the experience at Lilly and Amgen, the following are some general recommendations for companies that are considering its inclusion to their processes.

1. Review the main components of the PSAP (Section 4.2). Determine which portions already exist at the company with the appropriate timing. For example, Lilly already had a program-level data collection plan that was required by beginning of Phase 2, which led Lilly's PSAP to be focused on the analytical portions of the PSAP. This assessment can help make the decision of whether to do a complete PSAP following the guidance in Section 4.2 (as Amgen implemented, including both data collection and analytic portions) or to do a partial implementation of recommendations from Section 4.2 (as Lilly implemented, with the focus on the analytical portion).

2. Determine whether a new SOP (or business practice document) is needed or if it can be included in an existing procedure.

3. For actual creation of the SOP, roles and responsibilities as well as timing for creating and completing the first version of PSAP will need to be defined. Typically, work on the PSAP will begin prior to Phase 2, particularly on the standard data collection plan. The first complete PSAP for a product will generally be created at some point during Phase 2 development and can be maintained throughout the product life cycle. PSAPs should be periodically reviewed and updated as needed.

4. Create an example PSAP and/or a PSAP template. This will be important since most people may not have a good concept of what the document should look like and it provides a means to facilitate consistency of standard safety analyses across the organization.

5. Create a robust communication plan around the changes to SOPs. This will likely require a small group of internal subject matter experts to proactively reach out to the teams that are approaching the specified timing in the SOP.

6. Continue communication. As PSAPs are at the product level, it takes time to build awareness and experience. The importance of keeping the PSAP up to date, especially at important points in time for the product (e.g., before protocol development of important studies or before an important safety review) should continuously be emphasized.

4.4 Concluding Remarks

The PSAP is part of a well-defined, coordinated, program-wide approach that may help facilitate the planning, collection, and assessment of safety data during drug development. It provides a useful means to facilitate standardization of data collection and data analysis among the studies of the product. This proactive approach can make data integration easier and may avoid problems with inconsistency in data collection and analysis. This can enable better detection of potential safety issues and better understanding of them.

In order to implement the PSAP at the company level, some of the important considerations include establishing a process (typically via an SOP), developing an example and/or template PSAP to help teams understand what constitutes a PSAP, and implementing a short- and long-term communication plan and appropriate timing of development and regular updates.

At the program level, work on the PSAP will typically begin prior to Phase 2, particularly on the standard data collection plan. The first complete PSAP for a product will generally be created at some point during Phase 2 development and may be maintained throughout the product life cycle. PSAPs should be periodically reviewed and updated as needed. The PSAP will eventually form the basis for the SAP for the SCS.

References

Chuang-Stein, C. and Xia, H.A. 2013. The practice of pre-marketing safety assessment in drug development. *Journal of Biopharmaceutical Statistics*, 23, 3–25.

Cioms Working Group VI. 2005. Management of safety information from clinical trials. Geneva, Switzerland: Council for International Organizations of Medical Sciences.

Clinical Data Interchange Standards Consortium. Available: www.cdisc.org. Accessed August 1, 2014.

Cooper, A.J., Lettis, S., Chapman, C.L., Evans, S.J., Waller, P.C., Shakir, S., Payvandi, N., and Murray, A.B. 2008. Developing tools for the safety specification in risk management plans: Lessons learned from a pilot project. *Pharmacoepidemiol Drug Safety*, 17, 445–454.

Crowe, B., Brueckner, A., Beasley, C., and Kulkarni, P. 2013. Current practices, challenges, and statistical issues with product safety labeling. *Statistics in Biopharmaceutical Research*, 5, 180–193.

Crowe, B.J., Xia, H.A., Berlin, J.A., Watson, D.J., Shi, H., Lin, S.L., Kuebler, J. et al. 2009. Recommendations for safety planning, data collection, evaluation and reporting during drug, biologic and vaccine development: A report of the safety planning, evaluation, and reporting team. *Clinical Trials*, 6, 430–440.

MedDRA. *Medical Dictionary for Regulatory Activities*. Available: http://www.meddra. org. Accessed August 1, 2014.

Nilsson, M.E. and Koke, S.C. 2001. Defining treatment-emergent adverse events with the medical dictionary for regulatory activities. *Drug Information Journal*, 35, 1289–1299.

Pharmaceutical Users Software Exchange. 2013. PhUSE standard scripts working group site. Available: http://www.phusewiki.org/wiki/index.php?title= Standard_Scripts. Accessed August 1, 2014.

Rochester, C. G. 2009. Lifecycle planning for safety evaluation. *FDA/Industry Statistics Workshop*, Washington, DC.

U.S. Food and Drug Administration. 2009. Guidance for industry: Drug-induced liver injury: Premarketing clinical evaluation. Available: http://www.fda.gov/ downloads/Drugs/GuidanceComplianceRegulatoryInformation/Guidances/ UCM174090.pdf. Accessed August 1, 2014.

Xia, H.A., Crowe, B.J., Schriver, R.C., Oster, M., and Hall, D.B. 2011. Planning and core analyses for periodic aggregate safety data reviews. *Clinical Trials*, 8, 175–182.

5

Why a DMC Safety Report Differs from a Safety Section Written at the End of the Trial

Mark Schactman and Janet Wittes

CONTENTS

5.1 Introduction

Most Phase 2 or 3 clinical trials are designed to study the efficacy of a new treatment or, for a treatment already in use, its utility for a new application. People rarely enroll in a trial designed to demonstrate that an intervention under study is not harmful; rather, rational participants who understand the informed consent document they have signed trust that the investigator will tell them if the ongoing data show more risks than anticipated at the time they agreed to enter the study. A well-designed protocol specifies the outcomes for measuring efficacy. In safety monitoring, on the other hand, important outcomes are often unknown. A taxonomy of harms provides a structure for thinking about how to monitor for safety. A drug or other intervention has some risks that are known, some that are unexpected but nonserious, others that are unexpected but serious, those that are unexpected and life threatening, and some that though not biologically credible are frightening, if true (Proschan et al. 2006; Crowe et al. 2009; Xia et al. 2011).

Many trials, especially long-term Phase 3 trials and trials that study a serious disease or an especially vulnerable population, have a Data Monitoring

Committee (DMC) attached to them. The DMC, a committee of experts in the disease under study and in the methods of clinical trials, is ideally independent of the sponsor and of the investigators. The DMC is charged with monitoring the safety of the participants during the course of the trial and the integrity of the trial itself (Ellenberg et al. 2002; FDA 2006; Herson 2009). In trials where the investigators are masked to the ongoing data and therefore remain ignorant of emerging benefits and risks, the DMC must evaluate both on an ongoing basis with the view toward informing the participants or recommending changing the protocol or stopping the trial, should the DMC become convinced that risks outweigh the benefits. At intervals during the trial, the typical DMC receives reports on safety and perhaps efficacy data accumulated to date in the trial. In the ideal scenario, a group, again independent of the sponsor and the investigators, prepares the reports for the DMC. We refer to this group as the reporting statistician; others call it the Statistical Data Analysis Center (SDAC) (Fisher et al. 2001) or the independent statistician (DeMets and Fleming 2004). While this structure is the model the FDA Guidance on Data Monitoring Committees (2006) most strongly recommends, other models are possible. Note that while in this chapter we generally use the term reporting statistician in the singular, we stress the importance of having a group, not a single person, preparing the report. Our preference for a group stems from practicalities: (1) the timeline to create the report is short; (2) the data are often dirty, adding to the burden on the programming staff; and (3) new trends may be emerging that are best caught by more than one pair of eyes.

Looking at harms during the course of a clinical trial differs considerably from looking at harms when the trial is over. The data at the end of a trial are complete (or, at least, as complete as they will ever be) with outstanding queries resolved. The team responsible for assessing safety of the experimental product can look at the totality of the evidence not only from the trial at hand but from the drug development program as a whole. They can review the set of adverse events along with their severity and assess whether the event satisfies the regulatory definition of serious. They can summarize laboratory values, vital signs, and concomitant medications that may give more insight into the nature of the harms. Thus, at the end of at trial, the safety team should have the tools and data to make an informed assessment of the safety of the product being studied.

While a trial is ongoing, however, the data are by definition incomplete, and much of the data that are available are not clean. Coding of adverse events is often still in progress. The data arrive at different times; most of the data from wellperforming sites will be timely, but data from other sites will be delayed, with the delay sometimes protracted. In long-term studies, later protocol-defined visits often occur less frequently than earlier visits. For trials where the patients are seen, say, every 6 months, the lag in entering adverse events into the database occurs not because of a lag in data entry, but simply because many events have not yet been reported. Putting together the entire story of an event—the event itself as well as the concurrent laboratory values and concomitant medications—is therefore difficult. It is not usually realistic for the reporting statistician to look at concomitant medications and certain other clues to safety on an ongoing basis. At the end of the trial, the safety team has time to review data and to prepare presentations; during the trial, sponsors often give reporting statisticians a very short time to generate the report for the DMC under the assumption that nothing surprising will have occurred. If the results are surprising, there is often too little time to explore the data fully enough for the DMC to make an informed judgment about safety. The assessment the DMC is making could be based on biases in data collection. For example, events by arm may enter the database at differential rates even if the trial is double masked.

The teams involved in assessing the data differ considerably at the end of the trial from during the trial. At the end of the trial, the safety team should have at its fingertips a complete set of analyses, which it often has not seen previously. The DMC, on the other hand, sees versions of the report at regular or irregular intervals. Moreover, if the reporting statistician is thoughtful, this report is evolving in structure. When the study is young and few adverse events have occurred, the data in the report may be comprised of listings. As the study matures and the quantity of data increases, the listings give way to tables and graphs. A single, simple table may give rise to multiple, complex tables as potential signals emerge. The timing of that transformation depends on the rate of recruitment and the speed of entering the data into the database. The DMC then reviews the same report, with additional data, over and over. If the report is clear, seeing updated versions can give the DMC increasing insight into the safety profile of the drug, but a poorly constructed report can lead the members' eyes to glaze over and not notice changes from one report to another. One of the easiest ways to bury a signal is to include reams of unnecessary and uninformative presentations. A DMC that must review a poorly constructed massive report over and over again may have neither the time nor energy to wade through seemingly endless pages.

At the end of the trial, the study team can assess risks and benefits because the available data speak to both. Early on in the trial, on the other hand, data on efficacy may not have yet emerged, so that judgment about risks and benefits are related to observed risks and anticipated, but as yet unseen, benefits. Many industry sponsors structure their DMCs purely as safety committees. In such cases, DMCs see only adverse events, data from safety laboratory parameters, and concomitant medications. Not even a glimpse at emerging efficacy data is provided. Such a structure complicates assessment of the importance of emerging harms because of the lack of information about benefits.

Still another difference between the data at the end of the trial and the data during it relates to the different intimacy of knowledge of both the drug and the study among those involved at the end of the trial and those during the trial. If the reporting statistician fails to learn about the drug, the report may be insufficiently helpful to the members of the DMC. Similarly, the members of the DMC often review the data quickly, just before the meeting, and they may miss some important aspects of the safety profile. Moreover, what is presented to the DMC is often a small subset of the data collected in the case report forms. A report that contains all the data may overwhelm the DMC members, but the reporting statistician who fails to present relevant data hampers the DMC's ability to make informed judgments about safety.

Unlike final analyses where most of the analyses are prespecified, the reporting statisticians should review their analyses with an eye toward questions that are likely to arise at the DMC meeting. They should prepare additional analyses in anticipation of such questions. Therefore, the analysis plan described in the DMC charter cannot be fixed but is simply a framework for initial reports; the DMCs charter must allow the reporting statistician and DMC flexibility to modify the reports as needed to meet the committee's responsibility for monitoring patient safety.

Many trials have both open and closed reports. The open report, which goes to the sponsor and perhaps the executive committee, includes all, or some of, the tables but with the treatment groups combined. Many open reports present only baseline data, some data on dosing, information on adherence to study drug, and a summary of loss to follow up. Such reports should not provide any information that can give the reader a sense of what is happening either in safety or efficacy. Some open reports, however, present safety data pooled by treatment group. If an adverse event, or cluster of adverse events, occurs that is not typically seen in the control group, the readers of the open report may glean some preliminary information about the presumed safety of the experimental product. Some pooled data may even give a hint (perhaps an incorrect hint) as to the efficacy.

The remainder of this chapter amplifies the discussion of issues relevant to safety reporting for DMCs. We first describe what safety data typically look like, and then we discuss what we consider requirements for the safety section of the DMC report, including recommendations about the style of the report. A separate section includes some suggestions for taming Medical Dictionary for Regulatory Affairs (MedDRA®), with particular reference to reporting for a DMC. We discuss statistics—for example, the use of p-values in these reports and formal boundaries for safety. We then deal with the relationship between the reporting statistician and the DMC. We end with a mention of the FDA's Final Rule on expedited safety reporting and conclusions.

5.2 What Safety Data Look Like during the Course of a Trial

Most information on adverse events in clinical trials arrives as semistructured data. The adverse events case report form has a line for the investigator to write a brief description of the event (see Figure 5.1). On each line, the investigator reports the severity of the event and its relationship to the study drug. The form arrives at the data center where a person or a computer algorithm

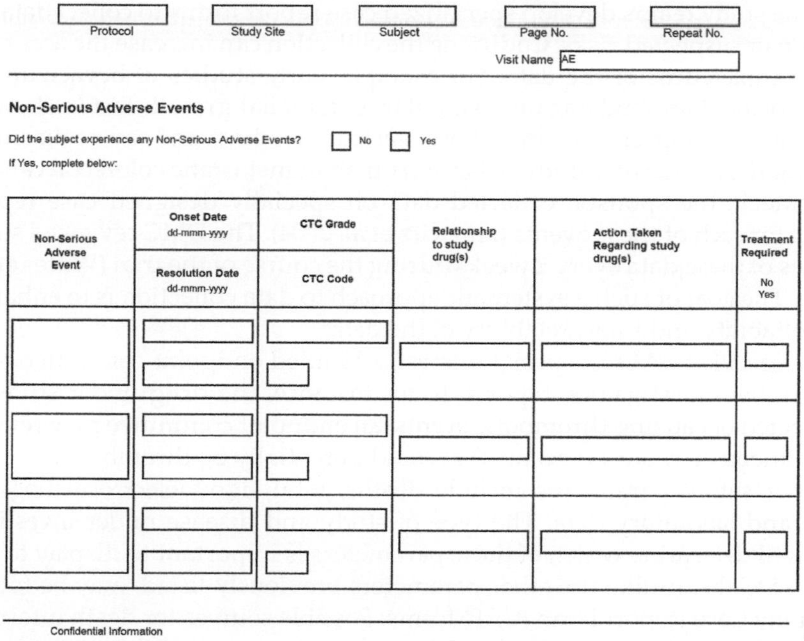

Protocol	Study Site	Subject	Page No.	Repeat No.

Visit Name AE

Non-Serious Adverse Events

Did the subject experience any Non-Serious Adverse Events? ☐ No ☐ Yes

If Yes, complete below:

Non-Serious Adverse Event	Onset Date dd-mmm-yyyy	CTC Grade	Relationship to study drug(s)	Action Taken Regarding study drug(s)	Treatment Required
	Resolution Date dd-mmm-yyyy	CTC Code			No Yes
					☐
					☐
					☐

Confidential Information

FIGURE 5.1
Sample AE CRF.

uses a standard dictionary to code the events. (The system currently accepted by drug regulatory authorities, the MedDRA, has thousands of unique terms.)

These individual reports become the basis for a summary table showing the number of adverse events by body system in the study groups. The typical table simply lists the events and the number of participants who have experienced the event. Usually, the table does not show the relationship or the time of the event with respect to the initiation of therapy. A standard report to the DMC has a table of serious adverse events (SAEs) structured in a way analogous to the adverse event table. For each SAE, the investigator prepares a narrative that describes the event in detail. Some members of DMCs read each of these narratives; some read only those that describe unexpected SAEs; some members do not read any of the narratives. During the course of the trial, the DMC is responsible for looking such tables to identify potential problems the drug may be causing.

One cannot expect classification systems for events as complicated, varied, and idiosyncratic as the possible collective of adverse events to be ideal, but some methods can enhance the quality of these data. For certain events, diaries provide a systematic approach to collection. For example, in trials of vaccines, the participant fills out a daily diary card identifying events they experience in the immediate days after immunization from a list of expected adverse events that often occur at such times.

Some study teams develop specialized case report forms to collect data on known or suspected risks. Codifying the collection can increase the accuracy and completeness of the data. For example, early studies of bevacizumab, a monoclonal antibody against vascular endothelial growth factor, showed evidence of hypertension, thrombosis, serious bleeding, and severe diarrhea. During the course of a study of bevacizumab in metastatic colorectal cancer, Genentech, the sponsor, collected data on specially designed case report forms for each of these events (Hurwitz et al. 2004). The DMC reviewed summaries of these data every 2 weeks during the course of the trial (Wittes et al. 2006). The goal of such a systematic approach to data collection is to enhance the reliability and interpretability of the data.

For specified SAEs of special interest, a blinded endpoint committee may review the investigators' reports. If, for instance, the drug under study is suspected of causing thrombotic events, an endpoint committee may review spontaneously reported events that could potentially be thrombotic.

Other safety parameters include deaths, vital signs, electrocardiogram data, and laboratory data. The type of study and disease under investigation will determine which of these parameters is important to display to the DMC. Deaths, unlike the other parameters previously listed, may be found on many Case Report Form (CRF) forms. Possible sources for death information include a dedicated death CRF, the adverse event CRF, and the treatment or study discontinuation CRF. While interim data are often messy and inconsistent, deaths are unlikely to be reported accidentally (especially if the data include a date of death and a cause). Therefore, the reporting statistician should use all sources of data on deaths. In our experience, many reporting statisticians rely only on the death form, thus undercounting fatalities. Many sponsors specifically tell the reporting statisticians to do just that. Here is an important example where analysis at the end of the trial differs from analysis during the trial: at the end of the trial, the clean data will have a death form for every death, but during the trial, the death forms may be delayed.

5.3 Requirements for the Safety Section of the DMC Report

When preparing a report for a DMC, reporting statisticians should put themselves in the mind of the DMC members—a hard task for many of us statisticians given that most DMC members are clinicians. The report should tell a story; it must be inviting so that the members will actually read it thoroughly. The order of materials needs to have a logical flow. Long tables with little information of interest, if included at all, should be relegated to appendices. Remember that if the report is for an in-person meeting, the members are likely to read the report on the plane or train to the meeting; therefore, reporting statisticians should structure the report in a way that allows the person with the shortest trip to grasp the important aspects of the report. When the meeting is a teleconference, some of the members will not have read the report at all prior to the meeting. Thus, it is essential that the report be extremely clear. We hear people sighing when they read this. Some of you are saying, "But I always spend as much time as I need to review the report." We know that you, our careful readers, always spend adequate time reviewing material sent to you, but we are speaking from experience being the reporting statisticians for many trials. While many members of DMCs do in fact read the reports carefully and come to meetings prepared to discuss what they have seen, others are less conscientious. Reporting statisticians can encourage full participation by preparing a report that is easy to read. If we are wrong, and all the members spend an adequate amount of time on the report, members will be grateful for a report structured logically with the important material in the first pages.

We know that recently many companies acting as reporting statisticians put the report on a protected website, rather than sending the report to the members. Our own procedure is to send hard copies of reports because when we have not, we have noticed that some members of committees have not even looked at the web-based materials.

Even if the DMC is charged with reviewing only safety, the report should include information on recruitment, baseline characteristics of the study population, adherence to study medication, follow-up rates, deaths, and adverse events with particular emphasis on SAEs.

The reporting statistician should be aware that some members of DMCs will not understand regulatory definitions of SAEs and will find the distinction between seriousness and severity confusing. Also, members who have been involved in oncology trials or trials studying HIV/AIDS will be more familiar reviewing adverse events by Common Terminology Criteria for Adverse Events (CTCAE, version 4.03, NIH 2010). Reporting statisticians should learn the preferences of the members and prepare data in a way that will facilitate the members' review.

Presentations of baseline data provide the DMC with a snapshot of the patient population and inform their interpretation of the safety data. Recruitment, drug adherence, and follow-up presentations allow the DMC to assess whether the trial is well run. For example, if recruitment is extremely slow and does not accelerate, the trial may never come to completion. In such a trial, patients are potentially being harmed by exposure to a study drug without the likelihood of the trial ever assessing the effectiveness of the treatment.

The reporting statistician must decide whether the report is going to be primarily graphical, primarily tabular, or a mixture of the two. We prefer reports with a mixture of types of presentations; certain kinds of questions lend themselves to graphs while others lend themselves to tables. Readers of reports have individual preferences; some prefer graphs and some prefer tables. Remember that the purpose of the report is to facilitate the ability of the DMC members to review the data effectively and efficiently. We recommend that reporting statisticians adjust their presentations to the particular committee's and particular members' preferences. Often, clinicians have a hard time interpreting graphs and tables that appear transparent to us statisticians. Volcano plots (described in the next paragraph), shift tables, and cumulative incidence plots are three examples of presentations that many members of DMCs will not understand at first exposure to them. Reporting statisticians must be able and willing to explain these presentations clearly and must understand what denominators are appropriate.

The heart of the typical DMC report is a set of tables of adverse events by system organ class (SOC) and preferred term (PT). These tables, especially in large-scale trials or in trials of the elderly or the very ill, can be so lengthy that they obscure signals. Presenting too much data, or presenting data with false precision, is a method for hiding signals in plain sight. The reporting statistician has the duty to review these tables before sending the report to the DMC. The first task is to group terms that represent the same, or nearly indistinguishable, outcomes. In looking at harms, the reporting statistician, and the DMC, may identify a specific event that requires further attention or a group of such events. Grouping AEs to construct a meaningful medical concept is not part of most statistician's expertise, so the exercise usually requires some clinician input. Sponsors may help facilitate this for prespecified adverse events of interest, but for unexpected events during the trial, reporting statisticians may need help from clinical and safety disciplines; either from their organization or from the clinicians on the DMC. In some trials comparing an active treatment to control, no particular event or cluster stands out, but overall the treated group appears to have more adverse events than the control. The volcano plot is a display that is often helpful as an adjunct to the standard adverse event table. This type of plot, originally developed for microarray data (Cui and Churchill 2003; Li 2012), can display risk differences on the x-axis and p-values on the y-axis. (We like to label the y-axis with p-values but scale it as $-\log_{10}$(p-value).) The figure highlights specific differences that may require more investigation and an overall picture of the adverse event profile. In Figure 5.2, we have added a label to any higher-level group term (HLGT) that is worse in Group B (right side of the graph) and with a p-value of < 0.01 (to partition the graph into areas of interest). The cutoff of 0.01 for the p-value is arbitrary; if the study were small, one might choose a cutoff of 0.05, 0.1 or even 0.2. In the case at hand, the distribution of adverse events appears to be similar between the groups as the number of events (dots) and the dispersion (risk differences) below the reference line of $p = 0.01$ both appear similar; however, a few conditions occur convincingly more frequently in Group B. Such a graph pinpoints areas that demand more careful and thorough investigation from the reporting statistician, ideally prior to the DMC meeting.

The traditional safety report presents the tables in terms of the proportion of participants who experience a specific event or cluster of events.

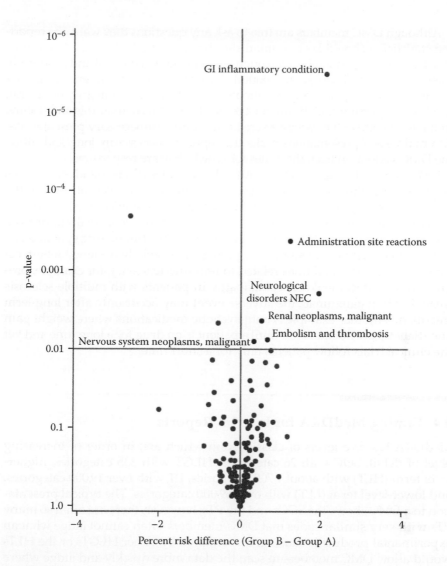

FIGURE 5.2
Volcano plot of adverse events by HLGT.

In long-term studies, however, if a drug is effective, people may stay on it longer than they remain on the control arm. In such cases, comparing annualized rates relative to time of exposure may be more relevant. In case where the adverse events occur most frequently at the beginning of exposure, piecewise rates (e.g., with the first 30 days and after 30 days) might be more informative. Depending on the type of metric chosen, the volcano plots may present not risk difference on the x-axis, but another measure such as hazard ratio, risk ratio, or difference between annualized event rates.

Although DMC members are free to ask any questions they want, the reporting statistician should keep in mind the charge of the DMC as outlined in its charter. Committee members are often very engaged in the trial and pose many interesting scientific questions about it. That said, not all the questions are relevant to protecting the safety of patients in the study or the integrity of the trial itself. The reporting statistician must remind the DMC of their role and decline (politely, of course) the request to create additional unnecessary presentations. Too many such presentations make the report unnecessarily long and divert the DMC's attention from the issues for which they are responsible.

DMCs cannot be expected to identify all types of SAEs. Some adverse events do not emerge during the clinical program. Several reasons contribute to the delay in recognizing these harms. Many times each clinical trial, or indeed the entire development program, is too small for rare events to occur or to be noticed. In those cases, the events may emerge in postmarketing spontaneous reports, in observational studies, or perhaps in specially designed long-term safety studies. For conditions related to immunodeficiency, for example, progressive multifocal leukoencephalopathy in patients with multiple sclerosis treated with natalizumab, the adverse event may occur only after long-term treatment. Another example is antipsychotic medications where weight gain and diabetes are not manifest until a patient is on drug for a long time and yet the clinical trials follow patients only for a short time.

5.4 Taming MedDRA for Interim Reports

MedDRA has five levels of classification which are, in order of increasing level of detail, SOC with 26 categories, HLGT with 335 categories, higher-level term (HLT) with about 1,700 categories, PT with over 19,000 categories, and lower-level term (LLT) with over 69,000 categories. The typical presentation to a DMC shows the SOCs and the PTs; however, there are often so many PTs with very similar codes that DMC members often cannot judge what an experimental product is causing. Presentation by either HLGTs or the HLTs would allow DMC members to scan the data more quickly and judge where imbalances exist.

MedDRA also provides Standardized MedDRA Queries (SMQs) to help summarize the data. SMQs may pull PTs from one or more SOCs. One example is cardiac failure, whose codes may occur both in the cardiac SOC as "cardiac failure" or some variant thereof and the respiratory SOC as "pulmonary edema." Bleeding is another important SMQ because people may experience bleeds in a variety of organ classes. Many SMQs have narrow or broad definitions with the latter necessarily including more PTs.

Sometimes, a drug causes not a single type of adverse event but many different types. All in all, the patients in the treated group seem to have more adverse experiences than the patients in the control group. This is hardly surprising, for a drug may affect many organs. Suppose in a trial of prevention of renal failure in type 2 diabetics, no particular adverse event occurred statistically significantly more frequently in the treated group, but overall, the treated group experienced more pulmonary, renal, hepatic, and coronary adverse events. A DMC might informally weigh the overall adverse impact of the drug against the potential gain and recommended stopping the trial viewing that a drug leading to so many adverse events would not be useful.

MedDRA also classifies what it calls "multiaxiality," or secondary SOCs. This system links a specific PT to more than one SOC. For example, the primary SOC for the PT "enterocolitis infectious" is "infections and infestations," but its secondary SOC is "gastrointestinal disorders." On the other hand, the primary SOC of the PT "enterocolitis" is "gastrointestinal disorders." Thus, if a drug caused enterocolitis, a reporting statistician who neglected the secondary SOC would undercount the number of events. In the absence of a classification committee, it is impossible to know whether an event the investigator calls "enterocolitis" is infectious or not. We contend that the reporting statistician who lumps such categories is more likely to identify a true signal than the statistician who relies on the narrow coding of the investigator's report.

Standard presentations of adverse events provide the number and percentage of patients experiencing each adverse events at the PT and SOC levels. This approach is often not useful as the PT may be too narrow and the SOC too broad. Consider the following portion of the neoplasm section of a SAE table. A reviewer of such a presentation would have a hard time determining if the data suggest harms. Overall, the two groups experience approximately the same number of events (85 in the Husky group and 91 in the Collie group), but for some PTs, the numbers differ considerably. Similar terms appear on different rows (e.g., colon cancer and colon neoplasm), and it is unclear whether these represent the same patients. (Note that we used the terms "Husky" and "Collie" rather than A and B to identify groups. In our experience, people do not remember from meeting to meeting what Groups A and B looked like, but the use of a label with a shape or color or place enhances the ability of the members to recall what they are seeing from table to table and what they had seen in previous reports.)

Now, consider a revised version of the table that uses MedDRA's HLGT. The picture here is clearer. Classifying the terms in a logical structure (e.g., terms grouping colon cancer and colon neoplasm under the same header) allows the reader to identify potentially important imbalances by treatment group.

A DMC should not rely solely on coding systems; rather, it should feel free to classify and reclassify to identify clusters of similar events. Conventional categorization by body system may underemphasize certain events that occur across systems, for example, bleeding or thrombotic events or pain, and SMQs might not capture all such events. Consider the following restructured table. Removing the known benign neoplasms (e.g., the uterine leiomyomas) and the basal cell carcinomas and then lumping together like terms make the table simpler, shorter, and more informative than the tables based on either PT or HLGT. A DMC looking at the SOC/PT version of the table, the type it typically reviews, would likely conclude that there is no difference in cancer between the Husky and Collie groups (Table 5.1). The DMC with the HLGT/PT table might worry about the apparent excess of gastrointestinal cancers in the Collie group but would fret about the splitting of terms like "colon cancer" from "colon neoplasm" (Table 5.2). Further, it might not notice that the equality in the skin category was due in part to the 9–6 split for basal cell carcinoma. A DMC whose report included Table 5.3 would likely decide to monitor colon and rectal cancer closely.

TABLE 5.1

Neoplasms by PT

PT	Husky N = 284 n (%)	Collie N = 292 n (%)
Neoplasms	85 (30)	91 (31)
Basal cell carcinoma	9 (3)	6 (2)
Prostate cancer	7 (2)	3 (1)
Thyroid neoplasm	6 (2)	4 (1)
Breast cancer	2 (1)	5 (2)
Colon adenoma	5 (2)	2 (1)
Lipoma	2 (1)	4 (1)
Malignant melanoma	2 (1)	4 (1)
Colon cancer	2 (1)	3 (1)
Rectal cancer	1 (< 1)	3 (1)
Renal neoplasm	2 (1)	2 (1)
Squamous cell carcinoma	3 (1)	1 (< 1)
Thyroid adenoma	0 (0)	4 (1)
Transitional cell carcinoma	3 (1)	1 (< 1)
Benign neoplasm of thyroid gland	1 (< 1)	2 (1)
Colon neoplasm	1 (< 1)	2 (1)
Lung neoplasm	1 (< 1)	2 (1)
Multiple myeloma	3 (1)	0 (0)
Non-Hodgkin's lymphoma	2 (1)	1 (< 1)
Non–small cell lung cancer	1 (< 1)	2 (1)
Pancreatic carcinoma	0 (0)	3 (1)
Uterine leiomyoma	1 (< 1)	2 (1)
Adrenal adenoma	1 (< 1)	1 (< 1)
Adrenal neoplasm	1 (< 1)	1 (< 1)
Benign colonic neoplasm	1 (< 1)	1 (< 1)
Bowen's disease	0 (0)	2 (1)
Hepatic neoplasm malignant	0 (0)	2 (1)
Lung adenocarcinoma	1 (< 1)	1 (< 1)
Lung adenocarcinoma metastatic	2 (1)	0 (0)

TABLE 5.2

Neoplasms by HLGT and PT v02

MedDRA HLGT/PT	Husky N = 284 n (%)	Collie N = 292 n (%)
Gastrointestinal Neoplasms, Malignant and Unspecified	**9 (3)**	**19 (7)**
Colon cancer	2 (1)	3 (1)
Rectal cancer	1 (< 1)	3 (1)
Colon neoplasm	1 (< 1)	2 (1)
Pancreatic carcinoma	0 (0)	3 (1)
Pancreatic neoplasm	0 (0)	2 (1)
Rectal neoplasm	1 (< 1)	1 (< 1)
Adenocarcinoma pancreas	1 (< 1)	0 (0)
Colon cancer stage II	0 (0)	1 (< 1)
Gastric neoplasm	1 (< 1)	0 (0)
Esophageal adenocarcinoma	0 (0)	1 (< 1)
Esophageal adenocarcinoma metastatic	0 (0)	1 (< 1)
Esophageal carcinoma	0 (0)	1 (< 1)
Pancreatic carcinoma metastatic	1 (< 1)	0 (0)
Rectal cancer metastatic	0 (0)	1 (< 1)
Rectal cancer stage II	0 (0)	1 (< 1)
Tongue neoplasm malignant stage unspecified	1 (< 1)	0 (0)
Skin Neoplasms, Malignant and Unspecified	**11 (4)**	**12 (4)**
Basal cell carcinoma	9 (3)	6 (2)
Malignant melanoma	2 (1)	4 (1)
Bowen's disease	0 (0)	2 (1)
Lentigo maligna	0 (0)	1 (< 1)
Skin cancer	0 (0)	1 (< 1)
Respiratory and Mediastinal Neoplasms, Malignant and Unspecified	**10 (4)**	**6 (2)**
Lung neoplasm	1 (< 1)	2 (1)
Non–small cell lung cancer	1 (< 1)	2 (1)
Lung adenocarcinoma	1 (< 1)	1 (< 1)

The example earlier posed a relatively easy problem because the events that the PTs classified as separate events so clearly belong together. In other cases, the way to reclassify is less clear. One useful approach to lumping requires the reporting statistician to lump on the basis of the pooled data, not the treatment-specific data, in order that the lumping is performed in an unbiased way. When it becomes clear that data need to be lumped, the reporting statistician can send a list of terms to the member of the DMC most knowledgeable about that cluster of events. If there is no relevant expert on the DMC, then the statistician can send the list to an expert and ask

TABLE 5.3

Gastrointestinal Neoplasm by Study-Specific Coding

	Husky	Collie
	N = 284	N = 292
Study-Specific Classification	n (%)	n (%)
Gastrointestinal Neoplasms, Malignant and Unspecified	9 (3)	19 (7)
Colon and rectal cancer	5 (2)	12 (4)
Pancreas	2 (1)	5 (2)
Esophagus	0 (0)	3 (1)
Tongue	1 (< 1)	0 (0)

for a categorization. One cannot expect perfection in the categorizations; one is looking only to increase the ability to detect a signal if one is there.

Clearly, there are advantages to lumping and splitting: lumping loses specificity but splitting loses sensitivity. Our strong belief is that in the context of clinical trials, when investigators report adverse events on a form constructed as in Figure 5.1, they are not thinking about MedDRA classification. One person might write "myocardial infarction" while another might write "acute myocardial infarction." To MedDRA, these are two different PTs so the splitter would separate them. In the absence of an adjudication committee, we are convinced that the lumper more correctly categorizes most events.

5.5 What about Statistics, Formal, or Informal

Monitoring for safety presents statistically difficult problems. In looking for safety signals, the DMC searches for the unknown, the rare unexpected event. Problems of multiplicity abound—many outcomes and many looks conspire to muddle the sample space and therefore make probabilities ill defined. While one can, and should, specify precisely the number of outcomes to evaluate for efficacy, by definition one cannot specify the number of hypotheses relevant to safety. Instead, the DMC must remain alert to react to surprises, turning a fundamentally hypothesis-generating ("data dredging") exercise into somewhat of a hypothesis-testing framework.

While DMCs typically have a statistician who represents the statistical view of the DMC, the reporting statistician should ensure that all DMC members understand the analyses performed. The reporting statistician should point out imbalances in the data or areas of the report on which to focus, but should refrain, if possible, from expressing an opinion about what the data are saying. Think of Dorothy Parker's "You can take a horticulture, but you can't make her think" (as quoted by Keats 1970). Or, if you are the reporting statistician, think of bringing the DMC to water but letting them take the drink.

The DMC, for its part, should consider the totality of the data and not focus on a single result or p-value. DMC members often ask for a p-value on a specific piece of data. For example, in a trial of roughly 100 participants per group, the DMC notices that 15% of active subjects experience a severe headache while only 5% of placebo subjects do. If asked, should the reporting statistician provide the p-value? Refusal to calculate a p-value is silly; it is not a magic number. In a trial of 1:1 randomization, let r_t and r_c be the number of events in the treated and control group, respectively. The statistician on the committee as well as some of the more statistically sophisticated clinicians will know that the formula $(r_t - r_c) / \sqrt{(r_t + r_c)}$ will give a reasonable approximation to z. Thus, if the reporting statistician says something like "Because of multiplicity and lack of prespecification, the p-value cannot be interpreted so I won't present it," someone on the committee will say, "10 over the square root of 20 is over 2 so this p-value is less than 0.05." Far better for the reporting statistician to respond with the p-value and let the committee discuss the import of the data. Someone on the committee may point out the problems with p-values calculated with no thought about prespecification or multiplicity. If no committee member raises a concern, the reporting statistician should do so.

A significant p-value does not necessarily reflect a true harm. After all, if p-values are calculated for each line in an adverse event table, some imbalances are expected to be statistically significant by chance. If the events were independent, then 1 in 20 of all lines with at least 6 events total would be expected to have a two-sided p-value less than 0.05. (In a large two-group study randomized 1:1, if there are fewer than six patients experiencing a particular type of event, then no configuration, no matter how extreme the allocation, can have a p-value less than 0.05.) But of course not all events are independent, so it is impossible to calculate the expected number of nominally significant p-values.

Importantly, a nonsignificant p-value does not necessarily mean that the drug did not cause the event. To take an extreme example, consider a small trial comparing some protein-based product to a saline placebo. If one participant in the trial suffers an anaphylactic reaction, the one-sided p-value is 0.5, hardly an indication of product-caused harm. But here, biology trumps statistics—what is relevant is the mechanism of action, not the p-value.

The DMC has several tools for monitoring safety. For adverse events the drug under study is known to cause, the DMC can use the emerging data to estimate the rates with the purpose of ensuring that the rate is not unacceptably higher than previously thought. By analogy with the approach it takes for efficacy, it can set a statistical boundary for safety. Crossing the boundary means that, with respect to this particular adverse outcome, the data have shown convincing evidence that the incidence rate is higher than expected. Some DMCs use futility bounds for efficacy to act as safety bounds.

They view participation in a trial studying an inefficacious product unsafe because the participants are exposed to risks without benefit.

Some DMCs establish a boundary more extreme than futility but less extreme than the symmetric bound. Still, another attractive theoretical approach is to define an *a priori* balance of risk and benefit; if the ongoing data show that the balance has changed importantly in the direction of excess risk, the DMC may recommend stopping the trial for safety. If the emerging event was previously unreported, the DMC may hypothesize an excess and use the remainder of the trial to test whether the excess is real (Wittes et al. 2004).

To fix ideas, consider a two-armed trial in which the treatment group is showing a slight excess in pulmonary emboli, a blood clot that enters the lung and may cause death. If the excess is real, then at the very least, the DMC should inform the investigators of the increased risk; if not real, then a warning sends an unnecessarily worrisome message. In deciding how to act, the difference between type I and type II errors emerges starkly. Committing the type I error—declaring an adverse event so bad that the study should stop when in fact the excess occurred by chance—can destroy a promising drug. On the other hand, a DMC that commits a type II error—ignoring a SAE—can harm the participants in the trial. Confronted with what seems like a signal, in this case an excess rate of pulmonary embolism, the DMC should first take measures to enhance the signal. Summarizing all thrombotic (clotting) events would increase the number of events; observing more total thrombotic events in the treated group would provide suggestive evidence that the observed excess in pulmonary embolism reflects a real effect. On the other hand, if biologically related events occurred more frequently in the control group, then the increase in pulmonary emboli might be simply noise. Next, the DMC should consider any laboratory data, such as clotting factors, which provide insight into the event. Think of the observation as a mystery to solve in real time—the DMC's responsibility is to identify an adverse event if it is real and not identify one if it is not real.

Unknown, but serious, risks that emerge during the course of a study may lead to anxious discussion within the DMC. If the signal is real, the DMC asks itself whether the possible—but as yet undemonstrated—benefit outweighs the risks. Especially troubling are unexpected life-threatening SAEs. For example, in a trial aimed to prevent myocardial infarction, a DMC that observes a small excess of stroke in the treated arm may take immediate action. It may ask for an expert on stroke to join the committee; it may ask for a special data collection instrument to enhance the accuracy of spontaneously reported stroke; or, if the risk seems unacceptably high compared to the benefit seen or hypothesized for myocardial infarction, it may recommend stopping the trial.

Two artificial examples point to the problem and provide some suggestions of approaches to dealing with such data. The first comes from an unblinded trial on heart failure testing whether the new treatment decreased mortality relative to standard therapy. Early in the trial, the DMC observed excess mortality in the treated group. The committee, although scheduled to meet every 6 months, asked for a safety update 3 months after its first meeting. At that time, the excess mortality in the treated group was even more pronounced. Worried that the unblinded nature of the study might have led the investigators to follow the treated group more intensively and therefore be more quickly aware of the deaths in the treated group, the DMC asked the investigators to determine the vital status of each participant on a specific date. To preserve the integrity of the study, the DMC did not describe why it wanted the data—it simply said that it could not responsibly monitor data without accurate information on mortality. The investigators then reported many more deaths; in fact, more deaths occurred in the control group. The DMC reacted to an apparent risk but it recognized the potential bias in the design and rather than stop the trial prematurely, it asked for rapid collection of relevant data.

In another example, a DMC may be monitoring a study on quality of life in patients with breast cancer. The treatment might be showing clear benefit on symptoms but excess early mortality in the treated group (20 deaths in the treated group and 8 in the control group, nominal p-value, 0.025). The DMC, recognizing that this difference in mortality could have occurred by chance, might ask for more information to help explain the excess mortality. If the sponsor were unable to give the DMC rapid accurate information on baseline factors, the DMC might feel it had no choice but to recommend stopping the study.

Finally, sometimes, an event occurs that is medically not credible, but, if true, devastating. Even a very low rate (a single event, perhaps) of fatal liver failure may lead to stopping the trial (e.g., troglitazone [DCCT Research Group 1987]). Here, judgment of the clinician members of the committee often guides decision making.

Thus far, we have focused on individual events, but typically, the DMC is confronted with a list of adverse events or SAEs as described earlier. Assuming that the reporting statistician has presented the data in an interpretable manner, lumping and splitting where appropriate, the DMCs scan the tables and figures to identify events occurring at increased rates in the experimental arm. Faced with a long list of events, the DMC needs some tools to separate the wheat (i.e., the true signals) from the chaff. Statisticians' usual armamentarium, p-values and confidence intervals, are not always useful. Methods described by Mehrotra and Heyse (2004), Crowe et al. (2009), and Xia et al. (2011), while often helpful at the end of a trial, may be unrealistic to implement during the course of a trial, both because the sample size is too small and because of problem of multiplicity. Some DMCs like to see p-values at each line of a table of adverse events; we worry that small p-values may be arising by chance while large p-values may give false comfort. To fix ideas, suppose that three participants have severe liver failure in the treated group and no such event occurs in the control. Assuming 1:1 allocation to treated and control, the one-sided p-value is 1/8 or 0.125, hardly an extreme probability. If, however, severe liver failure is expected hardly ever to occur in a population of the size and composition of the study group, then three events should worry a DMC. Thus, the judgment of whether a DMC should become concerned, and the assessment of the level of that concern, needs to be based not only on statistical considerations from the trial itself, but on clinical and epidemiological knowledge as well. The DMC should remember that its recommendations need not be binary. If it seems something worrisome, it can express that worry without recommending stopping the trial.

5.6 Interaction between the DMC and Sponsor

Having reviewed the ongoing data from a trial and having come to a conclusion about whether or not the protocol needs to be changed, the DMC must convey its recommendations to the sponsor of the trial. (Often, the recommendations go to the executive committee or to both the sponsor and the executive committee; in this section, we use the term "sponsor" to refer to the person or group to whom the DMC reports.) For many trials, the reporting statistician serves as the link between the DMC and sponsor. Using the reporting statistician as the bridge between the two groups helps maintain the integrity of the trial. Sponsors often ask questions about the trial that would, if answered, provide inappropriate clues about the ongoing data, and many DMCs, in an effort to be cooperative, answer questions that come dangerously close to revealing information that should remain masked. The reporting statistician has generally developed a relationship with the sponsor because of the many interactions related to timing of data transmission and understanding specific variables. Thus, the topics that are off limits have been clearly delineated between them. Having the reporting statistician serve as the link makes unmasking data from the DMC less likely to reach the sponsor.

The chair of the DMC is usually responsible for a brief communication with the same designated person at the sponsor's company or with the chair of the executive committee. This communication is sometimes limited to reading from a checkbox on a predefined form (see Figure 5.3); however, often the DMC wishes to communicate other information. For instance, the DMC might want to comment on recruitment, adherence, speed of data reporting, quality of data, or other aspects related to the operations of the trial. Whether the chair of the DMC conveys these comments or whether the reporting statistician is the conduit for them will depend on the sociology of the particular DMC.

<SPONSOR> <protocol number>
Data Monitoring Committee Recommendations
Meeting date: <Date XX, 2014>
CONFIDENTIAL

The Data Monitoring Committee for the <protocol number> study convened on <Date XX, 2014>. After reviewing data for the study, which included baseline characteristics, adverse events, serious adverse events, deaths, and clinical laboratory parameters, the Committee agreed:

☐ The study should continue unchanged.

☐ The study should be modified or terminated (specify details and rationale below).

Details and rationale for changes in study conduct:

_____   _____

<DMC Chair>   Date signed

FIGURE 5.3
Sample recommendation form.

5.7 Final Rule for Safety Reporting and the Associated FDA Guidance

In 2010, the U.S. Congress amended the legislation concerning the reporting of serious, unexpected, related adverse events (U.S. Code of Federal Regulations) by promulgating a "Final Rule." In 2012, the FDA issued a new guidance related to such reporting (FDA CDER 2012). This section first describes the process of expedited safety reporting prior to the Final Rule and then discusses the rule's implications.

Although one of the most important roles of the DMC is to report its ongoing findings when the data are strong enough to warrant a change in protocol or a modification in the informed consent document, DMCs have traditionally been reluctant to report findings unless members are confident that the observed excess of SAEs reflects true harm. The threshold for reporting differs depending on whether the adverse event is already known, and the data suggest that the rate is higher than previously thought or whether the risk is one that is newly identified.

For known risks, the role of safety monitoring is to ensure that during the course of the trial, the balance of risk and benefit continues to favor benefit. Even if the known risks are quite serious, the DMC may well recommend continuation of the trial if it is convinced that the participants are being well cared for and if they have been adequately informed of the risks they are incurring. For example, in a trial studying cancer where the drug under study is known to cause hand–foot syndrome, the DMC may ask whether the rates of progression and mortality are higher in the treated or the control group. As long as the excess hand–foot syndrome does not translate into excess mortality, the DMC may not ask for any change in protocol or any modification to the process of informed consent. Appealing to statistical significance in this type of situation is rather silly, for the treatment is known to cause the events. We have been involved in trials studying drugs with known risk and seen DMCs not worry about an emerging risk because it was not "statistically significant." Remember that statistical significance is the probability, under the "null" hypothesis, that the event or something more extreme would have occurred by chance. If we already know that a treatment causes some adverse event, we do not need to accumulate data to reject an artificial null hypothesis.

Lack of statistical evidence of effect within a particular trial in the case of an adverse event known to be related to the activity of the drug is not convincing evidence of no effect; it may simply be the consequence of testing the wrong hypothesis (the null) with inadequate power. We have seen many DMCs lulling themselves into complacency by improperly monitoring against a null hypothesis. The danger is especially important when the event is rare, but serious. Suppose one is studying a drug known to increase the probability of developing a blood clot and imagine that after 2000 patients have been followed for 3 months, five strokes have occurred in the treatment arm and none in the control. In such a situation, we have seen DMCs argue as follows. "The two-sided probability of a five to zero split is 0.06, not even reaching the conventional level of statistical significance. Because we are observing the data on multiple occasions, a difference this large may well occur by chance. Therefore, while we shall continue to monitor stroke, this finding has not raised our level of concern." Instead, we believe that a five to zero split in the situation of a drug known to cause clots is strong evidence of harm. Whether the harm is sufficient to lead to action is a different question. If the physicians discount data on known adverse events because the excess in the treated group has not yet reached statistical significance, the statistician needs to remind the DMC that its role is to recommend changing the informed consent form, modifying the protocol, or even stopping the study if the evidence of risk is high enough, even if not formally statistically significant.

Thus, the question the committee often faces is, "What rate is 'too high'"? Standard methods can apply when the null hypothesis is that the difference between the two groups is Δ; in this case, we reject the hypothesis if the data show a statistically significant difference above Δ. To fix ideas, suppose a study of idiopathic pulmonary hypertension is comparing a new treatment to placebo; the endpoint is the number of meters walked on a treadmill at 6 months (dead people scored as zero). The investigators say they would be comfortable using the new therapy if the excess number of people experiencing serious bleeds is below, say, 5%. The committee may establish for itself a guideline that would reject the hypothesis if the data crossed the boundary defined by a null hypothesis of 5%. Note that the threshold for an acceptable Δ depends on the efficacy endpoint of the study. If the study were designed to show a decrease in mortality, the investigators might well allow a large Δ. Similarly, if the study were to prevent events in a healthy population, the DMC might select a smaller Δ.

While arguments from biological plausibility aid in the interpretation of known risks and of risks associated with known mechanisms of action of the drug, observed unexpected events in the treated arm are more difficult to interpret. From a regulatory standpoint, in the United States, a sponsor of a clinical trial is responsible to report quickly to the FDA every adverse event that has the three attributes of "serious," "related," and "unanticipated." "Serious" in the regulatory context means that the event led to hospitalization or to increased length of hospital stay or that the event was life threatening or caused death. "Related" in the past had meant that the investigator deemed the drug was at least probably related to the event (in some settings, the definition was at least possibly related; sometimes, the event must have been definitely related). Much of the decision was up to the judgment of the investigator. In masked trials, the typical investigator rates events in the placebo group related as frequently as they do events in the treated group. For that reason, DMCs usually disregard the attribution of relatedness and consider, instead, the totality of all reported events.

An unanticipated event is one not known to be related to the process of the disease; an unexpected event is one that is not known to be related to the drug. It is these latter that the DMC regards with particular scrutiny because they may lead to a change in protocol or to the informed consent form.

Under the new FDA guidance (FDA CDER 2012), no longer does the investigator have the primary responsibility for assessing relatedness; the responsibility is now the sponsor's. For a rare unexpected event, especially one that like Stevens–Johnson syndrome, anaphylaxis, or tendon rupture that are commonly associated with drugs, the FDA is asking the sponsor to unmask the cases. If the sponsor judges that it is reasonably likely that the event has been caused by the experimental product, the sponsor should report the event to the FDA in an expedited manner. While some DMCs may take offence at the fact that the sponsor will unmask certain events, there will be very few of such unmaskings so that the integrity of the trial will not be affected but relevant data about the safety of the product will become known.

A more difficult problem is the occurrence of an increased rate of an antici-pated adverse event. Consider a clinical trial that is studying a population of mean age 62 years. Strokes or heart attacks or prostate cancers that occur in such a population would be anticipated because these are events that typically occur in population of this age. During the course of a trial, the DMC is the only group that can identify if these events are likely related to the product under study. It would not be able to tell whether a specific event is attribut-able to the experimental product, but if the rate of these events is higher in the treated than in the control group, the DMC may attribute causality. In order for the sponsor to satisfy the new guidance, the DMC would have to report an excess to the sponsor. This new process will represent a major change to the way that most DMCs have operated in the past. One approach to recogniz-ing such an event without overreporting is to use the sentinel event methods described by Lachenbruch and Wittes (2007). This time, the process would be as follows. The DMC, on identifying a possible increase in the treated group of the rate of a serious, unexpected, adverse event, would define that event (or the cluster of similar events) as the sentinel. It would then monitor the future events under what has become prespecified hypothesis and, if the future data continue to show an excess, it would report that excess to the sponsor.

5.8 Conclusion

Statistical issues posed by monitoring for safety differ considerably from issues posed by monitoring for efficacy. In the latter case, prior hypotheses govern both the design of the study and the consequent boundaries for monitoring data. For safety monitoring, however, the hypotheses are often data driven. For expected adverse events, the DMC's role should be to estimate the rates of events ensuring that they are not unacceptably high. For unexpected events, a DMC risks reacting to a falsely "discovered" endpoint (Benyamini and Hochberg 1995; Mehrotra and Heyse 2004) for the event may have occurred in the treatment arm purely by chance. Therefore, DMCs typically wish to dampen a rush to judgment about the product or intervention. In prevention trials among healthy volunteers or in trials of diseases or conditions not usu-ally accompanied by many types of SAEs (e.g., relief of minor headache pain), one analysis might compare the total number of adverse events to a standard rate known from historical data or to the control. Identification of sentinel events during the course of the trial coupled with nimble statistical guidelines for subsequent monitoring afford a DMC the opportunity to react to unantici-pated events at the same time as protecting against overzealous worry.

Especially for safety, a DMC should not be bound by purely statistical considerations. It may recommend modification or termination of a study when no statistical difference in safety can be shown because of the medical implications of an adverse event. On the other hand, it may recommend continuation of a study, even when it is reasonably sure that the experimental treatment is causing SAEs, if it believes the benefit will ultimately outweigh the risk. Now that the FDA has promulgated a new guidance on expedited safety monitoring, DMCs should discuss with sponsors the mechanism for reporting excess rates of anticipated (i.e., known to occur to the population being studied) but serious unexpected (i.e., not known to occur to participants receiving the study drug) related adverse events.

Acknowledgment

MedDRA® the Medical Dictionary for Regulatory Activities terminology is the international medical terminology developed under the auspices of the International Conference on Harmonization (ICH) of Technical Requirements for Registration of Pharmaceuticals for Human Use 2012.

References

Benyamini Y, Hochberg Y. 1995. Controlling the false discovery rate: A practical and powerful approach to multiple testing. *Journal of the Royal Statistical Society Series B* 57: 289–300.

Crowe BJ, Xia HA, Berlin JA, Watson DJ, Shi H, Lin SL. 2009. Recommendations for safety planning, data collection, evaluation and reporting during drug, biologic and vaccine development: A report of the safety planning, evaluation, and reporting team. *Clinical Trials* 6(5): 430–440.

Cui X, Churchill GA. 2003. Statistical tests for differential expression in cDNA microarray experiments. *Genome Biology* 4(4): 210. doi: 10.1186/gb-2003-4-4-210.

DCCT Research Group. 1987. Diabetes control and complications trial (DCCT): Results of feasibility study. *Diabetes Care* 10: 1–19.

DeMets DL, Fleming TR. 2004. The independent statistician. *Statistics in Medicine* 23: 1513–1517.

Ellenberg S, Fleming T, DeMets DL. 2002. *Data Monitoring Committees in Clinical Trials: A Practical Perspective.* Wiley, West Sussex, U.K.

Fisher MR, Roecker EB, DeMets DL. 2001. The role of an independent statistical analysis center in the industry-modified National Institutes of Health model. *Drug Information Journal* 35: 115–129.

Herson J. 2009. *Data and Safety Monitoring Committees in Clinical Trials*. Chapman and Hall, Boca Raton, FL.

Hurwitz H, Fehrenbacher L, Novotny W et al. 2004. Bevacizumab plus irinotecan, fluorouracil, and leucovorin for metastatic colon cancer. *New England Journal of Medicine* 350: 2335–2342.

Keats J. 1970. *You Might as Well Live: The Life and Times of Dorothy Parker*. Paragon House, New York.

Lachenbruch PA, Wittes J. 2007. Sentinel event methods for monitoring unanticipated adverse events. In: *Advances in Statistical Methods for the Health Sciences*, J-L Auget, N Balakrishnan, M Mesbah, G Molenberghs (eds.). Birkhauser, Boston, MA, pp. 61–74.

Li W. 2012. Volcano plots in analyzing differential expression with mRNA microarrays. *Journal of Bioinformatics and Computational Biology* 10(6): 1231003. doi: 10.1142/S0219720012310038.

Mehrotra DV, Heyse J. 2004. Use of the false discovery rate for evaluating clinical safety data. *Statistical Methods in Medical Research* 13(3): 227–238.

Proschan M, Lan K, Wittes J. 2006. *Statistical Monitoring of Clinical Trials: A Unified Approach*. Springer, New York.

U.S. Code of Federal Regulations. 2014. US Government Printing Office.

U.S. Department of Health and Human Services, Food and Drug Administration, Center for Biologics Evaluation and Research (CBER), Center for Drug Evaluation and Research (CDER), Center for Devices and Radiological Health (CDRH). 2006. Guidance for clinical trial sponsors: Establishment and operation of clinical trial data monitoring committees. http://www.fda.gov/downloads/RegulatoryInformation/Guidances/ucm127073.pdf. Viewed February 20, 2013.

U.S. Department of Health and Human Services, Food and Drug Administration, Center for Drug Evaluation and Research (CDER), Center for Biologics Evaluation and Research (CBER). 2012. Guidance for industry and investigators: Safety reporting requirements for INDs and BA/BE studies. http://www.fda.gov/downloads/Drugs/GuidanceComplianceRegulatoryInformation/Guidances/UCM227351.pdf. Viewed February 20, 2013.

U.S. Department of Health and Human Services, National Institutes of Health, National Cancer Institute. 2010. Common terminology criteria for adverse events (CTCAE) v4.03. NIH Publication No. 09-541. http://evs.nci.nih.gov/ftp1/CTCAE/CTCAE_4.03_2010-06-14_QuickReference_5×7.pdf. Viewed February 20, 2013.

Wittes J, Barrett-Connor E, Braunwald E et al. 2004. Monitoring the randomized trials of the Women's Health Initiative: The experience of the Data and Safety Monitoring Board. *Clinical Trials* 4: 218–234.

Wittes J, Holmgren E, Christ-Schmidt H et al. 2006. Making independence work: Monitoring the bevacizumab colorectal cancer clinical trial. In: *Data Monitoring in Clinical Trials: A Case Studies Approach*, DL DeMets, CD Furberg, LM Friedman (eds.). Springer, New York, pp. 360–367.

Xia HA, Ma H, Carlin BP. 2011. Bayesian hierarchical modeling for detecting safety signals in clinical trials. *Journal of Biopharmaceutical Statistics* 21(5): 1006–1029.

6

Safety Surveillance and Signal Detection Process*

Atsuko Shibata and José M. Vega

CONTENTS

6.1 What Is Safety Surveillance?

Safety surveillance, also referred to as pharmacovigilance, is defined as "the science and activities relating to the detection, assessment, understanding and prevention of adverse effects or any drug-related problem" [1]. Medicinal products (drugs) are required to follow a rigorous and highly regulated development process before they are allowed to be brought to market. Controlled clinical trials conducted prior to the granting of market authorization involve systematic and organized collection and analysis of adverse event data, as well as other data relevant to drug safety (e.g., electrocardiograms, laboratory measurements for liver and renal function). Although the controlled clinical trials are considered a hallmark of demonstrating the efficacy of a drug, safety data available from those trials have

---

* An earlier version of this chapter was published in the *Proceedings of the 14th International Conference on Information Fusion* (July 5–8, 2011, Chicago, IL). The views expressed herein represent those of the author(s) and do not necessarily represent the views or practices of the author's employer or any other party.

generally well-recognized limitations, such as the limited number of study subjects included in the trials (compared with the size of patient populations that may be exposed to the drug once on the market), the limited duration of exposure to the drug per study subject (particularly relevant in the case of a therapy intended for long-term use), and limited or no data in potentially higher-risk patient subpopulations that are often excluded from controlled clinical trials (e.g., patients with organ impairment, pediatric and geriatric patients, and women of childbearing age who may be treated during pregnancy and lactation). These limitations make it necessary that the marketing authorization holder of a drug and regulatory authorities continue to collect, analyze, and interpret data relevant to patient safety that become available after the drug is approved for commercial use.

6.2 Recent Trends in Pharmacovigilance

Over the past few years, the relative emphasis of pharmacovigilance activities has shifted (Figure 6.1). In the past, most pharmacovigilance departments at biopharmaceutical companies focused on the handling of individual adverse event reports (called individual case safety reports [ICSRs]) with less attention paid to systematic analysis and review of aggregate adverse

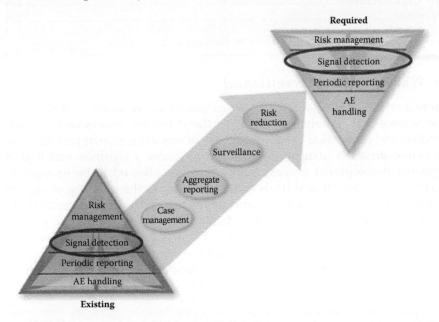

FIGURE 6.1
Recent trends in pharmacovigilance. (Reproduced from WCI. With permission.)

event data and risk management planning. While individual adverse event case handling remains highly regulated and timely reporting of ICSRs to regulatory authorities and other stakeholders continues to be an important compliance matter, the expectations for marketing authorization holders and clinical trial sponsors have increased in the areas of signal detection and risk management over the past several years. Some of these expectations have been delineated specifically in regulatory guidelines and requirements [2–5]. In addition, most sponsor companies (or marketing authorization holders) have internally committed to optimizing their ability to detect safety signals, as well as to managing safety risks proactively, and have developed corresponding processes. Because of the volume and complexity of adverse event data that must be evaluated for the detection of safety signals, various statistical methods, such as proportional reporting ratio (PRR) and its Bayesian variations, have been developed to aid routine monitoring of data, which are used in conjunction with more traditional pharmacovigilance approaches.

6.3 What Is a Safety Signal?

While the term "signal" has been used commonly and widely in the area of pharmacovigilance for years, its definition has evolved over the past few years. Clearly defined terminology is critical to ensure clarity and consistency in communication of drug safety information to patients, prescribers, manufacturers, and regulators [6], and therefore, establishing a common and clear definition of a safety signal is essential.

Most recently, Working Group VIII of the Council for International Organizations of Medical Sciences (CIOMS VIII) [7] defined a drug safety signal as follows, adapting the definition proposed by Hauben and Aronson [6]: "Information that arises from one or multiple sources (including observations and experiments), which suggests a new potentially causal association, or a new aspect of a known association, between an intervention and an event or set of related events, either adverse or beneficial, that is judged to be of sufficient likelihood to justify verificatory action."

Some of the key terms and concepts in the CIOMS VIII definition of a safety signal are worth further discussion:

- *Information that arises from one or multiple sources*: An earlier definition of a safety signal [8] referred to "report(s) of an event," implying that adverse event reports are the primary, if not the only, source of safety signals. The CIOMS VIII definition acknowledged that new information relevant to drug safety may arise from other sources, such as clinical and nonclinical experiments and published articles of clinical study results.

- *Suggests*: A safety signal is not synonymous with a confirmed safety issue. The information must be suggestive of something new that would be worthy of further investigation, after which the suggested association may or may not be confirmed.
- *New*: The concept of newness has always been an important part of safety signal detection. It should be noted that newness may result from emerging trends and from changes in the specificity, severity, and/or rate of occurrence (frequency) of a previously known adverse drug reaction.
- *Judged to be of sufficient likelihood to justify verificatory action*: As discussed earlier (regarding the word "suggests"), a safety signal by the CIOMS VIII definition precedes further investigation ("verificatory action"), not at the conclusion of such investigation. Importantly, this definitional element emphasizes the crucial role of clinical and scientific judgment in determining whether or not a possible association rises to the level warranting further assessment.

6.4 Natural History of Safety Signals

A safety signal has its life cycle; pharmacovigilance and signal detection are ongoing activities spanning the life cycle of a medicinal product, both before and after marketing authorization. Figure 6.2 provides a hypothetical depiction of the natural history that safety signals may follow.

When clinical trials involving human subjects are initiated, a potential therapeutic agent starts with zero adverse event cases in humans. As the drug proceeds through a clinical development program and then moves into the post-marketing phase, adverse event reports are received and accumulate over time and, for certain types of events, may begin to form discernible patterns. For the types of events that rarely occur in the patient populations being treated, a relatively small number of cases may require special attention and can serve as index cases. In some situations, the same or similar kinds of adverse events may be reported from the same clinical site or in the same geographic location, forming a cluster.

As the number of adverse event case reports increases, the relative reliance on different types of data and analytic techniques changes. When there are a relatively small number of cases, pharmacovigilance professionals can review the data, focusing on clinical characteristics of individual cases. Statistical and epidemiological methods become essential and more useful as large amounts of data are accumulated. However, irrespective of the size of the safety databases, clinical judgment and assessment remains important.

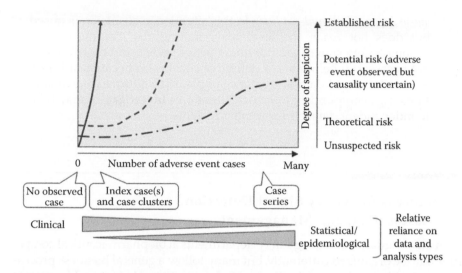

FIGURE 6.2
The natural history of safety signals. *Note*: This is a hypothetical depiction and not based on actual data.

Three hypothetical scenarios for how safety signals may evolve are illustrated in Figure 6.2.

Scenario 1 (*dash line*): Some suspicion about a safety risk exists at the time of starting a safety monitoring program. The suspicion could be theoretical based on the biological mechanism of action of the drug or the risk associated with other drugs in the same therapeutic class, or it may have been observed in animal studies. As the sample size of the clinical data (from clinical studies and post-market experience) grows, adverse event cases are reported. After a careful assessment of a case series, it is concluded that this drug is very likely causally associated with a particular adverse event. The conclusion may result in risk communication actions (e.g., updating the prescribing information) and/or additional risk assessment activities (e.g., investigations using independent data sources).

Scenario 2 (*solid line*): At the beginning of clinical development, no anticipation exists for a particular type of safety risk. However, after just a small number of adverse event cases start to be reported, particularly some of rare conditions that have high likelihood of being attributable to drug therapy, a causal relationship between the drug and the adverse event is suspected. After further investigation, the level of suspicion reaches the point where risk communication actions, such as a label change, are deemed necessary.

Scenario 3 (long-dash-and-dot line): In this situation, a suspicion is raised but does not reach the level for initiating risk communication actions. Safety signals are put under active review, but more data do not necessarily help make a definitive conclusion because of inherent limitations of data or existence of plausible alternative explanations (e.g., confounding). In such situations, investigators may turn to independent datasets for further insights.

6.5 Process for Safety Signal Detection, Assessment, and Management

Pharmacovigilance and drug safety departments at biopharmaceutical companies may be organized differently, but many follow a general business process similar to what is depicted in Figure 6.3. Information relevant and important to the safety monitoring of medicinal products, coming from multiple sources, not just adverse event reports, is reviewed, and observations are prioritized by the safety team responsible for a given product. Prioritized safety observations are then subjected to further investigation or assessment, and the team makes recommendations based on assessment results. After the team assessment (signal evaluation), recommendations are brought to a cross-functional safety committee, which makes decisions on risk communication and other actions to be taken. The level of urgency for risk communications and other actions is driven by clinical and public health impact of the new information on patient safety. As safety signal detection is an iterative process throughout the product's life cycle, observations and decisions at each of these workflow steps may lead to the adjustments to how routine monitoring is conducted.

A typical signal detection program (here, the term "program" refers to a set of planned activities and business procedures to be followed, not to a computer or software program) consists of a flow of sequential steps of signal detection, prioritization, and evaluation (Figure 6.4) as well as its linkage to risk management activities [7].

The term "signal detection" has been used by some authors and pharmacovigilance professionals synonymously with statistical data mining or disproportionality analyses of large adverse event databases. This is a misnomer, however, as signal detection is not limited to a narrow range of data sources or analytical approaches.

There is no single right approach to signal detection that would be optimal for all medicinal products in all situations. Just as diversification is important in financial investment, balancing the portfolio of various analytical methods aligned to a strategy and objectives is important in safety signal detection. Furthermore, no statistical methods or algorithms would replace the importance of medical and scientific judgment of trained

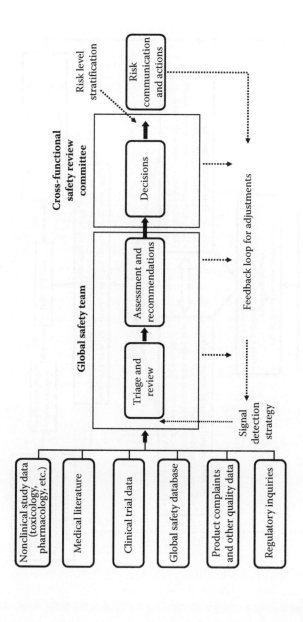

FIGURE 6.3
Signal detection assessment and management process.

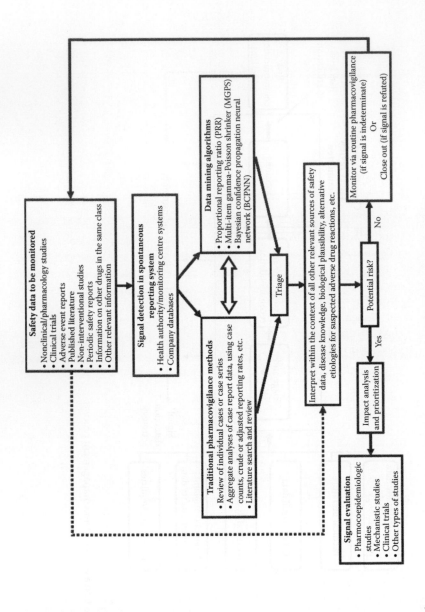

FIGURE 6.4

Signal detection and evaluation steps. (Reproduced from the Report of CIOMS Working Group VIII, *Practical Aspects of Signal Detection in Pharmacovigilance*, Council for International Organizations of Medical Sciences (CIOMS), Geneva, Switzerland, 2010. With permission, © CIOMS.)

pharmacovigilance professionals. At the same time, as the amount of information increases, codifying human experts' tacit reasoning and consistent application of sound pharmacovigilance logic become more and more important in supporting proactive and scalable safety surveillance.

6.6 Data Sources and Statistical Data Mining Methods That Are or May Be Used for Safety Signal Detection

Statistical data mining methods for application in pharmacovigilance emerged in the late 1990s, originally as a means of performing systematic signal detection in large databases from the spontaneous reporting systems (SRSs) of adverse event information maintained by health authorities and drug monitoring centers [7]. Some of the databases that can be used for signal detection in the post-authorization drug safety surveillance are listed in Table 6.1.

Data mining methods aim to examine a large dataset of adverse event report records by using statistical or mathematical tools. These methods are based on the concept of disproportionality. It is assumed that in the absence of disproportionality, the distributions of reported adverse events are the same across drugs. If a specific adverse event were associated with a given drug, however, this event would have higher reporting frequency and create reporting disproportionality. Statistics of disproportionate reporting are screened based on the ranking of drug–event combinations by the level of disproportionality or statistical "unexpectedness."

TABLE 6.1

Databases That Can Be Used for Signal Detection in Post-Authorization Drug Safety Surveillance

Type of Databases	Examples
Spontaneous reporting system (SRS) databases	VigiBase (WHO)
	EudraVigilance (EEA)
	AERS (United States)
	Sentinel (United Kingdom)
Prescription event monitoring databases	Drug Safety Research Unit (United Kingdom)
	Intensive Medicines Monitoring Program (New Zealand)
Large linked administrative databases or electronic health records (EHR) databases	Health-care insurance claims databases (managed by for-profit managed care organizations or government agencies)
Electronic medical records (EMR) databases	General Practice Research Database (United Kingdom)

Source: Adapted from the Report of CIOMS Working Group VIII, *Practical Aspects of Signal Detection in Pharmacovigilance*, Council for International Organizations of Medical Sciences (CIOMS), Geneva, Switzerland, 2010. With permission, © CIOMS.

TABLE 6.2

2 × 2 Table of Adverse Event Report Data for Disproportionality Analysis

	Reports for Event of Interest	Reports for All Other Events	Total
Reports for drug of interest	A	B	A + B
Reports for all other drugs	C	D	C + D
Total	A + C	B + D	A + B + C + D

Source: Reproduced from the Report of CIOMS Working Group VIII, *Practical Aspects of Signal Detection in Pharmacovigilance*, Council for International Organizations of Medical Sciences (CIOMS), Geneva, Switzerland, 2010. With permission, © CIOMS.

TABLE 6.3

Statistics for Disproportional Reporting

Statistic	Formula
Proportional reporting ratio (PRR)	$A(C + D)/C(A + B)$
Reporting odds ratio (ROR)	AD/CB
Relative reporting (RR)	$A(A + B + C + D)/(A + C)(A + B)$
Information component (IC)	$\log_2[A(A + B + C + D)/(A + C)(A + D)]$

Source: Adapted from the Report of CIOMS Working Group VIII, *Practical Aspects of Signal Detection in Pharmacovigilance*, Council for International Organizations of Medical Sciences (CIOMS), Geneva, Switzerland, 2010. With permission, © CIOMS.

Several statistical algorithms have been developed for disproportionality analysis. The data construct underlying all the disproportionality analysis methods is shown in Table 6.2. Commonly used statistical measures of association are listed in Table 6.3. Some of these statistics (e.g., relative reporting [RR] and information component [IC]) have been integrated into Bayesian approaches, such as Multi-Item Gamma-Poisson Shrinker (MGPS) and Bayesian confidence propagation neural network (BCPNN), which have been developed in part to account for the variability associated with small numbers of reports [7]. Figure 6.5 shows an example of disproportionality analysis using the proportional reporting ratio (PRR), demonstrating the usefulness of examining how the PRR for a given drug–event combination changes over time in detecting a safety signal.

While 2 × 2 table-based disproportionality analysis is the most commonly used in the contemporary statistical methods in pharmacovigilance, a variety of multivariate methods and sequential methods are being tested and applied for purposes of signal detection and preliminary evaluation. Disproportionality analysis is typically used to identify product–event combinations of unusually high frequency compared with the reference distributions based on data for other medicinal products represented in the database (interproduct analysis). The analytical methods can also be applied to examine time trends for a given drug, where reference distributions are established with historical data for the same drug (intraproduct analysis).

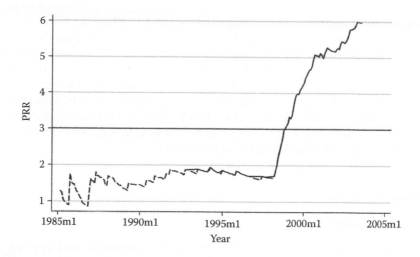

FIGURE 6.5
Time trend of PRRs for isotretinoin and reports of depression. (Reproduced from the Report of CIOMS Working Group VIII, *Practical Aspects of Signal Detection in Pharmacovigilance*, Council for International Organizations of Medical Sciences (CIOMS), Geneva, Switzerland, 2010. With permission, © CIOMS.)

Increasingly, observational research and pharmacoepidemiologic approaches are being applied to the assessment of safety risks, which may be difficult, if not impossible, to examine by analyzing spontaneously reported adverse events alone or by conducting interventional clinical studies.

6.7 Relationship of Signal Detection to Risk Communication, Risk Management, and Benefit–Risk Evaluation Processes

Results of signal evaluations (assessments) inform the sponsor and regulators and help determine the need for additional risk communication. Safety risks are typically communicated in the form of an investigator brochure (IB), informed consent form, and prescribing information (also referred to as product label). Additional forms of risk communication, such as Dear Investigator Letters (DILs) and Dear Healthcare Provider Letters (DHCP Letters), may be warranted for newly recognized important safety risks with significant public health and medical impact.

Certain safety risks of special interest may be deemed "important" for the purpose of risk management planning. Important identified risks and important potential risks are key components of the safety specification for a given medicinal product as per ICH guideline E2E (Pharmacovigilance Planning) [9]. Selection and assessment of important risks should be supported by the information that arises from safety signal detection and evaluation processes.

Regulators worldwide have recognized a need to place signal detection, risk management, and benefit–risk evaluation for medicinal products in a structured framework when marketing authorization holders communicate findings and their implications from those pharmacovigilance activities to regulatory authorities. One effort to operationalize this need is represented in the recently published ICH guideline E2C(R2) on the periodic benefit–risk evaluation reports (PBRER) [10]. The PBRER prepared following the guideline will include sections on safety signals (new, ongoing, and closed), risk evaluation, and integrated benefit–risk analysis.

6.8 Conclusions

Statistical data mining approaches have been developed and applied in the field of drug safety surveillance, adding to the toolkit of pharmacovigilance professionals. In addition, use of data sources other than spontaneous adverse event reporting systems has gained interest in recent years. Further development of statistical methods and technological solutions to analyze large amounts of data to detect signals for potential safety issues, while minimizing noise, would enhance the efficiency and effectiveness of pharmacovigilance activities. These technological and scientific advances must be placed in the context of well-documented business processes and combined with the medical and scientific judgment of trained pharmacovigilance professionals.

References

1. World Health Organization. The importance of pharmacovigilance—Safety monitoring of medicinal products. World Health Organization, Geneva, Switzerland, 2002.
2. U.S. Department of Health and Human Services, Food and Drug Administration, Center for Drug Evaluation and Research (CDER) and Center for Biologics Evaluation and Research (CBER). Guidance for industry: Good pharmacovigilance practices and pharmacoepidemiologic assessment. http://www.fda.gov/downloads/Drugs/GuidanceComplianceRegulatoryInformation/Guidances/ucm071696.pdf, March 2005.
3. U.S. Department of Health and Human Services, Food and Drug Administration, Center for Drug Evaluation and Research (CDER) and Center for Biologics Evaluation and Research (CBER). Draft guidance for industry format and content of proposed risk evaluation and mitigation strategies (REMS), REMS assessments, and proposed REMS modifications. http://www.fda.gov/downloads/Drugs/GuidanceComplianceRegulatoryInformation/Guidances/UCM184128.pdf, September 2009.

4. European Medicines Agency. Guideline on good pharmacovigilance practices (GVP). Module V—Risk management systems. http://www.ema.europa.eu/docs/en_GB/document_library/Scientific_guideline/2012/06/WC500129134.pdf, June 2012.

5. European Medicines Agency. Guideline on good pharmacovigilance practices (GVP) Module IX—Signal management. http://www.ema.europa.eu/docs/en_GB/document_library/Scientific_guideline/2012/06/WC500129138.pdf, June 2012.

6. Hauben M and Aronson JK. Defining 'signal' and its subtypes in pharmacovigilance based on a systematic review of previous definitions. *Drug Safety*, 32(2):99–110, 2009.

7. CIOMS Working Group VIII. *Practical Aspects of Signal Detection in Pharmacovigilance*. Council for International Organizations of Medical Sciences (CIOMS), Geneva, Switzerland, 2010.

8. CIOMS Working Group IV. *Benefit-Risk Balance for Marketed Products: Evaluating Safety Signals*. Council for International Organizations of Medical Sciences (CIOMS), Geneva, Switzerland, 1999.

9. International Conference on Harmonisation of Technical Requirements for Registration of Pharmaceuticals for Human Use. ICH harmonised tripartite guideline E2E: Pharmacovigilance planning. ICH, Geneva, Switzerland, November 2004.

10. International Conference on Harmonisation of Technical Requirements for Registration of Pharmaceuticals for Human Use. ICH harmonised tripartite guideline E2C(R2): Periodic benefit-risk evaluation report. ICH, Geneva, Switzerland, December 2012.

Section III

Evaluation/Analysis

7

Bayesian Adaptive Trials for Drug Safety

Jason T. Connor

CONTENTS

The U.S. Food and Drug Administration (FDA) has increased their focus on premarket confirmations of drug safety. Recent drug withdrawals include rofecoxib, valdecoxib, fen-phen, and tegaserod all for cardiovascular (CV) safety concerns that became apparent after these drugs were approved [1–3]. Other drugs such as rosiglitazone are prescribed with tighter restrictions after they were found to produce a higher CV risk [4]. These removals have led to requirements for evidence of drug safety for many new drug approvals. For instance, any diabetes drug must, in addition to efficacy, demonstrate that it is heart safe [5].

7.1 Premarket Demonstration

Demonstrating safety in the premarket setting, however, is difficult due to the often-rare nature of significant adverse events—one must observe CV events to show the rate of those events does not increase with a new treatment. Prospectively planning a premarket safety trial can be challenging due to the difficulty in predicting what that event rate is likely to be. When safety is required to be demonstrated via a bounded hazard ratio, as with the antidiabetic drug guidance, the power of the study increases with the event rate [6]. Therefore, in populations with low baseline CV event rates, such as would be the case for the population with a drug such as tegaserod, a drug for irritable bowel disease that was commonly prescribed for nonelderly females, premarket safety trials would require an enormous number of patient-years of exposure to produce sufficiently many events for adequate power. Furthermore, due to decreases in smoking rates, improvements in cholesterol drugs, and other coronary interventions (both drug and surgical), CV events and CV-related deaths are becoming increasingly less common [7,8].

Here, we discuss two alternatives: if primary assurance of safety must come from a bounded hazard ratio, such as in the FDA guidance for evaluating CV risk for new antidiabetic drugs, we illustrate an adaptive trial for choosing the optimal sample size when the true event rate is not well known, while also requiring type 1 error control necessary in a confirmatory study. Type 1 error control will be illustrated via simulation.

Second, we extend this design and illustrate how adding the risk difference to the risk (hazard) ratio can provide approximately equal power over the entire range of event rates [6]. We also discuss the benefits of jointly using the risk difference with the risk ratio where permissible.

7.2 Demonstrating Safety via a Bounded Risk Ratio

In December 2008, the FDA released its guidance document for antidiabetic drug safety [5]. In it, they suggest that any new drug application contains a prospective trial that demonstrates the 95% confidence interval (CI) for the hazard ratio for CV events be < 1.8 for approval. Furthermore, if the upper bound is > 1.3 (while being < 1.8), they suggest that further study is needed after approval until the upper bound can be shown to be < 1.3. A trial with a 95% CI entirely < 1.3 at the time of approval would require no further study.

One hundred twenty-five events would provide 90% power, and 95 events would provide 80% power to illustrate the 95% CI is < 1.8 based on a Cox proportional hazard model. Therefore, in a clinical trial with an expected event rate of 1 event per 100 patient-years and a 5% annual dropout rate, 90% power would be achieved by enrolling 200 patients per month for 24 months and following all enrolled patients for an additional 24 months. The total trial duration would be 48 months. If, however, the true event rate was actually 0.75%, the power would decrease to 80%—a doubling of the type 2 error. Furthermore, if the event rate were actually 0.5%, then the power would drop to 65%, a type 2 error 3.5 times as high. Because it is hard to predict CV event rates in noncardiac populations, such overestimates are not uncommon. Similarly, if the actual event rate is higher, then the trial is overpowered and enrolled more patients and took more time than was necessary.

Achieving the stronger noninferiority threshold of the upper bound < 1.3 requires observing even more events: 630 and 460 events are needed for 90% and 80% power, respectively. With 1 event for every 100 patient-years, this would require 63,000 patient-years of exposure or 12,600 patients followed for an average of 5 years for 90% power.

Often trial investigators make their best estimate of the event rate, then choose a sample size. Here we assume the same event rate in the treatment and control groups since this is a noninferiority trial starting from the assumption that both groups have the same adverse event rate. A blinded analysis of the event rate during the course of the trial may suggest the pretrial estimate was too high, leading to lower than expected study power. In this case, a sample size reestimation could be initiated to increase power to prespecified levels. Calculating a pooled estimate of event rates during the trial has advantages, maintaining the blind and not inflating type 1 error rate, but if the new drug (the treatment group) offers an improved safety profile seen by a lower event rate than the control group, then the sample size may be increased unnecessarily. Here, we present a trial that instead offers a sample size range, monitors the observed event rates in each arm, and uses predictive probabilities to stop accrual for success at the appropriate sample size—effectively using internal trial data that identify the sample size and ensure adequate power. Furthermore, it allows for the possibility of extending the trial if a superiority claim is possible, all while conserving type 1 error rates for both the noninferiority and superiority claims.

7.2.1 Design Goal

This is a two-arm randomized trial where patients are equally allocated to control or experimental treatment. The trial's goal is to satisfy the FDA anti-diabetic safety guidance: produce a 95% CI entirely below 1.3, which would exclude further study in the postmarket setting. Additionally, if the drug is so beneficial that superiority might be shown (CI for HR entirely < 1.0), then the trial may continue in hopes of establishing superiority. Thus, this is an adaptive noninferiority/superiority trial with an adaptive sample size. Multiple interim analyses are used to ensure the sample size is optimal to achieve the trial's goal, hence we call these Goldilocks trials [9]. We will refer to a CI < 1.8 as weak noninferiority and < 1.3 as strong noninferiority.

7.2.2 Statistical Model

The primary analysis will be a Cox proportional hazard model and the corresponding CIs for the treatment group hazard ratio. We use a Bayesian algorithm, based on the predictive probability of success at the final analysis, throughout the accrual stage to determine when to stop accrual. Thus, we use a Bayesian model to predict the likelihood of success for a frequentist test. A fully Bayesian primary analysis could also be performed.

The Bayesian model assumes CV events are exponentially distributed and uses a conjugate gamma prior for each treatment group. For randomization group G ($G = 1$ for control or $G = 2$ for treatment), we use the prior

$$\lambda_G \sim \Gamma(a = 0.5, b = 14.28).$$

This prior is vague as it contains one-half patient's worth of information for each group with a mean exposure time of 3.5 events per 100 patient-years of follow-up (0.5 events/14.28 patient-years = 0.035 events/patient-year). We use the same prior for each treatment group, and the total amount of prior information between both groups is one patient worth of information. The priors are used only in the predictive probability calculation, and as such, they affect the decision to stop accrual but not the primary analysis.

7.2.3 Design

The maximum sample size will be 5400 patients, and the final analysis will occur at most 30 months after the last patient is enrolled. Thus, if accrual is stopped early for predicted success, or if the trial continues to the maximum sample size, all patients are allowed to complete a 30-month follow-up, and the primary analysis is conducted at the completion of the follow-up period. For the primary analysis, we use two-sided 98% and 97% CIs to ensure overall 2.5% (one-sided) type 1 error rates for strong noninferiority and superiority, respectively.

Throughout the accrual stage, we calculate four predictive probabilities:

1. $P_{1.3,n,n}$: The predictive probability of strong noninferiority with the current sample size, that is, assuming accrual stops with the current n patients enrolled and follows patients an additional 30 months, the probability of showing the 98% CI is < 1.3

2. $P_{1.3,n,Nmax}$: The predictive probability of strong noninferiority at the maximum sample size, that is, assuming the trial enrolls to $Nmax$ and follows patients an additional 30 months, the probability of showing the 98% CI is < 1.3

3. $P_{1.0,n,n}$: The predictive probability of superiority at the current sample size, that is, assuming accrual stops with the current n patients enrolled and follows patients an additional 30 months, the probability of showing the 97% CI is < 1.0

4. $P_{1.0,n,Nmax}$: The predictive probability of superiority at the maximum sample size, that is, assuming the trial enrolls to $Nmax$ and follows patients an additional 30 months, the probability of showing the 97% CI is < 1.0

The trial has three stages. Stages 1 and 2 occur during accrual. In Stage 1, only early stopping for futility is allowed. If in Stage 1, a weak noninferiority (CI < 1.8) criterion is met, the trial transitions to Stage 2. In Stage 2, the sample size is optimized for either strong noninferiority or superiority. When accrual is stopped in Stage 2, or the trial accrues to the maximum sample size, the trial transitions to Stage 3, the follow-up phase.

Stage 1: Interim analyses are performed after the 60th CV event occurs, then after every 20th patient accrues. During Stage 1, the trial may stop for futility, or if there is evidence of noninferiority, the trial will transition to Stage 2. If, at a Stage 1 interim analysis, we observe (a) $P_{1.3,n,Nmax} < 0.05$, then even with continuing accrual to the maximum sample size and completing the 30 months of additional follow-up, there is little chance of demonstrating strong noninferiority, and so the trial would stop early for futility. On the other hand, if we observe (b) the 99% CI < 1.8, then the premarket criteria for weak noninferiority have been met. The 99% CI is used here to account for multiple interim analyses during the accrual phase. The trial continues accrual but transitions to Stage 2.

Stage 2: Stage 2 begins after an additional 100 events are observed. Interim analyses during Stage 2 are planned after every additional 20 events are observed. At each Stage 2 interim analysis, we calculate the four predictive probabilities described earlier. During Stage 2, the trial may stop for predicted success for either superiority or strong noninferiority, or it may enroll to the predetermined maximum of 5400 patients. One of four decisions can be made at each interim analysis during Stage 2:

a. Accrual will stop for predicted superiority success if $P_{1.3,n,n} > 0.95$ and $P_{1.0,n,n} > 0.80$. This says that if we stop accrual but continue to follow all enrolled patients for an additional 30 months, we will have a 95% or greater chance of showing strong noninferiority and an 80% or greater chance of showing superiority.

b. Accrual will stop for predicted strong noninferiority success if $P_{1.3,n,n} > 0.95$ and $P_{1.0,n,Nmax} < 0.60$. This says that if we stop accrual but continue to follow all enrolled patients for an additional 30 months, we will have 95% or greater chance of showing strong noninferiority, but we concede that demonstrating superiority is unlikely.

c. If $P_{1.3,n,n} > 0.95$ but $P_{1.0,n,n} < 0.80$ and $P_{1.0,n,Nmax} > 0.60$, then accrual will continue. While strong noninferiority is likely with the current sample size, additional data might also lead to a claim of superiority.

d. If the futility bound (condition a) in Stage 1 is not met and $P_{1.3,n,n} < 0.95$, then accrual will continue.

If conditions (a) or (b) are met, those conditions that lead to stopping accrual for predicted success, then the trial transitions to Stage 3. If ever the maximum sample size of 5400 patients is reached, accrual also stops and the trial transitions to Stage 3.

Decisions such as the condition in (c) are in part business decisions. Some may want to stop accrual and demonstrate noninferiority as soon as possible, whereas others may believe it to be in their best interest to extend the trial if it offered a reasonable chance to demonstrate superiority in addition to noninferiority.

Stage 3: In Stage 3, accrual has stopped, and all enrolled patients continue to be followed for a maximum of 30 months. Interim analyses occur in Stage 3 at 6, 12, 18, and 24 months after the final patient is enrolled. The first time the 98% CI for the hazard ratio for CV events is < 1.3, a new drug application will be submitted. Follow-up would continue in case the CI < 1.0 is ever observed, hence justifying a superiority claim for CV events for the new treatment.

7.2.4 Predictive Probability Calculation

The predictive probabilities used to stop the trial for futility or for predicted success are calculated using a combination of the current posterior probability, each individual patient's data (follow-up time and whether they've experienced an event), and simulated data from the current point forward along with the Cox proportional hazard model.

The current data include EV_G events observed in EXP_G patient-years of follow-up for the control ($G = 1$) and treatment ($G = 2$) groups, respectively. Thus, the posterior for group G becomes

$$\lambda_G \sim \Gamma(a = 0.5 + EV_G, b = 14.28 + EXP_G).$$

Furthermore, for each group, we have N_G patients where $N_G = EV_G + L_G + F_G$. N_G is the total number of patients in group G, EV_G is the number of patients with CV events, L_G is the number of patients who have been lost to follow-up, and F_G is the number of patients still being followed.

To calculate the predictive probabilities, we repeat the following process 100,000 times:

1. Draw $\lambda_G \mid$ Data $\sim \Gamma(0.5 + EV_G, 14.28 + EXP_G)$.
2. Simulate event times for the $i = 1 \dots F_G$ patients still being followed, $X^*_{i,G} \sim Exp(\lambda_G)$.
3. If $X^*_{i,G} < 30$ months, the event is observed and total time $= X_{i,G} + X^*_{i,G}$.

    Otherwise, the patient is censored and the total time $= X_{i,G} +$ 30 months.
4. This produces a full imputed dataset 30 months into the future.
5. Do Cox proportional hazard model on imputed dataset.
6. Track if 99% CI is < 1.8, 98% CI is < 1.3, or 97% CI is < 1.0.

The proportion of times out of 100,000 the 98% CI is < 1.3 is $P_{1.3,n,n}$. The proportion of times the 97% CI is < 1.0 is $P_{1.0,n,n}$.

The predictive probabilities for a trial running to the maximum sample size, $P_{B,n,Nmax}$ (where $B = 1.3$ or 1.0), are similarly calculated only now instead of all patients having 30 months of follow-up each future patient has a different follow-up time. All patients already enrolled will have a follow-up time equal to time needed to complete accrual + 30 months. The last patient enrolled will have a maximum of 30 months of follow-up. Patients enrolled in between will have a follow-up time that we assume is linear between these two times. The remaining follow-up time is a function of the time needed to complete accrual, and this may be chosen as a constant, the estimated rate before the trial started, or as an average of the last 6 or 12 months of patient per month from the trial.

7.2.5 Example Trial

In the example trial in Table 7.1, the first interim analysis occurs after 60 events. Here, we observe 33 events in 812 patient-years of follow-up in the control group and 27 events in 817 patient-years of follow-up in the treatment group. The predictive probability of showing CI < 1.3 at the maximum sample size is 0.920 > 0.05, so the trial does not stop for futility. The entire 99% CI was < 1.8, therefore, the trial transitions to Stage 2, and the next interim analysis occurs after 100 additional events are observed (Table 7.1 "Look 2").

At this interim analysis, $P_{1.3,n,n}$, the predictive probability for strong non-inferiority with the current n patients is 0.956 > 0.95, a signal to stop accrual early for predicted success. We then look at the predictive probabilities for

TABLE 7.1

Example Trial 1

Look	N	Months	Control Events, Patient-Years	Treatment Events, Patient-Years	$P_{1.8,n,n}$ $P_{1.8,n,\,N\,Max}$	$P_{1.3,n,n}$ $P_{1.3,n,\,N\,Max}$	$P_{1.0,n,n}$ $P_{1.0,n,\,N\,Max}$	Hazard Ratio 97% CI
1	2981	13.3	33	27	0.989	0.859	0.475	0.84
			812	817	0.999	0.920	0.580	(0.46, 1.42)
2	4696	20.9	85	75	> 0.999	0.956	0.391	0.90
			1995	2007	> 0.999	0.955	0.378	(0.63, 1.25)
3	4696	26.9	121	114				0.94
			3112	3129				(0.71, 1.25)
4	4696	32.9	162	151				0.93
			4218	4239				(0.72, 1.19)
5	4696	38.9	202	190				0.94
			5295	5325				(0.75, 1.17)
6	4696	44.9	241	228				0.94
			6360	6395				(0.77, 1.15)
7	4696	50.9	268	259				0.96
			7402	7443				(0.80, 1.16)

showing superiority. $P_{1.0,n,n} < 0.80$, so we do not predict showing superiority with the current sample size, and because $P_{1.0,n,N\,Max} = 0.378 < 0.60$, we concede that the trial is unlikely to demonstrate superiority even with 5400 patients. Therefore, accrual stops at the current sample size, 4696 patients, and trial transitions to Stage 3. At this time, the hazard ratio is 0.90, and the 97% CI is 0.63–1.25. Repeating the analysis every 6 months (Looks 3–7), superiority is never achieved.

This trial stops short of the maximum sample size of 5400 patients as it is seen to have high probability of demonstrating strong noninferiority, but not superiority. Then during the follow-up phase of Stage 3, this is precisely what the data bear out.

7.2.6 Power and Operating Characteristics

Multiple factors contribute to the study's power and average sample size: the maximum allowable sample size, accrual rate, CV event rate, stopping boundaries, and maximum follow-up length after the last patient is enrolled are key. The maximum sample size, follow-up times, and stopping boundaries are *a priori* defined design features. The accrual rate can be controlled to some extent by the number of sites and inclusion criteria. The hardest to predict or control is the CV event rate. One can attempt to project this based on the study's inclusion criteria and literature; however, it can be particularly difficult to predict in noncardiac populations (e.g., a trial for rofecoxib for arthritis) for which there is little literature on CV outcomes.

To determine operating characteristics, we start with various CV event rate scenarios (different control rates, then a range of treatment rates ranging from superiority to equivalence to inferiority) and then simulate 5000 trials per scenario. To simulate trials, we make the following assumption about the data-generating process. These are necessary to estimate operating characteristics but will not be used in the analysis of the actual trial:

- Times to event have an exponential distribution in both the treatment and placebo groups with the specified event rate.
- There is 1% annual loss to follow-up, sampled from an exponential distribution.
- Accrual is constant with the stated accrual rate per month.

Five thousand simulated trials give estimates of type 1 error with standard error < 0.2%, $(0.025(0.975)/5000)^{0.5} = 0.0022$. For powers on the order of 80%, the standard error is 0.56%, $(0.2(0.8)/5000)^{0.5} = 0.0056$.

In this trial, as suggested by the diabetes CV safety guidance, the trial population is enriched to patients at a higher risk for CV events. This elevates the baseline risk to attempt to ensure adequate CV events and therefore sufficient power. The predicted CV event rate is 5 events per 100 patient-years, though we also thoroughly explore the 3.5 events per 100 patient-years case,

TABLE 7.2

Operating Characteristics for the Primary Endpoint

Accrual	Control Rate	Treatment Rate	Hazard Ratio	Mean N	Mean Months	Power < 1.8	Power < 1.3	Power < 1.0
150	0.035	0.030	0.85	4588	37.1	0.99	0.99	0.19
patients	0.035	0.035	1.0	4909	43.8	0.92	0.84	0.017
per	0.035	0.045	1.3	4105	37.1	0.34	0.025	0.00
month	0.035	0.050	1.8	2482	16.6	0.00	0.00	0.00
	0.050	0.043	0.85	3980	32.7	0.99	0.99	0.22
	0.050	0.050	1.0	4561	39.0	0.94	0.92	0.020
100	0.035	0.030	1.3	3986	46.1	0.99	0.99	0.019
	0.035	0.035	1.8	4574	54.8	0.92	0.88	0.021

Accrual	Control Rate	Treatment Rate	Hazard Ratio	Accrual Looks	Follow-Up Looks	Stop Futility	Stop Max	Stop Early
150	0.035	0.030	0.85	4.1	1.1	0.01	0.25	0.74
patients	0.035	0.035	1.0	6.3	1.8	0.08	0.57	0.35
per	0.035	0.045	1.3	6.1	1.6	0.66	0.33	0.01
month	0.035	0.050	1.8	2.2	0.0	1.00	0.00	0.00
	0.050	0.043	0.85	4.8	1.0	0.01	0.08	0.92
	0.050	0.050	1.0	8.8	1.4	0.06	0.39	0.55
100	0.035	0.030	1.3	5.1	1.0	0.01	0.07	0.92
	0.035	0.035	1.8	9.5	1.5	0.08	0.41	0.52

Note: "Power < 1.8" represents the power to show the 99% CI is < 1.8. "Power < 1.3" represents the power to show the 98% CI is < 1.3. "Power < 1.0" represents the power to show the 97% CI is < 1.0.

in order to evaluate if the design is robust to decreases of up to 30% from the expected rate. We also show results for accrual rates of 100 and 150 patients per month for the primary scenario. Table 7.2 shows the operating characteristics for these scenarios.

Table 7.2 shows how the design enrolls more patients when the event rate is lower to ensure a sufficient number of observed CV events: in equivalent situations, the adaptive design enrolls more patients when the event rate is lower, 4909 patients for 3.5 events per 100 patient-years versus 4561 for 5.0 events per 100 patient-years. Furthermore, the sample size decreases when accrual decreases. This is because each enrolled patient produces greater exposure.

The overall type 1 error rates are controlled at the one-sided 0.025 level. When rates are equivalent between control and treatment groups, the probabilities of showing the HR is < 1 ("power < 1.0") are ≤0.025. Likewise, when the true hazard rate is 1.3, type 1 errors to demonstrate the HR is < 1.3 are ≤0.025.

The type 1 error rate is even lower to show HR < 1.8—mainly because the futility rule is determined by the likelihood of showing the hazard ratio is < 1.3 and if that is not achievable, the trial stops for futility.

Table 7.2 also shows the average number of analyses that occur during the accrual (Stages 1 and 2) and follow-up (Stage 3) stages and the distribution of why the study stops, for futility, reaching the maximum of 5400 patients, or stopping for predicted success. For instance, when the event rates are

5.0 events per 100 patient-years in both groups and the accrual rate is 150 patients per month, there tend to be 8.8 analyses during accrual and 1.4 during the follow-up period. Six percent of trials stop for futility, 39% enroll to the maximum of 5400 patients, and 55% stop early for predicted success. When the event rates are lower, 3.5 in both groups, 8% stop for futility, 35% stop early for predicted success, and 57% go to the maximum.

The key is that as event rates decrease or accrual rates increase, both of which lower power, the trial will enroll more patients in order to compensate and keep power high. This buffers against the trial being underpowered if the event rate is lower than expected and buffers against the trial running longer than necessary if, in fact, the event rate is higher than expected.

7.3 Risk Difference

The adaptive sample size can protect power over a range of event rates, but trials still need to be quite large when event rates are low. Another possibility is to introduce the risk difference in combination with the risk ratio [6].

Using the risk ratio is a particularly challenging way to show noninferiority when event rates are very low due to the mathematics defining the variability of a risk ratio. When there are few events in the control group, the CI for the control rate is wide and contains very small values—values in the denominator of the risk ratio—and therefore, the CI for the ratio may contain extremely large values even if the observed rates are similar.

Furthermore, using the risk ratio to ensure safety ignores the clinical relevance of the baseline risk and can lead to very different numbers needed to harm (NNH) to satisfy the definition of "safe." For instance, if the hazard ratio must be < 1.3 and the control rate is 0.5 events per 100 patient-years, then the NNH is $1/(1.3 \times 0.005 - 0.005) = 667$ patients. Whereas if the control rate is 3 events per 100 patient-years, then the NNH is $1/(0.039 - 0.030) = 111$. In one case, "safe" must have an NNH better than 667, and in the other, an additional CV event for every 111 treated patients might be deemed "safe." Clinically one could argue if the baseline CV event rate is very low, the allowed maximum hazard ratio should be a bit larger—a greater hazard ratio will still produce fewer additional events when the baseline risk is low. This can be done without increasing the NNH.

Therefore, we can extend the allowable NNH permitted by the hazard ratio (risk ratio) and add a risk difference component. For instance, if a lower bound of the NNH of 111 is permissible, which it is using a hazard ratio of 1.3 with a baseline risk of 3.0, it corresponds to a risk difference of 0.9 events per 100 patient-years. The benefit of using the risk difference is that power increases as the baseline risk decreases. Combining it with the risk ratio—for which power increases and the baseline risk increases—means that power can be higher over the range of event rates.

For example, in Table 7.2, when the event rates are 0.035 in both groups, the power is just 84%.

Continuing the earlier example, if we enroll 5400 patients in 3 years plus follow patients for another 2 years and use a joint definition of safety, 95% CI for risk ratio < 1.3% or 95% CI for risk difference < 0.009, then we maintain power of 90% across the entire range of event rates.

For a given control rate, Figure 7.1 shows the maximum allowed treatment rate using the higher of the risk ratio and risk difference. As the control rate goes to zero, maintaining the rule that the 95% CI for the hazard ratio must be < 1.3 means the lower bound for the NNH goes to infinity, making it impossible to show safety when the risk is zero and a large-fold increase in risk ratio would still result in very rare adverse events. Shifting to the risk difference below 0.03 means the allowable NNH is constant below that point (at 111).

Allowing safety to be declared one of two ways does slightly increase type 1 error at 0.03, particularly at the event rate that satisfies both the risk ratio and the risk difference criteria, the inflection point where 1.3 × control rate = control rate + 0.009. Using two-sided 95% CIs for the risk ratio and risk difference produces a type 1 error of 2.8% at that point because of the multiplicity—the 2.5% of type 1 errors via the risk difference are not entirely the same as the 2.5% of type 1 errors via the risk ratio. Using a two-sided 95.8% CI for both the risk ratio and risk difference ensures type 1 error is < 2.5% over the range of event rates. Figure 7.2 shows the power (in black, at the top) and type 1 errors (in gray at the bottom) for the risk ratio (small dashes), risk difference (long dashes), and combination (solid). The "either" situation overlaps the risk difference line for smaller event rates

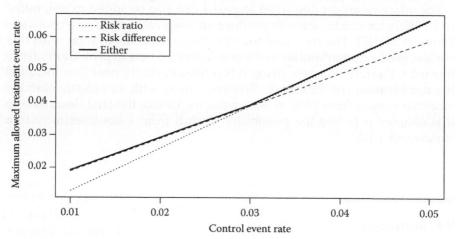

FIGURE 7.1

Maximum allowable treatment effect allowed by risk ratio (1.3 × control event rate), risk difference (0.009 + control event rate), or either. There is an inflection point where 1.3 × control rate = 0.009 + control rate, control rate = 0.03.

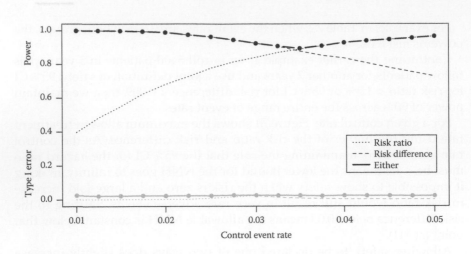

FIGURE 7.2
Power (black) and type 1 error (gray) using the risk ratio, risk difference, and either. Power increases as event rate increases for the risk ratio, while power increases as event rate decreases for the risk difference. Therefore, the combination of the two provides high power over the entire range of possible event rates.

(< 0.036 for power; < 0.030 for type 1 error) and overlaps the risk ratio line for larger event rates. Therefore, while the plot shows three lines, it may appear as only two. Therefore, assuming the treatment and control rates are equivalent, 5400 patients enrolled over 3 years and followed for 2 more years provide at least 90% power regardless of the underlying event rate, all while controlling type 1 error.

The adaptive aspect described in part 1 can also be added to risk ratio/risk difference combination to produce an adaptive sample size trial as in White et al. [17]. The trial designed by White, in collaboration with this author, used a design similar to the one described here to provide evidence toward CV safety of a new drug. It is a noninferiority trial that included the combination risk ratio/risk difference along with an adaptive sample size that ranged from 2500 to 4000 patients. Unlike the trial described in this chapter, it lacked the possibility to shift from a noninferiority to a superiority trial.

7.4 Summary

The last decade has seen numerous drug withdraws and increased FDA, congressional, and public scrutiny over drug safety. This has led to an increase in preapproval requirements, particularly the need to demonstrate CV safety for chronically administered drugs meant for non-CV use.

Meanwhile, the FDA Center for Devices and Radiological Health has been amenable to Bayesian adaptive designs for over a decade as illustrated by their guidance in 2010 [10] and numerous device approvals based on Bayesian adaptive designs. In 1997, the FDA Modernization Act included the "least burdensome mandate" with the goal to improve regulatory efficiency in medical devices [11]. Similarly in February 2010, the FDA Center for Drug Evaluation and Research (CDER) and Center for Biologics Evaluation and Research (CBER) jointly released draft guidance for adaptive clinical trial design [12]. Thus, there has been growing acceptance and even encouragement of Bayesian methods and adaptive clinical trials in all phases of drug and device development [13–16]. The 2010 FDA draft guidance on the use of adaptive designs for adequate and well-controlled drug trials addresses the key concerns of adaptive studies including the potential introduction of bias and the potential for adaptations to result in uninterpretable results or incorrect conclusions, with specific attention given to the type 1 error rate.

The Bayesian adaptive design detailed here meets the requirements for an adequate and well-controlled clinical trial. The sample size range is prespecified; however, rather than setting a fixed size with the possibility of stopping early as in group sequential designs, we offer a range and search for the optimal size within that range. Furthermore, the trial never stops immediately at a critical threshold; rather, the trial stops accrual for predicted success and allows for 30 additional months for follow-up. The *a priori* defined sample size range ensures a minimum amount of evidence, even if the effect size is large, and also a maximum sample size for planning purposes.

Group sequential designs, particularly in noninferiority safety studies where many events may occur after accrual during the follow-up phase, are unlikely to stop early. The design described here, instead, uses accruing information to predict how many events the current cohort is likely to provide with addition follow-up. Therefore, rather than having many patients with short exposure times, it relies on a smaller number of patients with greater exposure times—with that optimal number chosen adaptively and based upon internal trial data.

Furthermore, the delay between stopping accrual and performing the primary analysis spends less type 1 error than a group sequential trial with the sample number of interim analyses because success is not declared immediately. Rather the trial stops early for predicted success, and the final analysis is performed 30 months from the time the last patient is enrolled. Type 1 error is conserved here because if a trial stops for predicted success when in fact the drug is not heart safe, we may observe regression to the mean during the follow-up period and may not make a type 1 error. That is, during the extended follow-up period, we expect to observe more events in the treatment than control group, and the primary endpoint would not be met. If the treatment offers no increase in CV event versus the control, however, then the treatment will likely be demonstrated to be safe at the final analysis because we would continue to see a similar number of events in each group.

7.5 Conclusions

Here, we illustrate an adaptive trial that addresses the FDA guidance for antidiabetic medications. Furthermore, we introduce a second component, not explicitly discussed in the guidance, combining the risk ratio with the risk difference, which may be appropriate when expected event rates are very small. The risk ratio/risk difference combination has been used within the phase 3 adaptive safety drug trial [17,18.]

The trial is protected against loss in power due to misjudged *a priori* event rates while maintaining type 1 error control to meet the definition of adequate and well-controlled trials as described in the FDA CDER/CBER draft adaptive guidance.

References

1. Connolly HM, Crary JL, McGoon MD, Hensrud DD, Edwards BS, Edwards WD, and Schaff HV. Valvular heart disease associated with fenfluramine-phentermine. *The New England Journal of Medicine.* 1997; **337**: 581–588. doi: 10.1056/nejm199708283370901.
2. Karha J and Topol EJ. The sad story of Vioxx, and what we should learn from it. *Cleveland Clinic Journal of Medicine.* 2004; **71**: 933–939. doi: 10.3949/ccjm.71.12.933.
3. Loughlin J, Quinn S, and Rivero E. Tegaserod and the risk of cardiovascular ischemic events: An observational cohort study. *Journal of Cardiovascular Pharmacology and Therapeutics.* 2010; **15**: 151–157. doi: 10.1177/1074248409360357.
4. Nissen SE and Wolski K. Effect of rosiglitazone on the risk of myocardial infarction and death from cardiovascular causes. *The New England Journal of Medicine.* 1997; **356**: 2457–2471. doi: 10.1056/NEJMoa072761.
5. United States Food and Drug Administration Center for Drug Evaluation and Research. Guidance for industry: Diabetes mellitus—Evaluating cardiovascular risk in new antidiabetic therapies to treat type 2 diabetes, 2008. Available from: http://www.fda.gov/downloads/Drugs/GuidanceComplianceRegulatory Information/Guidances/ucm071627.pdf. Accessed June 26, 2012.
6. Broglio KR, Connor JT, and Berry SM. Combining risk difference and risk ratio in non-inferiority safety trials. *Journal of Biopharmaceutical Statistics.* 2013; **23**: 261–271.
7. Cushman WC, Evans GW, Byington RP et al. For the ACCORD study group. Effects of intensive blood pressure control in type 2 diabetes mellitus. *The New England Journal of Medicine.* 2010; **362**: 1575–1585. doi: 10.1056/NEJMoa1001286.
8. Yeh RW, Sidney S, Chandra M, Sorel M, Selby JV, and Go AS. Population trends in the Incidence and outcomes of acute myocardial infarction. *The New England Journal of Medicine.* 2010; **362**: 2155–2165. doi: 10.1056/NEJMoa0908610.
9. Broglio KB, Connor JT, and Berry SM. Not too big, not too small: A goldilocks approach to sample size selection. *Journal of Biopharmaceutical Statistics.* 2014; **24**(3): 685–705. doi: 10.1080/10543406.2014.888569.

10. United States Food and Drug Administration Center for Devices and Radiological Health. Guidance for industry and FDA staff: Guidance for the use of Bayesian statistics in medical device clinical trials, 2010. Available from: http://www.fda.gov/downloads/MedicalDevices/DeviceRegulationandGuidance/GuidanceDocuments/ucm071121.pdf. Accessed February 2013.

11. Food and Drug Administration Modernization Act (FDAMA) of 1997. Public Law 105–115. 105th Congress, 1997.

12. United States Food and Drug Administration Center for Drug Evaluation and Research and Center for Biologics Evaluation and Research. Guidance for industry: Adaptive design clinical trials for drugs and biologics (draft guidance), 2010. Available from: http://www.fda.gov/downloads/Drugs/GuidanceComplianceRegulatoryInformation/Guidances/ucm201790.pdf. Accessed February 2013.

13. Campbell G. The experience in the FDA's center for devices and radiological health with Bayesian strategies. *Clinical Trials.* 2005; **2**: 359–363. doi: 10.1191/1740774505cn093oa.

14. Campbell G. Statistics in the world of medical devices: The contrast with pharmaceuticals. *Journal of Biopharmaceutical Statistics.* 2008; **18**: 4–19. doi: 10.1080/10543400701668225.

15. Campbell G. Bayesian statistics in medical devices: Innovation sparked by the FDA. *Journal of Biopharmaceutical Statistics.* 2011; **21**: 871–887. doi: 10.1080/10543406.2011.589638.

16. Woodcock J, Temple R, Midthun K, Schultz D, and Sundlof S. FDA senior management perspectives. *Clinical Trials.* 2005; **2**: 373–378. doi: 10.1191/1740774505cn109oa.

17. White WB, Grady D, Giudice LC, Berry SM, Zborowski J, and Snabes MC. A cardiovascular safety study of LibiGel (testosterone gel) in postmenopausal women with elevated cardiovascular risk and hypoactive sexual desire disorder. *American Heart Journal.* 2012; **163**: 27–32. doi: 10.1016/j.ahj.2011.09.021.

18. BioSante Pharmaceuticals. Safety and efficacy of LibiGel® for treatment of hypoactive sexual desire disorder in postmenopausal women (BLOOM). ClinicalTrials.gov. Bethesda, MD: National Library of Medicine (US), 2000. Available from: http://clinicaltrials.gov/ct2/show/NCT00612742. NLM Identifier: NCT00612742. Cited February 21, 2013.

8

Observational Safety Study Design, Analysis, and Reporting

Hong Qiu, Jesse A. Berlin, and Paul E. Stang

CONTENTS

8.1 Applications of Observational Studies in Safety Evaluation

8.1.1 Where Observational Studies Can Fit

Drug safety is an important public health focus throughout the life cycle of a compound. The main data streams to monitor and examine drug safety are from randomized clinical trials (RCTs), spontaneous adverse event reports, and pharmacoepidemiologic safety studies.

RCTs are considered to be the gold standard to test causality, determine effect sizes, and assess risks and benefits of treatments. Randomization ensures that, on average, confounding factors (such as demographic status and comorbidities) are distributed equally across all treatment groups. The RCT design allows the objective observation of events in a controlled environment. However, the use of RCTs to fully characterize compound safety is limited by the fact that RCTs for product approval usually exclude important segments of the population who are at higher risk of adverse events, such as those who are elderly, women who are pregnant or lactating, those who have liver or kidney function impairment or other significant comorbidity, and those who are taking concomitant medications. Also, because of the short follow-up duration and limited sample size, most RCTs, and even entire development programs, are frequently underpowered to detect any increased risk of rare drug adverse events.

In the postmarketing period, drug adverse events are usually detected by routine pharmacovigilance practice from spontaneous reports. There is a substantial history of research that shows that these voluntary systems are plagued by substantial underreporting [1,2], so it is difficult to characterize safety issues until the product has been on the market for an extended period of time. The utility of the voluntary system is primarily to identify potential issues (i.e., signals) rather than help better understand and characterize specific drug event associations. Most of the statistical research using these data is focused on "signal-to-noise" ratio approaches or rates where the denominator represents the totality of reports rather than the actual exposure (proportional reporting rates) [3].

Observational studies of drug safety, conducted using large databases drawn from clinical practice, overcome some of the issues with RCTs and spontaneous reports. These databases reflect the experience of a potentially very large number of product users, thereby providing more generalizability, and facilitate the study of drug associations in the context of all of the comorbidities and diversity that one would expect when a product is marketed. Unlike with spontaneous reporting, data capture does not depend on a motivated reporter who suspects a causal relationship in a specific patient, but rather there is an analytic exercise focused on the strength of association between a medication and an outcome, while controlling for potential confounders that may also be captured in the same data. Furthermore, these databases offer a distinct, defined denominator of exposure, and descriptive analyses can help characterize the actual users of the product.

8.1.2 Study Design (Methods, Bias, Confounding by Indication)

8.1.2.1 *Choice of Study Design*

Multiple types of observational study design can be used in the quantitative measurement of risks from pharmaceutical products. Choice of a design depends on both the drug and the safety issue and should be tailored to address the specific hypotheses of interest.

The two basic types of observational study designs, cohort and case–control, are commonly used in pharmacoepidemiology to estimate the association between exposure and risk of events. Other designs, including self-controlled, case–cohort, or case–crossover designs, can be used depending on the study question of interest and what the nature is of the postulated relationship between drug exposure and the specific safety outcomes of interest. Choice of design can depend on a number of factors including (but not limited to) exposure persistence, duration of exposure effect period, strength of within-person confounding, strength of between-person confounding, and relative speed of onset of the outcome [4].

8.1.2.1.1 Cohort Study

In a cohort study, a group of subjects initially free of the study diseases (or events), with or without exposure to the drug of interest, is identified and followed for a certain period of time to ascertain the occurrence of the study diseases (or events). The rates of occurrence of the study diseases (or events) among two or more exposure groups are compared to estimate the relative risk. In some situations, cohorts are defined by a single exposure, for example, patients prescribed a particular medication and no comparator group is identified. Rates derived from such an approach can be compared to rates calculated from other data sources, possibly using different definitions of the outcome of interest and potentially introducing multiple additional potential sources of confounding.

In identifying the study cohort, one critical element is to define the exposure of the cohort. The exposed subjects may either be incident users, that is, patients who are newly exposed to the drug of interest at cohort entry, or prevalent users, that is, patients who may have started exposure to the drug of interest prior to cohort entry, as well as those recently initiating treatment. Although the prevalent-user design provides a more robust sample of those using the medication, it also may introduce multiple biases in assessing the relationship between exposure and outcomes [5] because the population of those who remain on a medication may be different from those at the start of treatment.

Because drug adverse events may occur at any time after the start of exposure, patients who experience early adverse events from the therapy may drop out and be underrepresented in the study cohort in a prevalent-user design; this would bias the study result toward the null. Thus, the prevalent-user cohort will usually be comprised largely of "survivors" of the early period of therapy or those in whom the treatment is more successful. The prevalent-user design may also introduce bias due to the varying onset time of risk factors for the outcome after initial exposure to the drug of interest. In some situations, ascertainment of potential confounders is made at the start of the observation period for a given patient, which can occur at different times after the true start of exposure, when the decision to prescribe the drug was made. Care needs to be taken to consider whether one might be statistically adjusting for factors on the causal pathway rather than for true confounders. For example, dyspepsia might be an early indicator of gastro-intestinal (GI) bleeding. If a drug causes dyspepsia, which evolves into overt bleeding, then adjusting for dyspepsia could inappropriately attenuate the estimated association between drug and bleeding.

In an incident-user design, cohort entry is defined as patients' start of therapy and is the observational design most similar to a randomized trial design, although treatment is not randomly assigned. This design eliminates many of the potential biases described earlier as the start time for exposure is known for each cohort member and all early events can be included in the analysis. The potential confounders are measured at cohort entry (prior to the start of therapy).

The incident-user design also has some limitations. It may be difficult to find patients newly exposed to a therapy, especially for drugs that have been on the market for a long period of time (as most will be prevalent users) or for those products with slow market uptake. In addition, in study-ing adverse events with a long lag time after exposure (e.g., cancer), an incident-user design may require longer follow-up times, and the cohort may have a large loss to follow-up. However, the more frequent use of electronic health-care record (EHR) databases in pharmacoepidemiology research has the promise of addressing some of these limitations. EHR databases usually include detailed information on prescriptions written (claims databases contain those prescriptions filled), most have active problem lists, and many also capture social and physiological details like smoking, blood pressure recordings, height, and weight. In some cases, the EHR will reflect more history and follow-up for a given patient, com-pared with claims data, as patients may change insurers but maintain the same physician.

A cohort study provides a complete picture from exposure to event occur-rence and provides a direct estimate of risk (incidence rate for both the exposed and comparator group). This facilitates the calculation of the pro-portion of an event that is attributable to the exposure (etiologic fraction). For a given exposure, a cohort study can investigate the relationship to multiple outcomes; however, a cohort study requires large numbers of subjects when studying rare outcomes or may require long follow-up for diseases with lon-ger latency (e.g., cancer).

8.1.2.1.2 Case–Control Study

Instead of the approach used in a cohort study to work from exposure to event, the principle behind a case–control design is first to identify those experiencing the event, then to determine who was exposed. Subjects are selected for a case–control study on the basis of the presence or absence of the study event, and prior exposure status is compared between the case and control groups. The controls should be comparable to the cases regarding the potential for exposure to the study drug(s).

Selection of controls is critical for a case–control design. In pharmacoepidemiology studies, the requirement that controls be similar to the cases in regard to past potential for exposure implies that the most likely source of the controls should be from the patients with diseases from the same background. For example, in investigating the relationship between an antipsychotic treatment and cardiovascular disease, the cases might reasonably be confined to patients with psychiatric disease indications. Often, medications are used for multiple indications, and the impact of the underlying disease being treated can be somewhat "controlled" by restricting the cohort to a particular indication. Specifically, for this example, in order to make sure the controls have the same chance to be exposed to antipsychotic treatment and have a similar baseline risk for cardiovascular disease, controls should also be chosen from among psychiatric patients.

The case–control design is appropriate for studying diseases with multiple possible causes. This design is especially useful for the study of rare diseases, since it does not depend on recruitment or identification of a large study sample in order to locate sufficient numbers of the cases. A case–control design, as well as a retrospective cohort study, can provide the opportunity to study diseases with long latencies. At the beginning of the investigation, all the cases have already occurred, which is the hallmark of the efficiency of this design. Although relative risk cannot be directly estimated from a case–control study, a well-planned case–control study can approximate the results from a cohort study, particularly if the disease is rare. Due to their "retrospective" nature, case–control studies are potentially subject to various biases mainly based on selection of cases and controls and how prior exposure is captured.

The advantages of a case–control design are less apparent in the context of an administrative (e.g., insurance claims) database, where there may already be large numbers of subjects with the necessary data to perform the analyses. When data collection is labor intensive, as it is with primary data collection directly from patients, caregivers, or paper medical records, the advantages of a case–control design are clearer.

8.1.2.1.3 Nested Case–Control Study

When cases are identified from a well-defined cohort, a nested case–control study can be carried out. In this design, all cases or a random sample of cases in the cohort will be identified as a case group. Controls are typically a random sample of the subjects remaining in the cohort at the time each case occurs (incidence density sampling). In this design, some subjects might appear in both case and control groups: cases occurring later in the follow-up can be potential controls for earlier cases. The nested case–control design has the advantages of both cohort and case–control design. Since both cases and controls are selected from the same source population, the likelihood of selection bias tends to be reduced compared to a regular case–control design, and the results are more credibly generalized to the population reflected in the cohort. As noted previously, this design is particularly efficient when extra information needs to be collected from participants in addition to the data available in the cohort, for example, when chart abstraction is required for subjects from an administrative database. In theory, all case–control studies can be considered as nested within a cohort, although the source cohort may not always be well-defined or easily identifiable.

8.1.2.1.4 Case–Crossover Study

Another special case–control design worth noting is the case–crossover design, which was introduced by Maclure [6] in 1991. This design only involves cases and compares the exposure just prior to the occurrence of the case (case period) to exposure in an earlier period of time (control period) for the same subject. In this way, each case serves as his or her own control, and the analytic units are not the subjects themselves, but the periods of time for the same subject. This self-matching approach minimizes selection bias and the individual characteristics that do not change over time and that may be potential confounding factors are "self-controlled." The case–crossover design is appropriate when the exposures are transient, vary over time, and have few carry-over effects; the effects on risk are immediate; and the outcomes are acute with minimal latency time. This design is attractive for many researchers because of the smaller sample sizes needed, making them more efficient and cost-effective. As may be expected, the selection of the control period and the length and timing of both the case period and control period are critical decisions. A major limitation of this design is the central assumption of independence between the baseline hazard and the study periods, which means that the baseline hazard should be invariably consistent over time. When this independence is not clear, using a case–crossover design will introduce significant confounding and bias. This design would not work, for example, for a study of hormonal contraceptives and risk of venous thromboembolism (VTE), where the risk is believed to be highest with initiation of the drugs [7,8]. The self-controlled case series [9,10] is a similar design that has gained much attention lately in pharmacoepidemiology.

8.1.2.2 Bias and Confounding

Identifying and controlling systematic error (bias) in observational studies is a critical concern in assuring the validity of the study results. Bias can arise from a number of sources and can be classified into three types: inappropriately selected study subjects (selection bias), errors during data collection/definition (information bias), or confounding. Most prominent in pharmacoepidemiology is confounding that arises when the indication for treatment is an independent risk factor for the outcome of interest (known as "confounding by indication"). For example, a treatment for diabetes may be under investigation for cardiovascular events, but diabetes itself confers an increased risk of the same events. Proton pump inhibitors are given to patients with GI symptoms that may represent early signs of GI bleeding, so the medication may appear to be associated with increased risk of bleeding, despite having demonstrated efficacy at preventing GI bleeding in randomized trials [11]. A full discussion of the various sources of bias in epidemiologic studies is beyond the scope of this chapter. For a more thorough understanding of these issues, a number of standard textbooks would provide a useful starting point [12,13].

8.1.3 Electronic Health-Care Databases

Electronic health-care databases, so-called automated databases, are computerized databases containing medical care or health-related data. There are a number of different types of databases, including claims data (coded transactions for reimbursement), electronic medical record (EMR), laboratory/diagnostic testing, facility data (hospital data, nursing home data), pharmacy data (Medco, IMS), surveillance data (SEER), prospective cohorts (Framingham), government surveys (e.g., NHIS, NHANES), registries (cancer, pregnancy, disease, exposure), or linked national databases (Sweden, Denmark). Ideally, databases used in this research should have the following characteristics: have large population coverage, contain complete information regarding a patient's medical/health condition, be representative of the general population from which it is drawn, be easily linkable by means of a patient's unique identifier, be updated on a regular basis, have records that are verifiable and accurate, and have the ability to conduct medical chart review to confirm outcomes. However, this is the ideal and is not generally what we find in the real world.

These databases have been utilized in pharmacoepidemiology research for several decades in North America and Europe. Most reflect the underlying qualities of the health-care delivery system. For example, EHR databases for a particular physician in the United States may not capture health care provided to the same patient outside of that practice, while databases from general practitioner (GP) gatekeeper systems (as in the United Kingdom) generally capture most of the patient's interactions because they must be referred and have care coordinated through the GP.

In the United States, claims databases are maintained by health insurance companies, health maintenance or managed care organizations, and the government (Medicaid and Medicare). Diagnosis (usually coded using ICD-9-CM codes) and prescribed treatment (usually prescriptions filled at the pharmacy) are usually well coded, and capture is fairly complete, as all those providing care to the patient will file for reimbursement. However, data not required for reimbursement (e.g., health behaviors, symptoms, signs, and over-the-counter medication use) are not usually captured. Furthermore, it can be difficult to distinguish between a true diagnosis and a "rule-out" diagnosis in these claim databases, as the clinician must provide a diagnosis for reimbursement. This ambiguous coding may cause misclassification.

EHRs were developed more recently, and their use is growing fast with increased computerization in medical care and the existence of government incentives. Clinicians use EHRs, instead of paper charts, to track patients to manage care. In the United States, EHRs may be inclusive of all care delivered within the managed care organization (e.g., Kaiser Permanente, Veterans Affairs) or be very fragmented, when the patient can be seen by any number of clinicians who may not be connected through a single EHR system. The strengths and limitations of these two types of electronic health-care databases are listed in the Table 8.1.

TABLE 8.1

Characteristics of Electronic Health-Care Databases

Feature	Medical Claims Database (from Insurance Company)	Electronic Health Record
Medical diagnoses	Coded and may be misleading because it is used for reimbursement.	May be coded or in text form; may also be part of active "problem" list.
Follow patient over time	In the United States, patients change insurance companies often (average every 2 years).	In the United States, patients change doctors less often than insurance companies.
Patient ID	May have different IDs for the same patient if the patient changes insurance company.	Usually can keep the same identifier for the same patient (within the same health-care system).
Medical history	Available as coded data only	Usually available as complete narrative
Signs, symptoms	Not captured	Captured in most systems
Surgical procedures	Coded	May be in a separate EHR if a surgery is performed by a different health-care provider
Prescriptions	Filled at pharmacy captured	Written by physician (so contained in EHR) but may not reflect dispensed medications
Diagnostic procedures	Yes	May not be captured if outside provider
Nonprescription medication and recreational drug (e.g., tobacco, alcohol) use	Not captured	May be captured in clinician notes
Clinician insight regarding causality between event and drug	Not captured	May be captured in clinician notes
Laboratory testing and results	Will capture that test was performed but not results	Captures tests performed and results for a given clinician
Data format	Standardized	Not standardized

8.1.4 Recent Empirical Findings Informing Study Design Choices: Lessons from the Observational Medical Outcomes Partnership

The most widespread use of observational research has been in medication safety research with most of the current attention being focused on the use of these large databases and methods for active surveillance. The concept has led to congressional action as the 2007 U.S. Food and Drug Administration (FDA) Amendments Act mandated that the FDA develop a system for using automated health-care data to identify risks of marketed drugs and other medical products. Under the direction of this act, the FDA launched in 2008 the Sentinel Initiative (http://www.fda.gov/safety/FDAsSentinelInitiative/ucm2007250.htm), which aims to develop and implement a proactive system that will complement existing systems. FDA's "Mini-Sentinel" pilot program, the FDA's initial step toward building a nationwide rapid-response electronic safety surveillance system for drugs and other medical products, includes 17 data partners across the United States and encompasses the data of nearly 100 million patients.

The Observational Medical Outcomes Partnership (OMOP; http://omop.org) is a public–private partnership among the FDA, academia, data owners, and the pharmaceutical industry that is focused on advancing the science of active medical product safety surveillance using existing observational databases. The OMOP's transparent, open innovation approach is designed to systematically and empirically study critical governance, data resource, and methodological issues and their interrelationships to inform active drug safety surveillance in observational data [14].

The OMOP program of research is extensive and has produced a steady output of interesting methodological and statistical insights. OMOP is focused on a set of "positive" and "negative" associations as exemplars against which a set of "experiments" have been implemented. The 165 positive associations were drawn from a systematic approach of identifying drug–outcome pairs with credible evidence of an association, while the 234 negative controls, based on the same medications, were selected based on the absence of such evidence. A series of methods (e.g., cohort, case–control, self-controlled, observational screening, and disproportionality) and the myriad study design decisions, or parameters, were systematically explored across five databases in the United States (total > 75 million lives) to determine the extent to which a method correctly identified the positive associations while failing to "signal" for the negative associations. The methods represent a spectrum of approaches to confounding adjustment from crude/unadjusted incidence rate ratios, to age-by-gender stratification, to the more advanced high-dimensional propensity score adjustment and Bayesian model to condition on all concomitant drug exposures. Their performance was characterized using receiver operating characteristic (ROC) curves, which plot sensitivity versus false-positive rate, the area which provides the probability that method will score a randomly chosen true-positive drug–outcome pair higher than a random unrelated drug–outcome pair. Using the OMOP approach, a risk identification system can perform at AUC > 0.80. The OMOP website provides extensive documentation on the performance of the methods across databases, outcome definitions, and parameter settings.

In general, the OMOP experiments have found differences by method, outcome definition, and database. A risk identification system should confidently discriminate positive effects with RR > 2 from negative controls in order to inform rational decision making. The findings included that database heterogeneity is common, as we have seen from a number of published studies, and the OMOP results suggest that replication has value but may result in conflicting results. Among the methods tested, self-controlled designs generally appeared to consistently perform well, but each method has parameters that are more sensitive than others and can substantially shift some drug–outcome estimates. Finally, broader outcome definitions have better coverage and comparable performance to more specific definitions.

OMOP also explored a number of statistical issues, particularly around significance levels in observational studies. Comparing the empirical distribution of negative controls with that of positive controls provided complementary information beyond the traditional p-value, leading to the conclusion that p < 0.05 did not guarantee that a true effect existed nor did a p > 0.05 guarantee that no effect was present. Their work suggests that using adjusted (calibrated) p-values to account for bias will provide a more appropriate assessment of whether an observed estimate is different from "no effect." In addition, the traditional interpretation of the 95% confidence interval as covering the true effect size 95% of the time may be misleading in the context of observational database studies. Specifically, OMOP found that coverage probability is much lower across all methods and all outcomes that were evaluated; these results were consistent across real data and simulated data. They suggested empirically adjusting confidence intervals to derive more robust coverage probabilities across method–outcome pairs.

8.2 Analysis in Observational Studies

8.2.1 Prespecified Analysis Plan

As noted earlier, studies should have a protocol in which all details of study design, conduct, and analysis are described. The protocol should include a prespecified analysis plan that addresses each of the specific study objectives. The importance of a prespecified analysis plan cannot be overemphasized. The potential for different analytic approaches to yield different results opens the analytic process to potential manipulation, the possibility for which can be greatly reduced by prespecification. It will generally be important to prespecify both primary and secondary analyses. If any preliminary analyses are to be conducted, for example, early looks at the data for reassessing feasibility or reevaluating sample size requirements, these should be prespecified, as well [15–17].

The protocol should present statistical power calculations, estimates of anticipated precision, or both. This is particularly relevant to studies of rare events, for which the interpretation of elevated relative risks that are not statistically significant creates challenges.

As noted in the ISPE guidelines [15], the analysis of nonrandomized studies should aim for the unbiased estimation of the parameters of interest (e.g., risk or rate differences, risk or rate ratios). Representing the precision of effect estimates is also crucial and should be quantified separately using confidence intervals. Typically, it will be helpful for both unadjusted and adjusted results to be presented. In studies using large databases, in particular, precision may be so great (i.e., confidence intervals may be so narrow) that statistical significance might be achieved, even for relatively small increases in risk or rate which may not be clinically meaningful. Regardless of the statistical significance, though, adjustment for potential confounders is an essential part of any analysis.

Having made the point that adjustment for confounders is necessary, it will seldom, if ever, guarantee complete adjustment. Consequently, residual confounding is almost always a possibility in any epidemiologic investigation. Thus, the impact of unmeasured confounders should always be considered. The inclusion of both adjusted and unadjusted results makes it possible to observe the effects of adjustment, but doesn't necessarily predict the effect that unmeasured confounders could have on the results.

8.2.2 General Points

In general, the nature of the analysis needs to match the question of interest. This is an obvious point in principle but requires thought in any particular instance. For example, if follow-up time varies across individuals, then an analysis that takes duration of follow-up into account will be appropriate. This might be an analysis of rates, with person-time in the "denominator," or it might be a time-to-event analysis, such as a Cox proportional hazard model. If duration of follow-up is relatively constant across individuals, then an analysis of risks would be appropriate, for example, a logistic regression analysis. Logistic regression is also (usually) an appropriate tool for the analysis of case–control studies because of unique properties of the odds ratio. When matching is used in the design phase of a study, matched analyses should be used, especially in case–control studies, where bias can actually be introduced by failure to retain the matching.

Regardless of the choice of analytic framework, the assumptions underlying any modeling need to be understood and examined. For example, models of rates, for example, Poisson regression models, require the assumption of constant hazards. This assumption implies that following 5000 people for 1 year (yielding 5000 person-years of follow-up) is equivalent to following 1000 people for 5 years. For outcomes with a long latency, such as cancer, this assumption may clearly be inappropriate. Cox models require the assumption (as the name implies) of proportional hazards, that is, the increase or decrease in the hazard (rate) of the event is constant over the duration of follow-up.

8.2.3 Use of Specific Statistical Techniques (e.g., to Minimize Confounding)

Appropriate statistical techniques to adjust for confounding should be clearly specified in the analysis plan, with justification and articulation of assumptions and limitations. *Post hoc* analyses, when they are considered appropriate, should be clearly indicated as such in any study reports. The same principles apply to assessment of effect modification, with the additional requirement that results should always be presented stratified by the specific effect modifier of interest.

Adjustment methods for confounding fall into several categories: stratification, multivariable models (such as logistic regression), and more advanced methods such as use of propensity scores or instrumental variables.

It will almost always be appropriate to do some kind of stratified analysis, that is, presenting contingency tables of exposure against outcome, stratified by the potential confounder or effect modifier, in order to get a better understanding of the data, particularly the sparseness of the data in specific strata. Proceeding straight to a multivariable model without a thorough understanding of the limitations of the data could lead to inappropriate analyses and interpretations of results. Choice of category definitions should be made in advance. Categories that are too sparse can lead to data analysis problems, but categories that are too broad risk residual confounding within those categories. Changes to definitions might be required once an analysis is underway. This is acceptable as long as the reasons for the revision are justified and these *post hoc* decisions are clearly indicated as such in study reports. When at all possible, the results of both the originally planned stratification and the revised analysis should be presented, to allow evaluation of the effects of the change.

Modeling is a more general and flexible approach to control for confounding, but the flexibility also introduces a number of challenges. In particular, model specification is crucial. The goal is to include all of the measured and appropriate variables and to ensure that the form of those variables is correctly specified. For example, adjustment for age can be done using predefined age categories or using age as continuous. Including age as a continuous variable assumes a linear relationship between age and outcome. This may not be a valid assumption, so investigation of nonlinearity and the possible use of quadratic terms or splines might be required. Using categories avoids the linearity assumption but generates the same challenges as stratification noted earlier. Prespecifying the choice of variables to be included in models is ideal, but in any case, a confounder selection algorithm should be specified. A popular such algorithm is the change-in-estimate criterion, which essentially includes variables in models that produce a change between the unadjusted and adjusted estimates of association [18]. A key point is that these modeling choices can affect results, again highlighting the importance of prespecification.

A detailed examination of propensity scoring is beyond the scope of this chapter. In brief, the propensity score is a model (often a logistic regression) of exposure status, that is, one is modeling exposure to the study drug of interest versus the comparator drug. This model yields a probability of exposure to study drug, based on observed characteristics (potential confounders). Stratification on propensity score quantile (e.g., quintile or decile), or matching on propensity score, produces matched sets or strata in which the probability of exposure is equal. In principle, this similarity of probabilities mimics randomized assignment to treatment and allows a valid comparison between exposed and unexposed groups.

Instrumental variables are a technique originally developed in economics. An instrumental variable is one that is related to exposure status but unrelated to outcome, except through exposure. Randomization is an ideal instrument, and use of randomization as an instrumental variable has been advocated in the context of pragmatic, randomized studies in which there might be reduced compliance with treatment assignment or loss to follow-up [19,20].

8.2.4 Sensitivity Analyses

For any study, sensitivity analyses should be performed to assess the impact of various study decisions relating to design, exposure definition, outcome definition, or choice of analytic approach. Sensitivity analyses, by demonstrating the robustness (or lack thereof) of conclusions to these design and analysis choices, can contribute to a better interpretation of study results. As should be evident by now, it is important to describe in detail any sensitivity analyses that are performed.

8.2.5 Assessment and Handling of Missing and Uninterpretable Data

The analysis plan should also specify how missing data will be handled (e.g., weight or smoking status is only available on a subset of the data). At a minimum, one should always describe the percentage of missing data for variables of interest, including potential confounders. A common practice is to assume that lack of positive information on the occurrence of an event (such as dialysis) or a risk factor (such as morbid obesity) indicates that the event or risk factor did not occur in that individual. Such an assumption may not always be valid, although in some instances, it might be justified.

More sophisticated approaches to missing data, such as multiple imputation [21,22], can be used to "fill in" the missing values in an unbiased manner, taking appropriate account of the fact that the replacement data were imputed, not actually observed, which has implications for estimating variability.

8.2.6 Quality Assurance and Quality Control

The FDA guidance on pharmacoepidemiologic studies states: "The quality control (QC) and quality assurance (QA) plan for the construction of the analytical data set(s) and analysis of data should be clearly described. QC consists of the steps taken during the analysis to ensure that it meets pre-specified requirements and that it is reproducible." [17] In addition, steps taken to ensure confidentiality and security of the data should be prespecified.

8.2.7 Transparency

In many situations, especially those in which a protocol is prepared as discussed earlier, there is a strong argument in favor of registration of the protocol, on a site such as clinicaltrials.gov, and a commitment needs to be made to disclosure of the study findings. Chavers and colleagues [23] recommend registration of protocols for observational studies, particularly prospective observational studies as they are "… nearly analogous to randomized control trial registration in terms of the benefits of registration." They also recommend registration of studies conducted in existing databases (such as claims data), when such studies have *"a priori* defined hypotheses and protocol-defined study conduct elements, such as studies conducted in response to regulatory requirements or funded though research grants." An option that is now available is the registration of prospective studies (registries) in the AHRQ "Registry of Patient Registries." [24] In addition, the ISPE guidance calls for clear documentation of all data management and statistical analysis programs and the archiving of these programs [15].

8.3 Reporting of Observational Study Results

Pharmacoepidemiology study reports should contain detailed study elements, including background, study objectives, study design, statistical analyses, detailed study results, interpretation of the findings, and discussion of study limitations. Transparency in reporting is essential. A brief description of the study drug(s) and safety concern(s) under investigation provides the study context. Study population and time period, data sources used, and methods to control for bias and confounding are important components of study design.

Statistical methods need to be described in enough detail to allow an appropriately trained reader to reproduce what was done. All study results, including sensitivity analysis and exploratory analysis results, need to be reported. Key results of the study should be interpreted without subjective speculation, and major limitations need to be discussed. If no association between the study drug and the safety outcome is found, investigators need to thoroughly discuss the statistical power and potential level of risk that can be ruled out, given the study findings. For studies that detect a statistically significant association between study drug and a safety outcome, investigators need to take into account the clinical significance of the findings, especially in large database studies where a small effect might be highly significant. The STROBE guidelines provide details to follow for reporting observational studies [25].

References

1. Hazell, L. and S.A. Shakir, Under-reporting of adverse drug reactions: A systematic review. *Drug Saf*, **29**(5): 385–396, 2006.
2. Alvarez-Requejo, A. et al., Under-reporting of adverse drug reactions. Estimate based on a spontaneous reporting scheme and a sentinel system. *Eur J Clin Pharmacol*, **54**(6): 483–488, 1998.
3. Evans, S.J., P.C. Waller, and S. Davis, Use of proportional reporting ratios (PRRs) for signal generation from spontaneous adverse drug reaction reports. *Pharmacoepidemiol Drug Saf*, **10**(6): 483–486, 2001.
4. Gagne, J.J. et al., Design considerations in an active medical product safety monitoring system. *Pharmacoepidemiol Drug Saf*, **21**(Suppl. 1): 32–40, 2012.
5. Ray, W.A., Evaluating medication effects outside of clinical trials: New-user designs. *Am J Epidemiol*, **158**(9): 915–920, 2003.
6. Maclure, M., The case-crossover design: A method for studying transient effects on the risk of acute events. *Am J Epidemiol*, **133**(2): 144–153, 1991.
7. Huerta, C. et al., Risk factors and short-term mortality of venous thromboembolism diagnosed in the primary care setting in the United Kingdom. *Arch Intern Med*, **167**(9): 935–943, 2007.
8. Lidegaard, O., B. Edstrom, and S. Kreiner, Oral contraceptives and venous thromboembolism: A five-year national case-control study. *Contraception*, **65**(3): 187–196, 2002.
9. Farrington, C.P., Relative incidence estimation from case series for vaccine safety evaluation. *Biometrics*, **51**(1): 228–235, 1995.
10. Whitaker, H.J. et al., Tutorial in biostatistics: The self-controlled case series method. *Stat Med*, **25**(10): 1768–1797, 2006.
11. Leontiadis, G.I. et al., Systematic reviews of the clinical effectiveness and cost-effectiveness of proton pump inhibitors in acute upper gastrointestinal bleeding. *Health Technol Assess*, **11**(51): iii–iv, 1–164, 2007.
12. Rothman, K.J., S. Greenland, and T.L. Lash, *Modern Epidemiology*, 3rd edn. Philadelphia, PA: Wolters Kluwer Health/Lippincott Williams & Wilkins, 2008, p. x, 758pp.

13. Strom, B.L., *Pharmacoepidemiology*. Chichester, England; Hoboken, NJ: John Wiley, 2005, p. xvii, 889pp.
14. Stang, P.E. et al., Advancing the science for active surveillance: Rationale and design for the Observational Medical Outcomes Partnership. *Ann Intern Med*, **153**(9): 600–606, 2010.
15. Andrews E.B. et al., Guidelines for good pharmacoepidemiology practices (GPP). *Pharmacoepidemiol Drug Saf*, **17**(2): 200–208, 2008.
16. Motheral, B. et al., A checklist for retrospective database studies—Report of the ISPOR Task Force on Retrospective Databases. *Value Health*, **6**(2): 90–97, 2003.
17. FDA. Guidance for industry and FDA staff: Best practices for conducting and reporting pharmacoepidemiologic safety studies using electronic healthcare data sets. http://www.fda.gov/downloads/Drugs/GuidanceCompliance RegulatoryInformation/Guidances/UCM243537. 2011. Last accessed: February 21, 2013.
18. Maldonado, G. and S. Greenland, Simulation study of confounder-selection strategies. *Am J Epidemiol*, **138**(11): 923–936, 1993.
19. Greenland, S., An introduction to instrumental variables for epidemiologists. *Int J Epidemiol*, **29**(6): 1102, 2000.
20. Greenland, S., S. Lanes, and M. Jara, Estimating effects from randomized trials with discontinuations: The need for intent-to-treat design and G-estimation. *Clin Trials*, **5**(1): 5–13, 2008.
21. Sterne, J.A. et al., Multiple imputation for missing data in epidemiological and clinical research: Potential and pitfalls. *BMJ*, **338**: b2393, 2009.
22. Allison, P.D., Multiple imputation for missing data. A cautionary tale. *Sociol Methods Res*, **28**: 301–309, 2000.
23. Chavers, S. et al., Registration of observational studies: Perspectives from an industry-based epidemiology group. *Pharmacoepidemiol Drug Saf*, **20**(10): 1009–1013, 2011.
24. AHRQ Registry of Registries. Registry of Patient Registries (RoPR) Policies and Procedures. http://effectivehealthcare.ahrq.gov/ehc/products/311/1115/ DEcIDE41_Registry-of-patient-registries-policies-procedures_FinalReport_ 20120531.pdf. 2012. Last accessed: August 5, 2014.
25. von Elm, E. et al., The strengthening the reporting of observational studies in epidemiology (STROBE) statement: Guidelines for reporting observational studies. *J Clin Epidemiol*, **61**(4): 344–349, 2008.

9

Emerging Role of Observational Health-Care Data in Pharmacovigilance

Patrick Ryan, David Madigan, and Martijn Schuemie

CONTENTS

9.1 Background

Prior to regulatory approval, while a drug is in development, randomized clinical trials represent the primary sources of safety information. Such experiments are generally regarded as the highest level of evidence, leading to an unbiased estimate of the average treatment effect (Atkins et al. 2004). Unfortunately, most trials suffer from insufficient sample size and lack of applicability to reliably estimate the risk of other potential safety concerns for the target population (Berlin et al. 2008; Waller and Evans 2003). Even if one leverages meta-analytic tools, rare side effects, long-term outcomes (both positive and negative), and effects in patients with comorbidities may still be unknown when a product is approved because of the relatively small size and short duration of clinical trials. For products intended to treat chronic, non-life-threatening conditions that occur in large populations, the International Conference on Harmonisation (Azoulay et al. 2012) recommends a baseline safety database that involves at least 1500 patients on average with at least a 6-month exposure time to reliably (i.e., 95% of the time) identify events happening at the 1% level (U.S. Food and Drug Admin 1999). In other words, events that occur less frequently than 1 in 100 patients are not expected to be detected under this recommendation.

Once a drug has been approved and is introduced on the market, the FDA's "postmarketing surveillance programs focus primarily on (1) identifying events that were not observed or recognized before approval, and (2) identifying adverse events that might be happening because a product is not being used as anticipated" (Furberg et al. 2006). Spontaneous adverse event reporting remains the cornerstone of pharmacovigilance activities. The FDA receives about 400,000 reports annually, primarily from drug manufacturers who are required to report serious, unexpected safety events within 15 days, with further reports coming directly from health-care providers and patients (Lasser et al. 2002). A primary use of spontaneous adverse event reports is to facilitate clinical review of case series. A case series of related events is often the first initial warning that a potential association may exist. Occasionally even a single reliable case report may be sufficient to provide definitive evidence about a rare, serious idiosyncratic event (Aronson and Hauben 2006). While some authors regard case series as occupying a low rank on the hierarchy of evidence, behind observational cohort studies and randomized trials (Horisberger and Dinkel 1989), the use of case series analysis certainly has a prominent role when other information is limited (Vandenbroucke 2001). With hundreds of thousands of reports submitted to FDA, WHO, and other organizations each year, clinical review of every report becomes impractical, and statistical data mining algorithms are becoming increasingly popular tools for safety reviewers (Burton et al. 2007; Noren and Edwards 2009). Various disproportionality (DP) analysis methods exist, but each approach attempts to answer the same question: which drug–event combinations are reported more frequently than we would have expected if the drug and event were truly independent (Hauben et al. 2005)? Experience gained internationally shows that spontaneous reporting is effective in providing information about a wide range of different adverse effects and other drug-related problems. It has been particularly helpful in detecting adverse events that are often allergic or idiosyncratic reactions, characteristically occurring in only a minority of patients and usually unrelated to dosage, and that are serious, unexpected, and unpredictable and unusual effects that are related to the pharmacological effects of the drug and are dosage related (Meyboom et al. 1997). On the other hand, some authors have suggested that spontaneous reporting is of less use for the study of adverse effects with a relatively high background frequency and occurring without a suggestive temporal relationship (Meyboom et al. 1999). However, recent work has suggested that DP analyses of spontaneous reports can achieve performance levels on a par with some standard observational study approaches (Harpaz et al. 2013).

While the spontaneous adverse event reporting system has value in generating hypotheses about potential associations, it has several limitations that make causal assessments difficult: voluntary reporting suffers from chronic underreporting and other biases, and the unknown nature of underlying population makes true reporting rates difficult to obtain and use for comparisons. It has been estimated that only about 1% of all adverse drug reactions and about 10% of all serious adverse drug reactions are reported (Furberg et al. 2006). Reports are "usually based on suspicion, and may be preliminary, ambiguous, doubtful or wrong" (Meyboom et al. 1999).

9.2 Opportunity for a Risk Identification System

Various stakeholders have recognized the need to improve current pharmacovigilance practice and the opportunities that exist to expand the use of observational data in that pursuit (Berlin 2008). In 2007, the U.S. Congress passed the FDA Amendment Act, which in part mandated the "establishment of a postmarket risk identification and analysis system" that leverages observational health-care data, including administrative claims and electronic health records (EHRs), to monitor approved medicines on a periodic basis. In response, FDA established the Sentinel Initiative, an effort to create and implement a national, integrated, electronic system for monitoring medical product safety (Platt et al. 2009).

While still in its infancy, there is much debate about the intended design and scope of a national medical product postmarket risk identification and analysis system. One vision of a risk identification system will involve a systematic process for analyzing multiple observational health-care data sources, including administrative claims and EHRs, to better understand the effects of medical products by estimating temporal relationships between exposure and outcomes. The risk identification system can be used to (1) characterize known side effects, (2) monitor preventable adverse events, and (3) explore remaining uncertainties. The goal of the risk identification system is to contribute supplemental information to other existing sources of safety information (including preclinical data, clinical trials, and spontaneous adverse event reporting) to support decision making about appropriate use of medical products for regulatory agencies, providers, and patients.

Figure 9.1 provides a conceptual framework for risk identification. There are various sources of risk of medical products that can result in injury or death, including known side effects, medication and device errors, product defects, and other remaining uncertainties. These risks are influenced by many factors, including patient characteristics (such as demographics, comorbidities, concomitant medications, and health service utilization), health system factors (such as utilization practice and provider behavior),

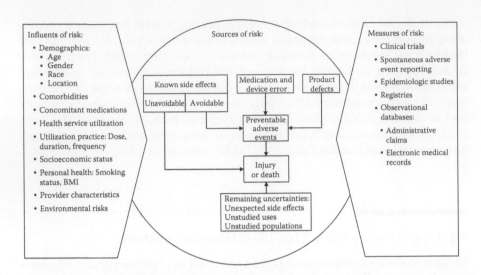

FIGURE 9.1
Conceptual framework for a risk identification system.

and other environmental sources. The discovery of how treatment effects vary by baseline risk is one of the important contributions of postmarketing surveillance of drugs (Sturmer et al. 2008). The current measures of risk include clinical trials, spontaneous adverse event reporting systems, epidemiologic studies, and registries. A risk identification system offers the opportunity for the systematic use of observational health-care databases, such as administrative claims and EHRs, to improve our measures of the sources of risks. Analyses against these data can account for the measurable influents of risk to provide robust, supplemental information that can be used to both identify and evaluate potential drug safety issues. While evaluation studies have been common practice for decades, use of these data in a formal exploratory analytic framework is new and requires further research to determine its relative contribution to such a system.

When considering drug safety in a causal inference framework, one can consider Hill's considerations of (1) strength, (2) consistency, (3) specificity, (4) temporality, (5) biologic gradient, (6) plausibility, (7) coherence, (8) experimental evidence, and (9) analogy (Hill 1965). The strength of association should be considered because stronger associations may be more compelling, but weak associations do not rule out causal connections (Rothman 2002). Consistency refers to the repeated observation of an association in different populations under different circumstances. Specificity relates to the number of causes that lead to a specific effect and the number of effects produced from a given cause. Temporality refers to the necessity that the cause precedes the effect. Biologic gradient addresses the degree to which there is a dose–response relationship, where the amount of response increased with increased exposure. Plausibility reflects the scientific rationale for the existence of an association, typically in drug safety, related to the mechanism of action and the biologic pathways that lead to the effect. Coherence is the degree to which the interpretation of the association does not conflict with the current understanding of the natural history and biology of the disease. Experimental evidence for drug safety analyses typically refers to evidence that comes from human randomized clinical trials, but can also include randomized preclinical experiments in animal models.

A risk identification system can apply Hill's considerations as part of its process for generating and evaluating hypotheses. Specifically, analyses conducted across a network of observational databases can be used to identify potential drug safety issues based on strength, consistency, specificity, and temporality. Methods produce estimates of the strength of temporal associations between exposure and subsequent outcomes. Applying the methods to multiple sources provides an assessment of consistency, as formal tests for heterogeneity can be used to measure differences between source populations. Evaluating multiple outcomes for each drug and multiple exposures for each outcome can provide insights into the specificity of any specific drug–outcome association. However, these exploratory analysis results will not be sufficient to address issues of biologic plausibility, and the use of observational data does not meet the same standards of evidence that come from a randomized experimental design. Methods for studying dose effects require further research, as the degree to which dose and amount of exposure can be accurately measured and used within a hypothesis generating framework is undetermined.

While hypothesis generating analyses are inherently exploratory in nature, basic principles of formal evaluation can be applied to raise the collective confidence in the reliability of the process. Research questions and statistical analyses should be specified in advance, with all methodological considerations addressed during study planning rather than after study completion. This includes decisions around definitions of exposure and outcome, inclusion/exclusion criteria imposed on the sample, and strategies for statistical adjustment (Horisberger and Dinkel 1989). Analysis processes should be fully transparent and reproducible and should minimize subjective assessment to improve the generalizability of the approach. Many of these principles are well defined in guidelines for conducting full evaluation studies (Schneeweiss et al. 2009) but have not yet been adopted for exploratory analyses. With these principles in place, hypothesis generation can play an important role in a risk identification system's contribution to causal inference of drug safety issues. These exploratory analyses can identify and prioritize areas that warrant further examination. Evaluation studies may be used to refine estimates of the strength of the association, but attention can particularly focus on biologic plausibility and coherence to put the preliminary results in proper clinical context with other evidence, including clinical trials, preclinical data, spontaneous adverse events, and other epidemiologic studies.

9.3 Review of Observational Databases

Administrative claims and EHR databases have been actively used in pharma-coepidemiology for more than 30 years (Strom 2005) but have seen increased use in the past decade owing to increased availability at lower costs and technological advances that made computational processing on large-scale data more feasible. Observational database studies provide empirical investigations of exposures and their effects but differ from experiments in that assignment of treatment to subjects is not controlled (Rosenbaum 2002). Observational database studies can be based on many forms of epidemiologic investigation, applying alternative study designs and employing a range of analysis strategies (Hartzema et al. 1999, 2008; Jewell 2004; Rothman 2002; Rothman et al. 2008; Szklo and Nieto 2007). Many such databases contain large numbers of patients that make possible the examination of rare events and specific subpopulations that previously could not be studied with sufficient power (Rockhill et al. 1998; Rodriguez et al. 2001). Because the data reflect health-care activity within a real-world population, they offer the potential to complement clinical trial results. Long-term longitudinal capture of data in these sources can also enable studies that monitor the performance of risk management programs or other interventions over time (Weatherby et al. 2002).

Administrative claims databases have been the most actively used observational health-care data source. These databases typically capture data elements used within the reimbursement process, as providers of health-care services (e.g., physicians, pharmacies, hospitals, and laboratories) must submit encounter information to enable payment for their services (Hennessy 2006). The submitted information commonly includes pharmacy claims for dispensing of prescription drugs (e.g., the drug dispensed, the dispensing date, and the number of days of drug supply) and medical claims (inpatient and outpatient) that detail the dates and types of services rendered. Medical claims typically contain diagnosis codes used to justify reimbursement for the services provided. Information on over-the-counter drug use and in-hospital medication is usually unavailable, and the patient's compliance with the prescription is generally unknown (Suissa and Garbe 2007).

EHRs generally contain data captured at the point of care, with the intention of supporting the process of clinical care as well as justifying reimbursement and providing data for quality measurement. A patient record may include demographics (birth date, gender, and race), height and weight, and family and medical history. Many EHR systems support provider entry of diagnoses, signs, and symptoms and also capture other clinical observations, such as vital signs, laboratory results, and imaging reports. Beyond those, EHRs may often contain findings of physical examinations and the results of diagnostic tests (Schneeweiss and Avorn 2005). EHR systems usually also have the capability to record other important health status indications, such as alcohol use and smoking status (Lewis and Brensinger 2004), but the data may be missing in many patient records (Hennessy 2006). As a result of discontinuous care within the U.S. health-care system, a patient may have multiple EHRs scattered across the providers the individual has seen, but rarely are those records integrated, nor can they usually be linked, so each EHR reflects a different and incomplete perspective of that person's health-care experience. Recent efforts to advance health information exchange aim to reduce this fragmentation.

Neither administrative claims nor EHRs represent the ideal information required to assess a particular effect. For example, diagnoses recorded on medical claims are used to support justification for the payment of a given visit or procedure; a given diagnosis could represent the condition that the procedure was used to rule out or could be an administrative artifact (e.g., the code used by a medical coder to maximize the reimbursement amount). Some diagnosis codes have been studied through source record verification and have demonstrated adequate performance characteristics (Donahue et al. 1997; García Rodríguez and Pérez Gutthann 1998; Hennessy et al. 2009; Lee et al. 2005; Miller et al. 2008; Pladevall et al. 1996; So et al. 2006; Tunstall-Pedoe 1997; Varas-Lorenzo et al. 2008; Wahl et al. 2010; Wilchesky et al. 2004), whereas other conditions and systems provide less certainty (Harrold et al. 2007; Leonard et al. 2008; Lewis et al. 2007; Strom 2001). Limitations exist in EHR systems as well, in which, apart from concerns about incomplete capture, data may be artificially manipulated to serve clinical care (e.g., an incorrect diagnosis recorded to justify a desired medical procedure). Most systems have insufficient processes to evaluate data quality a priori, requiring intensive work on the part of the researcher to prepare the data for analysis (Hennessy et al. 2007). To estimate potential drug exposures, for example, researchers can make inferences in administrative claims sources based on pharmacy dispensing records, whereas inferences for EHR systems rely on patient self-report and physician prescribing orders (Hennessy 2006). Neither approach reflects the timing, dose, or duration of drug ingested, so assumptions are required in interpretation of all study results.

9.4 Review of Observational Database Methods

The principal concern for all observational studies is the potential for systematic error or bias. Schneeweiss and Avorn illustrated some of the potential sources of bias that are introduced throughout the data capture process for both administrative claims and EHRs (Schneeweiss and Avorn 2005). An observational study is biased if the treated and control groups differ prior to treatment in ways that can influence the outcome under study (Rosenbaum 2002). Several forms of bias can arise in the design and conduct of an observational study. In the context of drug safety analyses, one of the most challenging issues is confounding by indication, that is, a situation in which the indication for the medical product is also an independent risk factor for the outcome (Walker 1996). Therefore, a medical product can spuriously appear to be associated with the outcome when no appropriate control for the underlying condition exists, and confounding may persist despite advanced methods for adjustment (Bosco et al. 2010). For example, proton pump inhibitors might produce an apparent increase in risk of gastrointestinal bleeding. This apparent increase could arise because that class of drugs is used to treat symptoms that might also be indicative of such bleeding. A predisposition for health-care utilization can also produce confounding, perhaps because of functional status or of access due to proximity, economic, and institutional factors (Brookhart et al. 2010b). An additional concern is immortal time bias, whereby outcomes are not observable within the defined time at risk (Rothman and Suissa 2008; Suissa 2007, 2008).

The prominent observational database study design used in most pharmacoepidemiology circles and under consideration for use within a national postmarket risk identification and analysis system is the cohort design. Within a cohort design, patients exposed to a target treatment on interest are compared with a different group of patients who are not exposed to the treatment, and the rates of events that occur during some period postexposure are compared between the two groups. Cohort studies allow many possible approaches to address confounding. One design strategy is to impose restrictions on the selected sample to increase validity, potentially at the expense of precision. These restrictions are quite analogous to clinical trials and include ensuring that only incident drug users are studied; the restrictions also ensure similar comparison groups, patients without contraindications, and comparable adherence, as demonstrated by Schneeweiss et al. (2007), who showed how bias was reduced at each stage of restriction using statin and 1-year mortality as an example.

The restriction to incident users deserves special attention, as implementation of a new-user design can eliminate prevalent user bias (Cadarette et al. 2009; Ray 2003; Schneeweiss 2010). Within a new-user design framework, measures of association focus on events occurring after the first initiation of treatment, thus allowing a more direct comparison with an analogous group using an alternative treatment. One can logically extend the design to study drug switching and add-on therapies, as long as incident use of the target drug is preserved (Schneeweiss 2010).

Comparator selection is also an important design consideration to reduce confounding by indication (Ryan et al. 2013a). The comparator definition should ideally yield patients in the same health circumstance as those eligible to be new users of the target medication. Frequently, when assessing a drug safety issue, the comparator is chosen to represent the standard of care that would have been provided to that patient had the person not been prescribed the target drug, such that relative effect estimates represent risk above and beyond what the patient could otherwise expect. However, a comparator may also be selected specifically to address a question about difference in risk stemming from the underlying biological mechanism (e.g., choosing a comparator drug with the same indication). A challenge in comparator selection arises when no standard of care exists or when significant channeling bias to a particular drug class is present. For example, this bias might occur when a particular class of drugs is reserved for the most severely ill patients, who might be at increased risk for an adverse event because of the increased severity of disease. In this regard, evaluation studies can be highly sensitive to the comparator selected, and a criticism of these studies is often the subjective nature by which the comparator was selected.

Once a design is established, researchers can further reduce bias through analysis strategies, such as matching, stratification, and statistical adjustment. Variables commonly considered for adjustment are those for which the distribution at baseline differs between the exposed and unexposed populations or those known to potentially influence treatment decisions. To produce confounding, these variables also need to be associated with outcome occurrence; they may include patient demographics (such as age, gender, and race) or patient comorbidities (expressed either as a set of binary classifiers of specific diseases or as a composite index of comorbidity). One commonly used measure is the Charlson index (Bravo et al. 2002; Charlson et al. 1987, 1994; Cleves et al. 1997; D'Hoore et al. 1993, 1996; Li et al. 2008; Needham et al. 2005; Quan et al. 2005; Southern et al. 2004; Zhang et al. 1999), which was originally developed to predict mortality but has also been shown to be related to health-care expenditures (Farley et al. 2006). Adjustment for a comorbidity index is useful for exploratory data analysis (Schneeweiss et al. 2001) but may not suffice to address all sources of confounding. Additional variables often cited include prior use of medications and markers for health service utilization, such as number of outpatient visits and inpatient stays. The specific definitions and applications of covariates are highly variable across drug safety evaluation studies. Covariate selection can influence the magnitude of effect measures, regardless of the modeling approach undertaken, particularly if effect modification exists (Lunt et al. 2009).

Once variables are identified, one can control for them through direct matching or stratification, whereby the target and comparator groups are logically divided by the attributes of the covariates. However, in a multivariable context, the data may be too sparse to provide adequate sample size to allow matching on all covariates or to provide subpopulations within each covariate-defined stratum (i.e., there may be empty cells defined by combinations of covariates). A popular tool to overcome this limitation is propensity score analysis (Rosenbaum 2002; Rubin 1997).

As with other approaches, the propensity score model is only as good as the covariates selected to provide the adjustment. A propensity score is a single metric that is intended to account for all of the explanatory variables that predict who will receive treatment. Propensity scores generally balance observed confounders but do not necessarily produce balance on factors not incorporated into the model. Such imbalances represent a particular problem for analysis of electronic health-care databases, in which many important covariates, such as smoking status, alcohol consumption, body mass index, and lifestyle and cultural attitudes regarding health, are not captured. Sturmer et al. (2007) demonstrated that further adjustment could be achieved by conducting supplemental validation studies to collect additional information on previously unmeasured confounders. Schneeweiss et al. (2005) showed how unmeasured confounders biased estimates of COX-2 inhibitors and myocardial infarction. Seeger et al. (2003, 2005, 2007) highlighted how a model without the appropriate variables included could yield a biased estimate in a case study that explored association of statin therapy and myocardial infarction. Strategies for automated selection of large sets of covariates have been proposed as potential solutions to reduce the possibility of missing an empiric confounder (Schneeweiss et al. 2009). Sensitivity analysis has been proposed as an additional approach to assess the potential consequences of unobserved confounding (Schneeweiss 2006) but is unfortunately rarely reported in published studies. For example, Rosenbaum (2002) posits the existence of a latent confounder and explores the magnitude of the confounding that would be required to explain away the observed effect.

Instrumental variable (IV) analysis presents another potential solution to adjusting for confounding through control of a factor that is related to exposure but unrelated to outcome (Brookhart et al. 2010a; Hogan and Lancaster 2004; Schneeweiss 2007). Several studies have shown how IV analysis can reduce bias (Dudl et al. 2009; Rassen et al. 2009, 2010; Schneeweiss et al. 2008). A challenge in IV analysis is identifying a covariate that satisfies the criteria of an IV, particularly with regard to having no association with the outcome. For active surveillance, in which one may explore multiple outcomes for a given exposure, the selection of a common IV becomes even harder.

One consideration for all statistical adjustment techniques in drug safety evaluation studies is the danger of the statistical adjustment itself introducing bias (Greenland and Morgenstern 2001). Statistical control can sometimes either increase bias or decrease precision without affecting bias and can thereby produce less reliable effect estimates (Schisterman et al. 2009). For example, bias can also be induced if an analysis improperly stratifies on a collider variable (Cole et al. 2010), that is, a variable that is itself directly influenced by two other variables. As a result, care is necessary in any evaluation study to develop a parsimonious model that achieves an appropriate balance between bias and variance.

Self-controlled study designs represent the other family of methods that are also commonly applied to observational databases, especially for vaccine studies. Self-controlled designs, such as self-controlled case series (Suchard et al. 2013), self-controlled cohort (Ryan et al. 2013b), and temporal pattern discovery (Noren et al. 2013), aim to address the threat of between-person confounding by comparing exposed and unexposed time at the individual level (Whitaker et al. 2006). Many variants and analysis choices within these designs attempt to address other sources of systematic error, including varying the time at risk and using time-varying exposures as additional covariates. Confounders (e.g., sex) that do not change over time within a person are inherently controlled in such designs.

Some believe each design may have potential applications for examining specific types of associations based on the attributes of the exposure and outcome (e.g., certain designs are presumed to be appropriate for short-term effects) (Gagne et al. 2012). Although researchers have made substantial progress to establish theoretical and conceptual arguments for design and analysis considerations in observational data analysis, very little empirical evidence exists to support best practice and determine how observational analyses should be properly interpreted when evaluating the potential effects of medical products (Ioannidis 2005).

9.5 Methodological Progress toward a Risk Identification System

The CDC has played a leading role in establishing public health surveillance programs to inform medical product safety issues. The National Electronic Injury Surveillance System-Cooperative Adverse Drug Event Surveillance System has enabled monitoring of adverse drug events leading to emergency department visits (Budnitz et al. 2006). Another successful project has been the CDC Vaccine Safety Datalink (VSD), which demonstrated the feasibility of establishing a distributed network of administrative claims sources and conducting systematic analyses to detect vaccine-related adverse events (Davis et al. 2005). The sequential probability ratio test was applied to detect increases in intussusception following introduction to the rotavirus vaccines as well as decreases in several events after the changeover from the whole-cell pertussis vaccine to the acellular pertussis vaccine. In these instances, as with spontaneous reporting, the public health surveillance objective is to identify cases of serious events following exposure that wouldn't otherwise be expected. A primary distinction between spontaneous reporting and the VSD is that the spontaneous reporting system captures all potential events of any origin, whereas studies designed for VSD focus on a restricted set of specific adverse events known to be potentially caused by vaccines. In that respect, the VSD approach still falls within an evaluation paradigm, whether the system must first be presented with a prior hypothesis of a specific drug–condition relationship and craft an analysis to assess the purported effect. The method was since enhanced (Kulldorff et al. 2011) and applied to drug safety surveillance as part of the HMO Research Network, which demonstrated the ability to identify known drug-safety issues, including acute myocardial infarction risk following exposure to rofecoxib (Brown et al. 2007, 2009). However, as the authors note, these studies suffer from several methodological limitations, notably failure to fully address confounding and length of exposure. Also, as with the applications to public health surveillance, these methods have not yet been applied in an exploratory framework to generate hypotheses but are instead applied to targeted drug–condition pairs. As such, "identification" of the rofecoxib myocardial infarction effect comes without regard to how many other false-positive cases may be detected when using the same approach.

As we move into active drug safety surveillance, the goal shifts from case detection to association detection. That is, the interest in the system expands beyond detecting rare, idiosyncratic events that would not be expected to be seen without exposure to detecting elevations in risks of conditions that occur in the background population. Where clinical trials may be sufficient for detecting strong associations with highly prevalent outcomes (such as nuisance side effects like headache and nausea), and spontaneous reporting and public health surveillance tools may serve the purposes for identifying cases of rare events (such as Stevens–Johnson syndrome, toxic epidermal necrolysis, and Guillain–Barré syndrome), the largest opportunity for an active surveillance system rests in complementing those systems in the gap in between. This may include adverse events that are less commonly observed in clinical trials, that have weaker associations to exposure, and that are observed sufficiently often in the general population that case series may not be sufficient to fully characterize the relationship. Several notable adverse events that fall within this category include acute myocardial infarction, fracture, gastrointestinal bleeding, suicidality, and renal and hepatic dysfunction.

While there is a lot of excitement for the potential of a risk identification system for hypothesis generation, it is widely recognized that significant methodological research is needed to inform the appropriate use of observational data and analysis methods before a national system can be reliably used. A common challenge across all methods is determining how to manage the potential false alarms when exploring such a large set of potential outcomes and determining when evidence is sufficiently compelling to warrant follow-up (Avorn and Schneeweiss 2009). Until recently, no empirical studies had demonstrated the performance characteristics of these methods across a large sample of drug–event pairs or quantitatively identified the incremental value in supplementing current pharmacovigilance practice with these new methods, either in terms of identifying new issues or faster time to detection.

The Observational Medical Outcomes Partnership (OMOP) was established in 2009 as a public–private partnership to conduct methodological research to inform the establishment of a national active surveillance system (Stang et al. 2010). OMOP was chaired by the FDA, managed by the Foundation for the National Institutes of Health (FNIH), and supported by the pharmaceutical industry, with broad participation from government, academia, payers, health-care systems, and patient groups, across multiple disciplines, including epidemiology, statistics, and medical informatics, and across the applied health sciences. The OMOP program of work involved a series of experiments to evaluate the feasibility and utility of alternative

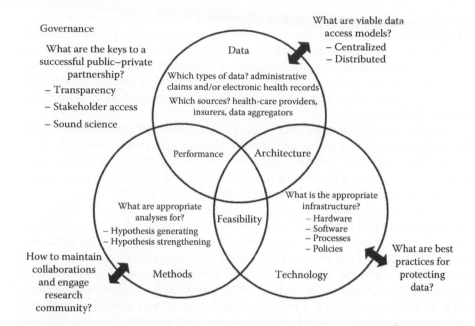

FIGURE 9.2
Interactions between data, analytic methods, technology, and governance that have driven OMOP's research agenda.

analysis methods and observational health-care databases for identifying and evaluating safety and benefit issues of drugs already on the market. OMOP established a data community of disparate data sources, comprising over 200 million lives in aggregate, and designed a large-scale methodological framework whereby a suite of commonly used epidemiology methods were systematically applied to each database and tested against a defined set of test cases (positive and negative controls) to empirically measure performance in identifying known safety issues and discerning from false-positive findings. The OMOP program had a number of unique features, including its public–private governance, its open innovation approach, and the transparency of its research processes and findings (Figure 9.2).

9.6 Findings from the Observational Medical Outcomes Partnership

Fundamentally, the OMOP studies focused on heterogeneity and its impact on observational studies: data capture and the populations they represent (Ryan et al. 2010), the definitions of exposures and outcomes, selection of comparators, study design decisions, and execution of the analysis using the methods and choices of parameter "settings" within a method. Characterizing heterogeneity and the performance of methods is intended to provide insights into how best to achieve more robust approximation of the true relationships between exposures and health outcomes of interest (HOIs) with the goal that observational research can credibly contribute another source of information to inform clinical decisions by identifying associations in larger defined patient populations.

Attempts to analyze data from multiple sources must contend with additional issues such as inconsistent data structures and vocabularies. A major aspect of the OMOP research program involved either minimizing such issues (e.g., creating a consistent data structure and vocabularies) or understanding them and how they affect data analysis and results. To identify and address issues related to inconsistent data structures, access, and vocabularies, we established a network of disparate data sources that included both central (housed in the OMOP research lab) and distributed (data remained with collaborators) databases. These databases consisted of claims data and EHRs. Each database was converted to a standards-based common data model (CDM) (Overhage et al. 2012), with a standard structure and vocabulary to which we applied a set of epidemiological and statistical methods that were identified in searches, discussions, and contests. A set of "known associations" were identified and served as our "positive control" test cases; similarly, a set of associations believed to represent "negative controls" were also identified and together formed the associations used to test the performance characteristics of the methods.

A typical epidemiologic analysis uses a single data source and applies a specific analytic method to the data in a specific manner, using specific definitions for exposures and outcomes (possibly subject to a validation procedure). In practice, researchers usually consider myriad choices among data sources, analytic methods, and definitions of exposure and outcomes, then select a single choice for each based on the research question, the nature of the data sources, and other considerations. Considerable evidence suggests that these choices differ by investigator and results may not be reliably reproduced from other data sources or analytic methods that are equally valid.

Rather than choosing a single database or analytic method, OMOP attempted to empirically evaluate the methodological performance across databases, analytic methods, and drug–outcome scenarios. In OMOP's first experiment, the evaluation involved testing 14 analytic methods (Table 9.1), each with up to 162 parameter settings, against 53 drug–outcome pair test cases, across 10 databases. The output of this analysis consisted of quantitative estimates of association for 1.4 million drug–outcome association estimates. As a follow-up experiment, the suite of methods was expanded to include additional parameter settings and were executed across 5 U.S. databases and a network of European databases for 399 drug–outcome pairs. All results from these experiments were placed in the public domain (http://omop.org). This section highlights the key findings from these experiments.

TABLE 9.1

Fourteen Analytical Methods Applied across 10 Databases in the OMOP Experiment

Method Name	DP Analysis	Case-Based Methods	Exposure-Based Methods	Sequential Testing Methods	Contributor
Bayesian logistic regression		X			Rutgers University/ Columbia University
Bayesian multivariate self-controlled case series		X			Columbia University
Case–control surveillance		X			Eli Lilly and Company
Case–crossover		X			University of Utah
Conditional sequential sampling procedure				X	Harvard Pilgrim/ Group Health
DP analysis	X				Columbia University/Merck & Co., Inc.
High-dimensional propensity score			X		Columbia University
High-throughput screening			X		Regenstrief Institute/ Indiana University
ICTPD	X				Uppsala Monitoring Centre
Incident user design			X		University of North Carolina
LGPS and longitudinal evaluation of observational profiles of adverse events related to drugs (LEOPARD)	X				Erasmus University Medical Center Rotterdam
Maximized sequential probability ratio test				X	Harvard Pilgrim/ Group Health
Multiset case–control estimation		X			Columbia University/ GlaxoSmithKline
Observational screening			X		ProSanos/ GlaxoSmithKline

9.6.1 Performance

To gain insight into the ability of a method to distinguish between positive and negative controls, we used the effect estimates to compute the area under the curve (AUC), a measure of predictive accuracy: an AUC of 1 indicates a perfect prediction of which test cases are positive and which are not. An AUC of 0.5 is equivalent to random guessing.

Often we are not only interested in whether there is an effect or not but would also like to know the magnitude of the effect. However, in order to evaluate the accuracy of the effect size estimates for a particular analytic method, we must know the true effect size. This true effect size is never known, and so we restrict some of our analyses to the negative controls where we assume that the true log relative risk is zero. Using the negative controls in real data, we compute "bias," the average difference between the log relative risk and zero. An unbiased estimator would yield a bias of zero. The "mean square error (MSE)" is the average squared difference between the log relative risk and zero. Since zero is the true relative risk, smaller MSEs are desirable. "Coverage" is the fraction of the 95% intervals that include zero. In the case of an unbiased estimator with accurate confidence interval estimation, we would expect the coverage probability to be 95%.

In the initial experiment, the OMOP team found that several methods perform modestly well but still returned relatively high false-positive rates. This study suggested that all methods yield a false-positive rate of at least 15% at conventional levels of significance, and although specificity can be improved through additional restriction in threshold criteria, it also comes at a cost of substantially decreased sensitivity. Imposing a restriction to require the point estimate of RR > 2 decreases false-positive rates to < 10% for most methods, but consequently, sensitivity is reduced to < 33%. The performance of risk identification methods was much in line with operating characteristics observed for widely used clinical diagnostic tests, such as prostate-specific antigen screening for prostate cancer, mammography for breast cancer, and rapid strep tests (AUC range 0.74–0.80), but all methods failed to provide the definitive information that is desired (Ryan et al. 2012).

The second experiment was conducted to replicate these findings across a broader array of drug–outcome test cases and across an international data network to determine if results would be consistent. The experiment focused on four HOIs: acute myocardial infarction, acute liver injury, acute renal failure, and gastrointestinal bleeding. Performance estimates were obtained for each outcome to determine if the system behavior may be differential across events.

Table 9.2 presents the analytic method that provided the best AUC for each outcome–database combination as well as the associated AUC value. Each six-digit code specifies a particular set of design choices (for details, see http:// bit.ly/10iWgs2). For every database–outcome combination, self-controlled methods (self-controlled cohort [SCC], self-controlled case series [SCCS],

TABLE 9.2

AUC Optimal Analytic Method for Each Database and Outcome

Data Source	Acute Kidney Injury	Acute Liver Injury	Acute Myocardial Infarction	Upper Gastrointestinal Bleed
MDCR	OS: 401002 (0.92)	OS: 401002 (0.76)	OS: 407002 (0.84)	OS: 402002 (0.86)
	Study design: self-controlled cohort	Study design: self-controlled cohort	Study design: self-controlled cohort	Study design: self-controlled cohort
	Exposures to include: all occurrences	Exposures to include: all occurrences	Exposures to include: all occurrences	Exposures to include: all occurrences
	Outcomes to include: all occurrences	Outcomes to include: all occurrences	Outcomes to include: first occurrence	Outcomes to include: all occurrences
	Time at risk: length of exposure + 30 day	Time at risk: length of exposure + 30 days	Time at risk: length of exposure + 30 days	Time at risk: length of exposure + 30 days
	Include index date in time at risk: no	Include index date in time at risk: no	Include index date in time at risk: no	Include index date in time at risk: yes
	Control period: length of exposure + 30 day	Control period: length of exposure + 30 days	Control period: length of exposure + 30 days	Control period: length of exposure + 30 days
	Include index date in control period: no	Include index date in control period: no	Include index date in control period: no	Include index date in control period: yes
CCAE	OS: 404002 (0.89)	OS: 403002 (0.79)	OS: 408013 (0.85)	SCCS: 1931010 (0.82)
	Study design: self-controlled cohort	Study design: self-controlled cohort	Study design: self-controlled cohort	Outcomes to include: all occurrences
	Exposures to include: all occurrences	Exposures to include: all occurrences	Exposures to include: first occurrence	Prior distribution: normal
	Outcomes to include: all occurrences	Outcomes to include: all occurrences	Outcomes to include: first occurrence after exposure	Variance of the prior: determined through cross-validation
	Time at risk: length of exposure + 30 day	Time at risk: length of exposure + 30 days	Time at risk: all time postexposure start	Time at risk: all time postexposure start
	Include index date in time at risk: no	Include index date in time at risk: yes	Include index date in time at risk: no	Include index date in time at risk: no

(Continued)

TABLE 9.2 (*Continued*)

AUC Optimal Analytic Method for Each Database and Outcome

Data Source	Acute Kidney Injury	Acute Liver Injury	Acute Myocardial Infarction	Upper Gastrointestinal Bleed
	Control period: length of exposure + 30 day	Control period: length of exposure + 30 days	Control period: all time prior to exposure start	Apply multivariate adjustment on all drugs: no
	Include index date in control period: yes	Include index date in control period: no	Include index date in control period: no	Required observation time: 180 days
MDCD	OS: 408013 (0.82)	OS: 409013 (0.77)	OS: 407004 (0.80)	OS: 401004 (0.87)
	Study design: self-controlled cohort	Study design: self-controlled cohort	Study design: self-controlled cohort	Study design: self-controlled cohort
	Exposures to include: first occurrence	Exposures to include: first occurrence	Exposures to include: all occurrences	Exposures to include: all occurrences
	Outcomes to include: first occurrence after exposure	Outcomes to include: first occurrence	Outcomes to include: first occurrence	Outcomes to include: all occurrences
	Time at risk: all time postexposure start	Time at risk: all time postexposure start	Time at risk: length of exposure + 30 days	Time at risk: length of exposure + 30 days
	Include index date in time at risk: no	Include index date in time at risk: no	Include index date in time at risk: no	Include index date in time at risk: no
	Control period: all time prior to exposure start	Control period: all time prior to exposure start	Control period: 365 days prior to exposure start	Control period: 365 days prior to exposure start
	Include index date in control period: no	Include index date in control period: no	Include index date in control period: no	Include index date in control period: no
MSLR	SCCS: 1907010 (1.00)	OS: 406002 (0.84)	OS: 403002 (0.80)	OS: 403002 (0.83)
	Outcomes to include: all occurrences	Study design: self-controlled cohort	Study design: self-controlled cohort	Study design: self-controlled cohort
	Prior distribution: normal	Exposures to include: all occurrences	Exposures to include: all occurrences	Exposures to include: all occurrences
	Variance of the prior: determined through cross-validation	Outcomes to include: first occurrence after exposure	Outcomes to include: all occurrences	Outcomes to include: all occurrences

(*Continued*)

TABLE 9.2 (*Continued*)

AUC Optimal Analytic Method for Each Database and Outcome

Data Source	Acute Kidney Injury	Acute Liver Injury	Acute Myocardial Infarction	Upper Gastrointestinal Bleed
	Time at risk: all time postexposure start	Time at risk: length of exposure + 30 days	Time at risk: length of exposure + 30 days	Time at risk: length of exposure + 30 days
	Include index date in time at risk: no	Include index date in time at risk: no	Include index date in time at risk: yes	Include index date in time at risk: yes
	Apply multivariate adjustment on all drugs: yes	Control period: length of exposure + 30 days	Control period: length of exposure + 30 days	Control period: length of exposure + 30 days
	Required observation time: none	Include index date in control period: no	Include index date in control period: no	Include index date in control period: no
GE	SCCS: 1949010 (0.94)	OS: 409002 (0.77)	ICTPD: 3016001 (0.89)	ICTPD: 3034001 (0.89)
	Outcomes to include: all occurrences	Study design: self-controlled cohort	Control period: −1080 days to −361 days before exposure start	Control period: −810 days to −361 days before exposure start
	Prior distribution: normal	Exposures to include: first occurrence	Time at risk: 60 days from exposure start	Time at risk: 60 days from exposure start
	Variance of the prior: determined through cross-validation	Outcomes to include: first occurrence	Use control period in expected calculation: yes	Use control period in expected calculation: yes
	Time at risk: 30 days from exposure start	Time at risk: length of exposure + 30 days	Use 1 month prior to exposure in expected calculation: yes	Use 1 month prior to exposure in expected calculation: no
	Include index date in time at risk: yes	Include index date in time at risk: no	Use 1 day prior to exposure in expected calculation: no	Use 1 day prior to exposure in expected calculation: yes
	Apply multivariate adjustment on all drugs: yes	Control period: length of exposure + 30 days		
	Required observation time: 180 day	Include index date in control period: no		

Notes: MDCR, MarketScan Medicare Supplemental Beneficiaries; CCAE, MarketScan Commercial Claims and Encounters; MDCD, MarketScan Multistate Medicaid; MSLR, MarketScan Lab Supplemental; GE, GE Centricity.

and information component temporal pattern discovery [ICTPD]) provide the optimal performance, with AUCs ranging from a low of 0.77 for acute liver injury in the MDCD database to 1.00 for acute kidney injury in the MSLR database. In general, AUCs are highest for acute kidney injury and lowest for acute liver injury, with acute myocardial infarction and gastrointestinal bleeding in between. Performance across the five data sources is similar despite their substantial differences in patient populations (see Table 9.3).

Figure 9.3 presents the AUC value for all analytic methods, broken down by database and method. Several findings emerge from Figure 9.3:

- The case–control method, longitudinal gamma Poisson shrinker (LGPS), and DP consistently underperform other methods, often yielding AUCs close to 0.5.
- Within each method, the specific design choices that correspond to the global optimum generally perform well for all outcomes and databases. Consider, for example, the SCC design; with the exception of acute myocardial infarction in MDCD, performance of the database–outcome optimum design choices does not exceed the global optimum by more than 0.10 in AUC.
- The design choices within each method affect performance significantly. For the majority of drug–outcome–method triples, there are design choices that yield AUC values at or close to 0.5, despite the existence of design choices with quite high AUC values for the same drug–outcome pair.

The experiment further estimated bias, MSE, and 95% confidence interval coverage for database–outcome optimal analytic methods. Because the true relative risks are unavailable for the positive controls, the evaluation just draws on the negative control test cases. In our experiments, the case–control, SCC, and LGPS methods generally yield positively biased effect estimates, whereas the cohort method generally yields negatively biased estimates. The SCCS method yields estimates that are close to unbiased. All three self-controlled methods produce smaller MSEs than the other methods, with SCCS being especially close to zero. No method provides coverage probabilities that are close to the nominal 95%. On average, coverage probabilities for the cohort, DP, ICTPD, LGPS, and SCC methods are all below 50%. Average coverage for the case–control method is 63%, whereas for SCCS, the average coverage is 76% (Ryan et al. 2013).

9.6.2 Database Heterogeneity

One opportunity in developing a risk identification system is the ability to explore disparate data sources representing diverse populations and different data capture processes. From the OMOP experiment results,

TABLE 9.3

Data Sources Used in the Experiment

Abbreviation	Name	Description	Population	Observation Time
CCAE	MarketScan® Commercial Claims and Encounters	Represents privately insured population and captures administrative claims with patient-level de identified data from inpatient and outpatient visits and pharmacy claims of multiple insurance plans	Total: 46.5 m % male: 49% Mean age: 31.4 (18.1)	Pt-years: 97.6 m 2003–2009
MDCD	MarketScan Multistate Medicaid	Contains administrative claims data for Medicaid enrollees from multiple states, including inpatient, outpatient, and pharmacy services	Total: 10.8 m % male: 42% Mean age: 21.3 (21.5)	Pt-years: 20.7 m 2002–2007
MDCR	MarketScan Medicare Supplemental Beneficiaries	Captures administrative claims for retirees with Medicare supplemental insurance paid by employers, including services provided under Medicare-covered payment, employer-paid portion, and any out-of-pocket expenses	Total: 4.6 m % male: 44% Mean age: 73.5 (8.0)	Pt-years: 13.4 m 2003–2009
MSLR	MarketScan Lab Supplemental	Represents privately insured population that has at least one recorded laboratory value, with administrative claims from inpatient, outpatient, and pharmacy services supplemented by laboratory results	Total: 1.2 m % male: 35% Mean age: 37.6 (17.7)	Pt-years: 2.2 m 2003–2007
GE	GE Centricity™	Derived from data pooled by providers using GE Centricity office (an ambulatory electronic health record) into a data warehouse in a HIPAA-compliant manner	Total: 11.2 m % male: 42% Mean age: 39.6 (22.0)	Pt-years: 22.4 m 1996–2008

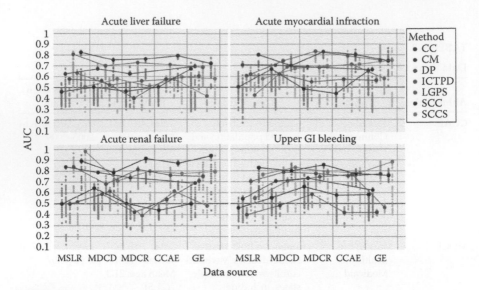

FIGURE 9.3
(See color insert.) Performance of methods across AUC values for all analytic methods, again broken down by database and method. The solid lines represent the AUCs for the set of design choices within each method that provided the best performance on average across all outcomes and databases.

we assessed the impact of data source heterogeneity of drug–outcome effect estimates. Our findings suggest that 20%–40% of observational database studies can swing from statistically significant in one direction to statistically significant in the opposite direction depending on the choice of database, despite holding study design constant. We also found that almost all studies can be statistically significant in some databases and statistically nonsignificant in others. Although variability in statistical power across the databases could explain this latter finding, it does not explain the former finding of contradictory statistical significance. These results are illustrated in Figure 9.4. We believe the findings of database heterogeneity have two immediate implications for drug safety research. First, when interpreting results from a single observational data source, more attention is needed to consider how the choice of data source may be affecting results. Second, where possible, studies should examine multiple sources to confirm that significant findings are consistently identified or that results are at least consistent across databases (although such consistency in and of itself does not establish causality). When interpreting results across multiple sources, it is important to characterize the observed heterogeneity and limit the use of composite estimates that could otherwise hide the uncertainty in effect estimates that is not driven by sampling variability. Forest plots provide especially useful insights into heterogeneity and should always be included (Madigan et al. 2013b).

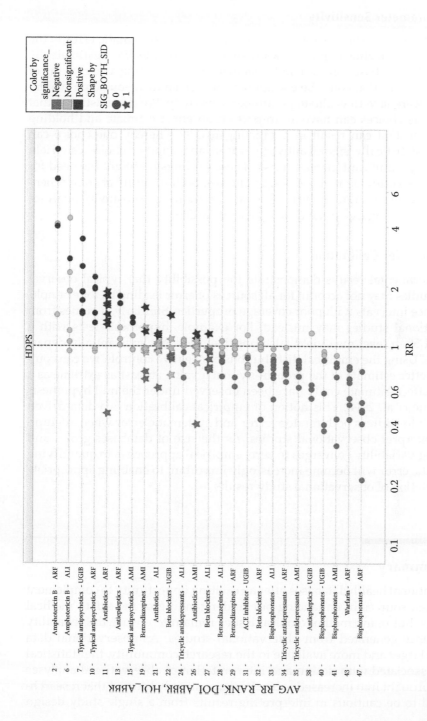

FIGURE 9.4

(See color insert.) Database heterogeneity. ARF, acute renal failure; ALI, acute liver injury; UGIB, upper gastrointestinal bleeding; AMI, acute myocardial infarction; HDPS, high-dimensional propensity score-adjusted cohort design.

9.6.3 Parameter Sensitivity

A complementary dimension to the identification of systematic error is the impact of design choices made within an analysis on the effect estimate. By holding the database constant and systematically evaluating a large array of analysis choices that could be considered for a given drug–outcome evaluation, we were able to evaluate parameter sensitivity. We demonstrated that all analysis choices can have an impact on an effect estimate and holding all else constant can result in conflicting results whereby one choice can produce statistically significantly decreased and another choice can yield statistically significant increased risk. These findings highlight the need for increased scrutiny of analysis design choices, both in terms of comprehensive prespecification in analysis plans and in thorough sensitivity analyses performed during execution (Madigan et al. 2013a).

9.6.4 Need for Calibration

The experimental results demonstrate the possibility that typical observational studies may not account for all sources of error leading to, for example, confidence intervals with poor coverage properties. Similarly, p-values from observational studies may mislead. For example, hypothesis tests with a nominal 5% α level may yield false-positive rates that substantially depart from 5%. Using the empirical distributions of negative controls, we can compute a better estimate of the probability that a value at least as extreme as a certain effect estimate could have been observed under the null hypothesis (Schuemie et al. 2014). This notion of empirical calibration is a data-driven approach to address both random error and systematic error when attempting to interpret observational studies. As the size of databases grows and sampling variability converges to zero, empirical approaches to quantifying systematic error will become increasingly important to enabling appropriate interpretation of observational study results.

9.7 Summary

Observational health databases offer an exciting opportunity to complement other data sources in enhancing our understanding of the safety medical products, but many methodological issues remain to ensure the reliability of evidence generated from observational studies. As observational data become larger and more available to the research community, the statistical issues associated with design and interpretation of observational analyses will be brought into increased focus. We have already learned that researchers need to be cautious in interpreting results from a single study design or single statistical method in a single database. Furthermore, replication of results using a different method and dataset cannot ensure the validity of the results. Further empirical research is needed to develop a reliable, automated process for systematic surveillance of drug safety using observational data.

References

Aronson JK, Hauben M. 2006. Anecdotes that provide definitive evidence. *BMJ* 333(7581):1267–1269.

Atkins D, Best D, Briss PA, Eccles M, Falck-Ytter Y et al. 2004. Grading quality of evidence and strength of recommendations. *BMJ* 328:1490.

Avorn J, Schneeweiss S. 2009. Managing drug-risk information—What to do with all those new numbers. *N. Engl. J. Med.* 361(7):647–649.

Azoulay L, Yin H, Filion KB, Assayag J, Majdan A et al. 2012. The use of pioglitazone and the risk of bladder cancer in people with type 2 diabetes: Nested case-control study. *BMJ* 344:e3645.

Berlin JA, Glasser SC, Ellenberg SS. 2008. Adverse event detection in drug development: Recommendations and obligations beyond phase 3. *Am. J. Public Health* 98:1366–1371.

Bosco JL, Silliman RA, Thwin SS, Geiger AM, Buist DS et al. 2010. A most stubborn bias: No adjustment method fully resolves confounding by indication in observational studies. *J. Clin. Epidemiol.* 63:64–74.

Bravo G, Dubois MF, Hébert R, De Wals P, Messier L. 2002. A prospective evaluation of the Charlson comorbidity Index for use in long-term care patients. *J. Am. Geriatr. Soc.* 50:740–745.

Brookhart MA, Rassen JA, Schneeweiss S. 2010a. Instrumental variable methods in comparative safety and effectiveness research. *Pharmacoepidemiol. Drug Saf.* 19:537–554.

Brookhart MA, Sturmer T, Glynn RJ, Rassen J, Schneeweiss S. 2010b. Confounding control in healthcare database research: Challenges and potential approaches. *Med. Care* 48:S114–S120.

Brown JS, Kulldorff M, Chan KA, Davis RL, Graham D et al. 2007. Early detection of adverse drug events within population-based health networks: Application of sequential testing methods. *Pharmacoepidemiol Drug Saf.* 16(12):1275–1284.

Brown JS, Kulldorff M, Petronis KR, Reynolds R, Chan KA et al. 2009. Early adverse drug event signal detection within population-based health networks using sequential methods: Key methodologic considerations. *Pharmacoepidemiol. Drug Saf.* 18(3):226–234.

Budnitz DS, Pollock DA, Weidenbach KN, Mendelsohn AB, Schroeder TJ, Annest JL. 2006. National surveillance of emergency department visits for outpatient adverse drug events. *JAMA* 296(15):1858–1866.

Burton MM, Hope C, Murray MD, Hui S, Overhage JM. 2007. The cost of adverse drug events in ambulatory care. *AMIA Annu. Symp. Proc.* 2007:90–93.

Cadarette SM, Katz JN, Brookhart MA, Sturmer T, Stedman MR et al. 2009. Comparative gastrointestinal safety of weekly oral bisphosphonates. *Osteoporos. Int.* 20:1735–1747.

Charlson ME, Pompei P, Ales KL, MacKenzie CR. 1987. A new method of classifying prognostic comorbidity in longitudinal studies: Development and validation. *J. Chronic Dis.* 40:373–383.

Charlson ME, Szatrowski TP, Peterson J, Gold J. 1994. Validation of a combined comorbidity index. *J. Clin. Epidemiol.* 47:1245–1251.

Cleves MA, Sanchez N, Draheim M. 1997. Evaluation of two competing methods for calculating Charlson's comorbidity index when analyzing short-term mortality using administrative data. *J. Clin. Epidemiol.* 50:903–908.

Cole SR, Platt RW, Schisterman EF, Chu H, Westreich D et al. 2010. Illustrating bias due to conditioning on a collider. *Int. J. Epidemiol.* 39:417–420.

Davis RL, Kolczak M, Lewis E, Nordin J, Goodman M et al. 2005. Active surveillance of vaccine safety: A system to detect early signs of adverse events. *Epidemiology* 16(3):336–341.

D'Hoore W, Bouckaert A, Tilquin C. 1996. Practical considerations on the use of the Charlson comorbidity index with administrative data bases. *J. Clin. Epidemiol.* 49:1429–1433.

D'Hoore W, Sicotte C, Tilquin C. 1993. Risk adjustment in outcome assessment: The Charlson comorbidity index. *Methods Inf. Med.* 32:382–387.

Donahue JG, Weiss ST, Goetsch MA, Livingston JM, Greineder DK, Platt R. 1997. Assessment of asthma using automated and full-text medical records. *J. Asthma* 34:273–281.

Dudl RJ, Wang MC, Wong M, Bellows J. 2009. Preventing myocardial infarction and stroke with a simplified bundle of cardioprotective medications. *Am. J. Manag. Care* 15:e88–e94.

Farley JF, Harley CR, Devine JW. 2006. A comparison of comorbidity measurements to predict healthcare expenditures. *Am. J. Manag. Care* 12:110–119.

Furberg CD, Levin AA, Gross PA, Shapiro RS, Strom BL. 2006. The FDA and drug safety: A proposal for sweeping changes. *Arch. Intern. Med.* 166(18):1938–1942.

Gagne JJ, Fireman B, Ryan PB, Maclure M, Gerhard T et al. 2012. Design considerations in an active medical product safety monitoring system. *Pharmacoepidemiol. Drug Saf.* 21(Suppl. 1):32–40.

García Rodríguez LA, Pérez Gutthann S. 1998. Use of the UK general practice research database for pharmacoepidemiology. *Br. J. Clin. Pharmacol.* 45:419–425.

Greenland S, Morgenstern H. 2001. Confounding in health research. *Annu. Rev. Public Health* 22(1), 189–212.

Harpaz R, DuMouhcel W, LePendu P, Bauer-Mehren A, Ryan P, Shah NH. 2013. Performance of pharmacovigilance signal-detection algorithms for the FDA adverse event reporting system. *Clin. Pharmacol. Ther.* 93:539–546.

Harrold LR, Saag KG, Yood RA, Mikuls TR, Andrade SE et al. 2007. Validity of gout diagnoses in administrative data. *Arthritis Rheum.* 57:103–108.

Hartzema AG, Porta MS, Tilson HH. 1999. *Pharmacoepidemiology: An Introduction.* Cincinnati, OH: Harvey Whitney Books.

Hartzema AG, Tilson HH, Chan KA. 2008. *Pharmacoepidemiology and Therapeutic Risk Management.* Cincinnati, OH: Harvey Whitney Books.

Hauben M, Madigan D, Gerrits CM, Walsh L, Van Puijenbroek EP. 2005. The role of data mining in pharmacovigilance. *Expert Opin. Drug Saf.* 4(5):929–948.

Hennessy S. 2006. Use of health care databases in pharmacoepidemiology. *Basic Clin. Pharmacol. Toxicol.* 98:311–313.

Hennessy S, Leonard CE, Freeman CP, Deo R, Newcomb C et al. 2009. Validation of diagnostic codes for outpatient-originating sudden cardiac death and ventricular arrhythmia in Medicaid and Medicare claims data. *Pharmacoepidemiol. Drug Saf.* 19:555–562.

Hennessy S, Leonard CE, Palumbo CM, Newcomb C, Bilker WB. 2007. Quality of Medicaid and Medicare data obtained through Centers for Medicare and Medicaid Services (CMS). *Med. Care* 45:1216–1220.

Hill AB. 1965. The environment and disease: Association or causation? *Proc. R. Soc. Med.* 58:295–300.

Hogan JW, Lancaster T. 2004. Instrumental variables and inverse probability weighting for causal inference from longitudinal observational studies. *Stat. Methods Med. Res.* 13:17–48.

Horisberger B, Dinkel R. 1989. *The Perception and Management of Drug Safety Risks.* Berlin, Germany: Springer-Verlag.

Ioannidis JP. 2005. Why most published research findings are false. *PLoS Med.* 2:e124.

Jewell N. 2004. *Statistics for Epidemiology.* Boca Raton, FL: Chapman & Hall.

Kulldorff M, Davis RL, Kolczak M. Lewis E, Lieu T, Platt R. 2011. A maximized sequential probability ratio test for drug and vaccine safety surveillance. *Sequential Analysis: Design, Methods and Applications.* 30(1):58–78.

Lasser KE, Allen PD, Woolhandler SJ, Himmelstein DU, Wolfe SM, Bor DH. 2002. Timing of new black box warnings and withdrawals for prescription medications. *JAMA* 287(17):2215–2220.

Lee DS, Donovan L, Austin PC, Gong Y, Liu PP et al. 2005. Comparison of coding of heart failure and comorbidities in administrative and clinical data for use in outcomes research. *Med. Care* 43:182–188.

Leonard CE, Haynes K, Localio AR, Hennessy S, Tjia J et al. 2008. Diagnostic E-codes for commonly used, narrow therapeutic index medications poorly predict adverse drug events. *J. Clin. Epidemiol.* 61:561–571.

Lewis JD, Brensinger C. 2004. Agreement between GPRD smoking data: A survey of general practitioners and a population-based survey. *Pharmacoepidemiol. Drug Saf.* 13:437–441.

Lewis JD, Schinnar R, Bilker WB, Wang X, Strom BL. 2007. Validation studies of the health improvement network (THIN) database for pharmacoepidemiology research. *Pharmacoepidemiol. Drug Saf.* 16:393–401.

Li B, Evans D, Faris P, Dean S, Quan H. 2008. Risk adjustment performance of Charlson and Elixhauser comorbidities in ICD-9 and ICD-10 administrative databases. *BMC Health Serv. Res.* 8:12.

Lunt M, Solomon D, Rothman K, Glynn R, Hyrich K et al. 2009. Different methods of balancing covariates leading to different effect estimates in the presence of effect modification. *Am. J. Epidemiol.* 169:909–917.

Madigan D, Ryan PB, Schuemie M. 2013a. Does design matter? Systematic evaluation of the impact of analytical choices on effect estimates in observational studies. *Ther. Adv. Drug Saf.* 4(2):53–62.

Madigan D, Ryan PB, Schuemie M, Stang PE, Overhage JM et al. 2013b. Evaluating the impact of database heterogeneity on observational study results. *Am J Epidemiol.* 178(4):645–651. doi: 10.1093/aje/kwt010. Epub 2013 May 5.

Meyboom RH, Egberts AC, Edwards IR, Hekster YA, de Koning FH, Gribnau FW. 1997. Principles of signal detection in pharmacovigilance. *Drug Saf.* 16(6):355–365.

Meyboom RH, Egberts AC, Gribnau FW, Hekster YA. 1999. Pharmacovigilance in perspective. *Drug Saf.* 21(6):429–447.

Miller DR, Oliveria SA, Berlowitz DR, Fincke BG, Stang P, Lillienfeld DE. 2008. Angioedema incidence in US veterans initiating angiotensin-converting enzyme inhibitors. *Hypertension* 51:1624–1630.

Needham DM, Scales DC, Laupacis A, Pronovost PJ. 2005. A systematic review of the Charlson comorbidity index using Canadian administrative databases: A perspective on risk adjustment in critical care research. *J. Crit. Care* 20:12–19.

Noren GN, Bergvall T, Ryan PB, Juhlin K, Schuemie MJ, Madigan D. 2013. Empirical performance of the calibrated self-controlled cohort analysis within temporal pattern discovery: Lessons for developing a risk identification and analysis system. *Drug Saf.* 36(Suppl. 1):S107–S121. doi: 10.1007/s40264-013-0095-x.

Noren GN, Edwards IR. 2009. Modern methods of pharmacovigilance: Detecting adverse effects of drugs. *Clin. Med.* 9(5):486–489.

Overhage JM, Ryan PB, Reich CG, Hartzema AG, Stang PE. 2012. Validation of a common data model for active safety surveillance research. *J. Am. Med. Inform. Assoc.* 19:54–60.

Pladevall M, Goff DC, Nichaman MZ, Chan F, Ramsey D et al. 1996. An assessment of the validity of ICD Code 410 to identify hospital admissions for myocardial infarction: The Corpus Christi Heart Project. *Int. J. Epidemiol.* 25:948–952.

Platt R, Wilson M, Chan KA, Benner JS, Marchibroda J, McClellan M. 2009. The new Sentinel Network—Improving the evidence of medical-product safety. *N. Engl. J. Med.* 361(7):645–647. doi: 10.1056/NEJMp0905338. Epub 2009 Jul 27.

Public Law 110-85: Food and Drug Administration Amendments Act of 2007; 2007.

Quan H, Sundararajan V, Halfon P, Fong A, Burnand B et al. 2005. Coding algorithms for defining comorbidities in ICD-9-CM and ICD-10 administrative data. *Med. Care* 43:1130–1139.

Rassen JA, Brookhart MA, Glynn RJ, Mittleman MA, Schneeweiss S. 2009. Instrumental variables I: Instrumental variables exploit natural variation in nonexperimental data to estimate causal relationships. *J. Clin. Epidemiol.* 62:1226–1232.

Rassen JA, Mittleman MA, Glynn RJ, Brookhart MA, Schneeweiss S. 2010. Safety and effectiveness of bivalirudin in routine care of patients undergoing percutaneous coronary intervention. *Eur. Heart J.* 31:561–572.

Ray WA. 2003. Evaluating medication effects outside of clinical trials: New-user designs. *Am. J. Epidemiol.* 158:915–920.

Rockhill B, Newman B, Weinberg C. 1998. Use and misuse of population attributable fractions. *Am. J. Public Health* 88:15–19.

Rodriguez EM, Staffa JA, Graham DJ. 2001. The role of databases in drug postmarketing surveillance. *Pharmacoepidemiol. Drug Saf.* 10:407–410.

Rosenbaum PR. 2002. *Observational Studies*, 2nd edn. New York: Springer.

Rothman KJ. 2002. *Epidemiology: An Introduction*. Oxford, U.K.: Oxford University Press.

Rothman KJ, Greenland S, Lash T. 2008. *Modern Epidemiology*. Philadelphia, PA: Lippincott Williams & Wilkins.

Rothman KJ, Suissa S. 2008. Exclusion of immortal person-time. *Pharmacoepidemiol. Drug Saf.* 17:1036.

Rubin DB. 1997. Estimating causal effects from large data sets using propensity scores. *Ann. Intern. Med.* 127:757–763.

Ryan PB, Madigan D, Stang PE, Marc Overhage J, Racoosin JA, Hartzema AG. 2012. Empirical assessment of methods for risk identification in healthcare data: Results from the experiments of the Observational Medical Outcomes Partnership. *Stat. Med.* 31(30):4401–4415.

Ryan PB, Schuemie MJ, Gruber S, Zorych I, Madigan D. 2013a. Empirical performance of a new user cohort method: Lessons for developing a risk identification and analysis system. *Drug Saf.* 36(Suppl. 1):S59–S72. doi: 10.1007/s40264-013-0099-6.

Ryan PB, Schuemie MJ, Madigan D. 2013b. Empirical performance of a self-controlled cohort method: Lessons for developing a risk identification and analysis system. *Drug Saf.* 36(Suppl. 1):S95–S106. doi: 10.1007/s40264-013-0101-3.

Ryan PB, Schuemie MJ, Welebob E, Duke J, Valentine S, Hartzema AG. 2013c. Defining a reference set to support methodological research in drug safety. *Drug Saf.* 36(Suppl. 1):S33–S47.

Ryan PB, Stang PE, Overhage JM, Suchard MA, Hartzema AG et al. 2013d. A comparison of the empirical performance of methods for a risk identification system. *Drug Saf.* 36(Suppl. 1):S143–S158. doi: 10.1007/s40264-013-0108-9.

Ryan PB, Welebob E, Hartzema AG, Stang PE, Overhage JM. 2010. Surveying US observational data sources and characteristics for drug safety needs. *Pharm. Med.* 24(4):231–238.

Schisterman EF, Cole SR, Platt RW. 2009. Over adjustment bias and unnecessary adjustment in epidemiologic studies. *Epidemiology* 20:488–495.

Schneeweiss S. 2006. Sensitivity analysis and external adjustment for unmeasured confounders in epidemiologic database studies of therapeutics. *Pharmacoepidemiol. Drug Saf.* 15:291–303.

Schneeweiss S. 2007. Developments in post-marketing comparative effectiveness research. *Clin. Pharmacol. Ther.* 82:143–156.

Schneeweiss S. 2010. A basic study design for expedited safety signal evaluation based on electronic healthcare data. *Pharmacoepidemiol. Drug Saf.* 19:858–868.

Schneeweiss S, Avorn J. 2005. A review of uses of health care utilization databases for epidemiologic research on therapeutics. *J. Clin. Epidemiol.* 58:323–337.

Schneeweiss S, Glynn RJ, Tsai EH, Avorn J, Solomon DH. 2005. Adjusting for unmeasured confounders in pharmacoepidemiologic claims data using external information: The example of COX2 inhibitors and myocardial infarction. *Epidemiology* 16:17–24.

Schneeweiss S, Patrick AR, Sturmer T, Brookhart MA, Avorn J et al. 2007. Increasing levels of restriction in pharmacoepidemiologic database studies of elderly and comparison with randomized trial results. *Med. Care* 45:S131–S142.

Schneeweiss S, Rassen JA, Glynn RJ, Avorn J, Mogun H, Brookhart MA. 2009. High-dimensional propensity score adjustment in studies of treatment effects using health care claims data. *Epidemiology* 20:512–522.

Schneeweiss S, Seeger JD, Landon J, Walker AM. 2008. Aprotinin during coronary-artery bypass grafting and risk of death. *N. Engl. J. Med.* 358:771–783.

Schneeweiss S, Seeger JD, Maclure M, Wang PS, Avorn J, Glynn RJ. 2001. Performance of comorbidity scores to control for confounding in epidemiologic studies using claims data. *Am. J. Epidemiol.* 154:854–864.

Schuemie MJ, Ryan PB, DuMouchel W, Suchard MA, Madigan D. 2014. Interpreting observational studies: Why empirical calibration is needed to correct p-values. *Stat. Med.* 33(2):209–218.

Seeger JD, Kurth T, Walker AM. 2007. Use of propensity score technique to account for exposure-related covariates: An example and lesson. *Med. Care* 45:S143–S148.

Seeger JD, Walker AM, Williams PL, Saperia GM, Sacks FM. 2003. A propensity score-matched cohort study of the effect of statins, mainly fluvastatin, on the occurrence of acute myocardial infarction. *Am. J. Cardiol.* 92:1447–1451.

Seeger JD, Williams PL, Walker AM. 2005. An application of propensity score matching using claims data. *Pharmacoepidemiol. Drug Saf.* 14:465–476.

So L, Evans D, Quan H. 2006. ICD-10 coding algorithms for defining comorbidities of acute myocardial infarction. *BMC Health Serv. Res.* 6:161.

Southern DA, Quan H, Ghali WA. 2004. Comparison of the Elixhauser and Charlson/ Deyo methods of comorbidity measurement in administrative data. *Med. Care* 42:355–360.

Stang PE, Ryan PB, Racoosin JA, Overhage JM, Hartzema AG et al. 2010. Advancing the science for active surveillance: Rationale and design for the Observational Medical Outcomes Partnership. *Ann. Intern. Med.* 153:600–606.

Strom BL. 2001. Data validity issues in using claims data. *Pharmacoepidemiol. Drug Saf.* 10:389–392.

Strom BL. 2005. *Pharmacoepidemiology*. Chichester, U.K.: Wiley.

Sturmer T, Glynn RJ, Rothman KJ, Avorn J, Schneeweiss S. 2007. Adjustments for unmeasured confounders in pharmacoepidemiologic database studies using external information. *Med. Care* 45:S158– S165.

Sturmer T, Rothman KJ, Avorn J. 2008. Pharmacoepidemiology and "in silico" drug evaluation: Is there common ground? *J. Clin. Epidemiol.* 61(3):205–206.

Suchard MA, Zorych I, Simpson SE, Schuemie MJ, Ryan PB, Madigan D. 2013. Empirical performance of the self-controlled case series design: Lessons for developing a risk identification and analysis system. *Drug Saf.* 36(Suppl. 1):S83–S93. doi: 10.1007/s40264-013-0100-4.

Suissa S. 2007. Immortal time bias in observational studies of drug effects. *Pharmacoepidemiol. Drug Saf.* 16:241–249.

Suissa S. 2008. Immortal time bias in pharmacoepidemiology. *Am. J. Epidemiol.* 167:492–499.

Suissa S, Garbe E. 2007. Primer: Administrative health databases in observational studies of drug effects—Advantages and disadvantages. *Nat. Clin. Pract. Rheumatol.* 3:725–732.

Szklo M, Nieto FJ. 2007. *Epidemiology: Beyond the Basics*. Burlington, MA: Jones & Bartlett.

Tunstall-Pedoe H. 1997. Validity of ICD code 410 to identify hospital admission for myocardial infarction. *Int. J. Epidemiol.* 26:461–462.

U.S. Food and Drug Admin. 1999. *Managing the Risks from Medical Product Use: Creating a Risk Management Framework*. Silver Springs, MD: U.S. Food Drug Admin. http://www.fda.gov/downloads/Safety/SafetyofSpecificProducts/UCM180520.pdf. Accessed August 14, 2014.

Vandenbroucke JP. 2001. In defense of case reports and case series. *Ann. Intern. Med.* 134(4):330–334.

Varas-Lorenzo C, Castellsague J, Stang MR, Tomas L, Aguado J, Perez-Gutthann S. 2008. Positive predictive value of ICD-9 codes 410 and 411 in the identification of cases of acute coronary syndromes in the Saskatchewan Hospital automated database. *Pharmacoepidemiol. Drug Saf.* 17:842–852.

Wahl PM, Rodgers K, Schneeweiss S, Gage BF, Butler J et al. 2010. Validation of claims-based diagnostic and procedure codes for cardiovascular and gastrointestinal serious adverse events in a commercially-insured population. *Pharmacoepidemiol. Drug Saf.* 19:596–603.

Walker AM. 1996. Confounding by indication. *Epidemiology* 7:335–336.

Waller PC, Evans SJ. 2003. A model for the future conduct of pharmacovigilance. *Pharmacoepidemiol. Drug Saf.* 12:17–29.

Weatherby LB, Nordstrom BL, Fife D, Walker AM. 2002. The impact of wording in "Dear doctor" letters and in black box labels. *Clin. Pharmacol. Ther.* 72:735–742.

Whitaker HJ, Farrington CP, Spiessens B, Musonda P. 2006. Tutorial in biostatistics: The self-controlled case series method. *Stat. Med.* 25:1768–1797.

Wilchesky M, Tamblyn RM, Huang A. 2004. Validation of diagnostic codes within medical services claims. *J. Clin. Epidemiol.* 57:131–141.

Zhang JX, Iwashyna TJ, Christakis NA. 1999. The performance of different lookback periods and sources of information for Charlson comorbidity adjustment in Medicare claims. *Med. Care* 37:1128–1139.

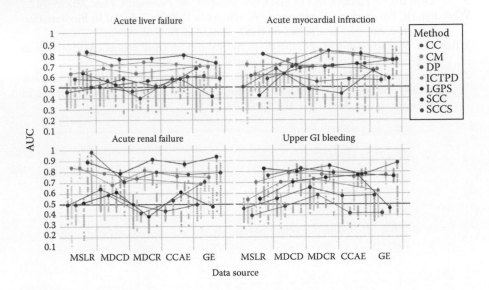

FIGURE 9.3

Performance of methods across AUC values for all analytic methods, again broken down by database and method. The solid lines represent the AUCs for the set of design choices within each method that provided the best performance on average across all outcomes and databases.

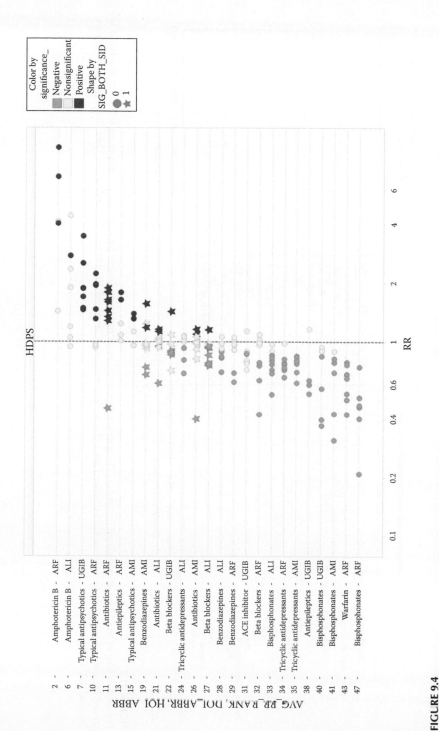

FIGURE 9.4

Database heterogeneity. ARF, acute renal failure; ALI, acute liver injury; UGIB, upper gastrointestinal bleeding; AMI, acute myocardial infarction; HDPS, high-dimensional propensity score-adjusted cohort design.

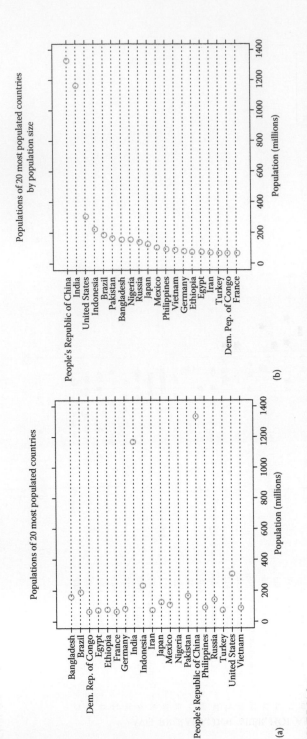

FIGURE 11.2
Population size of the 20 most populated countries. The graphical perception techniques of table lookup and pattern recognition are displayed in (a) and (b), respectively. Graphs constructed by Susan Duke. (From concepts from Cleveland, W., *The Elements of Graphing Data*, 2nd edn., Hobart Press, Summit, NJ, 1994.)

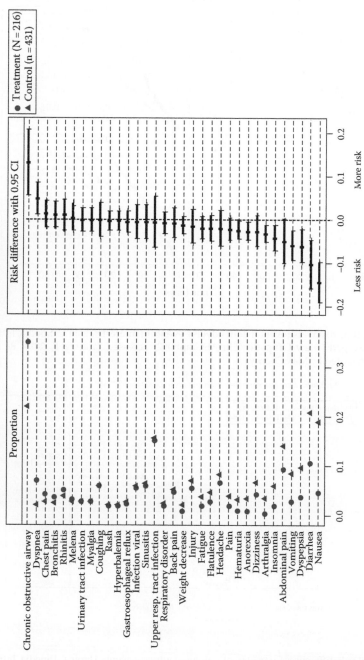

FIGURE 11.3

Adverse event double dot plot for most frequent adverse events. (From CTSpedia, http://www.ctspedia.org/do/view/CTSpedia/ClinAEGraph000, accessed January 8, 2013. R-code contributed by Frank Harrell, SAS code contributed by Sanjay Matange.)

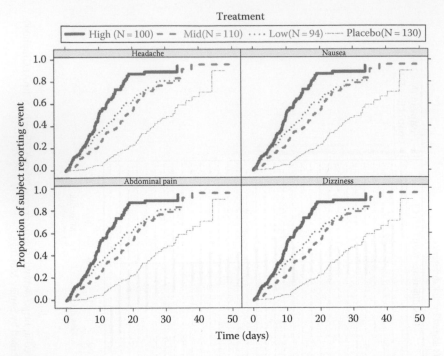

FIGURE 11.5
Trellis plot for time to adverse event in four adverse events of interest. (From CTSpedia, http://www.ctspedia.org/do/view/CTSpedia/ClinAEGraph002, accessed January 8, 2013. R- and SAS-code contributed by Max Cherny.)

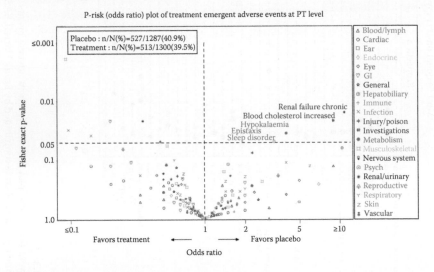

FIGURE 11.6
Volcano plot of risk differences for treatment-emergent adverse events. (From CTSpedia, http://www.ctspedia.org/do/view/CTSpedia/ClinAEGraph003, accessed January 8, 2013. SAS code contributed by Qi Jiang, Haijun Ma and Jun Wei.)

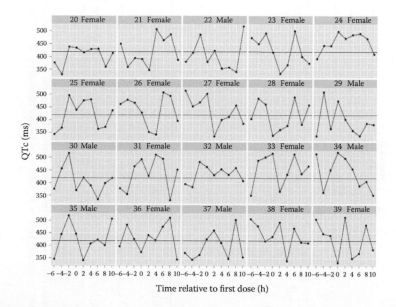

FIGURE 11.7
Individual QTc values relative to first dose (based on simulated data). The 4 h time point presented here represents T_{max} of the experimental drug. All QTc values corrected by the Fridericia method. (From CTSpedia, http://www.ctspedia.org/do/view/CTSpedia/ClinECGGraph003, accessed January 8, 2013. R-code contributed by Max Cherny.)

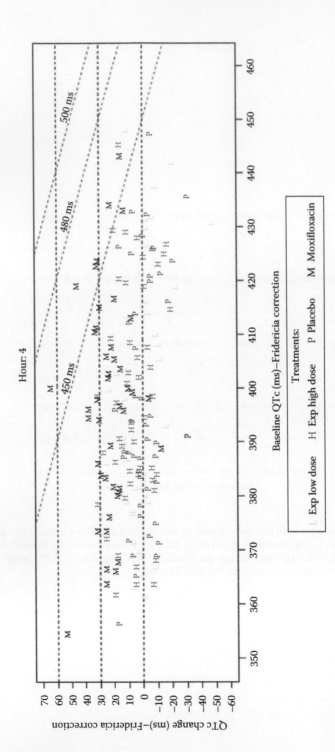

FIGURE 11.8

QTc change from baseline. In this example, the 4 h time point is shown (representative of T_{max} for the experimental drug). (From CTSpedia, http://www.ctspedia.org/do/view/CTSpedia/ClinECGGraph000, accessed January 8, 2013. R-code contributed by Sanjay Matange.)

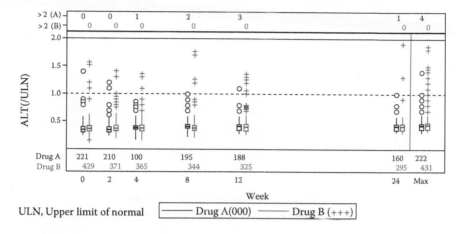

FIGURE 11.9
Distribution of alanine aminotransferase (ALT) by time and treatment. Maximum value for each subject over the course of the 24 weeks is displayed in the right panel. (From CTSpedia, http://www.ctspedia.org/do/view/CTSpedia/ClinLFTGraph007, accessed January 8, 2013. Contributed by Robert Gordon; original plot cited in Amit, O. et al., *Pharm. Stat.*, 7, 20, 2008.)

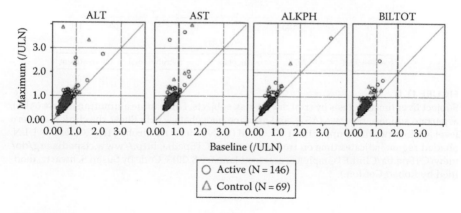

FIGURE 11.10
Lab shift plots: baseline versus maximum. ALT, alanine aminotransferase; AST, aspartate aminotransferase; ALKPH, alkaline phosphatase; BILTOT, total bilirubin. The clinical concern level is 2× ULN (two times upper limit of normal) for BILTOT and ALKPH. It is usually 3× ULN for AST and ALT, but for some therapeutic areas, 5× ULN, 10× ULN, and 20× ULN are used to assess toxicity grades. (From CTSpedia, http://www.ctspedia.org/do/view/CTSpedia/ClinLFTGraph002, accessed January 8, 2013. Code contributed by Sunil J Mistry and Mark Jones, modified by Sanjay Matange, Robert Gordon, and Max Cherny.)

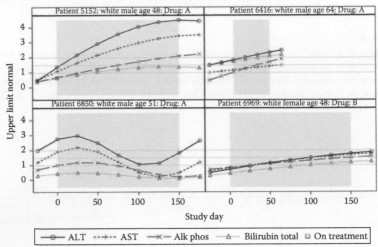

Liver function tests by study day: At-risk subjects

FIGURE 11.11

Subject liver function tests by trial day: at-risk subjects. For alanine aminotransferase (ALT), aspartate aminotransferase (AST), and alkaline phosphatase (Alk Phos), the clinical concern level (CCL) is two times upper limit of normal (ULN); for total bilirubin, the CCL is 1.5× ULN. Shaded region indicates time on treatment. (From CTSpedia, http://www.ctspedia.org/do/ view/CTSpedia/ClinLFTGraph005, accessed January 8, 2013. Code by Susan Schwartz, modified by Robert Gordon.)

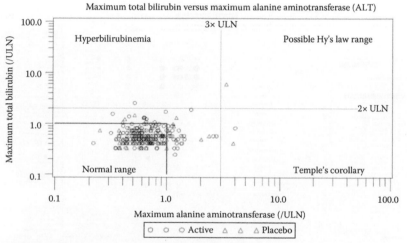

FIGURE 11.12

Scatter plot of maximum transaminase versus maximum total bilirubin. (From CTSpedia, http://www.ctspedia.org/do/view/CTSpedia/ClinLFTGraph000, accessed January 8, 2013. SAS code contributed by Robert Gordon.)

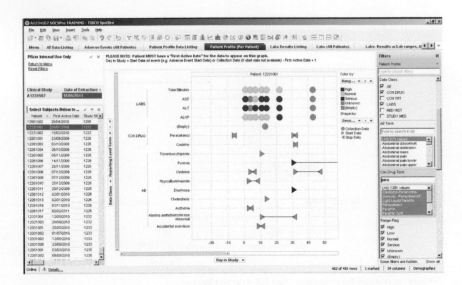

FIGURE 11.13

Example of patient profile interactive graph, in Spotfire. (This is discussed in detail in Section 11.4.) (Courtesy of Jones, D., Timely safety surveillance of clinical study safety data, Tibco Software, Atlanta, GA; Pfizer, La Jolla, CA. http://www.tibco.com/company/news/releases/2012/tibco-spotfire-to-showcase-graphical-representation-of-safety-data-for-timely-safety-surveillance-during-clinical-trials.)

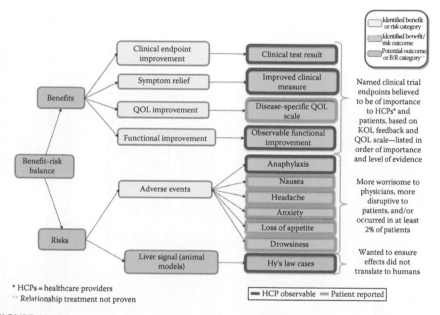

FIGURE 11.14

A completed benefit–risk value tree for a hypothetical compound (from PhRMA BRAT framework; EMA has successfully field tested a similar "effects tree").

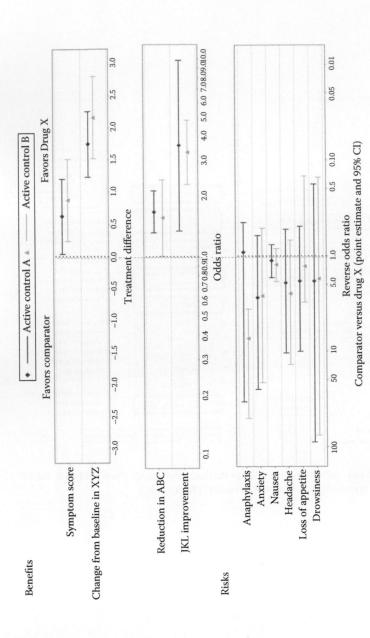

Benefits are from year 1 and year 2: two phase 3 studies (12–24 weeks) (n = 600)

Risks are from primary safety population: two phase 3 studies (12–54 weeks) and one phase 2 study (24 weeks) (n = 670)

FIGURE 11.15
Benefit–risk interval plot. Parameters displayed are based on value tree in Figure 11.14. (Graph created by Susan Duke, based on hypothetical data.)

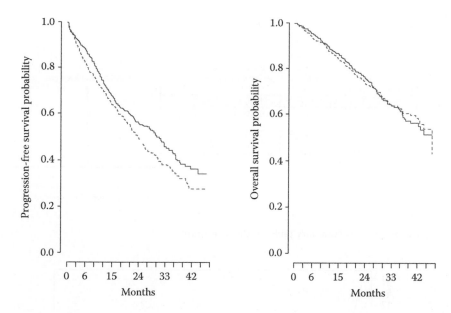

FIGURE 15.1
KM estimate of the survival probability for progression or death and overall death (red solid lines, treatment; blue dotted lines, placebo).

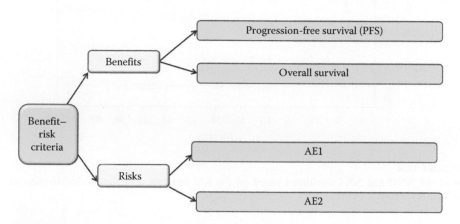

FIGURE 15.2
Example of a value tree for the case study.

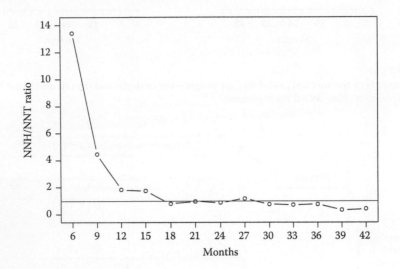

	Outcome	Placebo risk	Treatment risk	Risk difference: per 100 person-years (95% CI)	Rank	Risk difference forest plot
Benefits	PFS	35.4	29.0	−6.4 (−12.1, −0.7)	1	
	OS	15.2	15.1	−0.2 (−3.6, 3.2)	2	
Risks	AE1	7.2	10.0	2.8 (−0.3, 6.1)	4	
	AE2	0.0	4.0	4.0 (2.7, 5.4)	3	

Favors treatment   Favors control

FIGURE 15.3
Example of benefit–risk summary table for the case study.

FIGURE 15.4
Ratio of NNH and NNT calculated based on the KM estimate over time (reference line of NNH/NNT ratio = 1 in red).

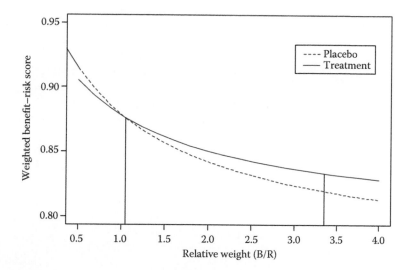

FIGURE 15.5
Weighted benefit–risk scores versus ratio of total weights for benefits and risks.

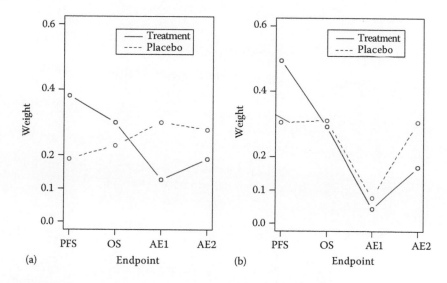

FIGURE 15.6
Central weight vectors for uniform (a) and Dirichlet (b) distributions.

10

Roadmap for Causal Inference in Safety Analysis

Jordan C. Brooks, Alan S. Go, Daniel E. Singer, and Mark J. van der Laan

CONTENTS

10.1 Introduction and Motivation

Safety analysis involves assessing the causal effect of a treatment or exposure regimen on one or more outcomes of interest based on observing a sample of subjects over time. The observed data might constitute an observational study or a randomized controlled trial or even a collection of observational or randomized studies. Since randomized trials are typically not powered for safety outcomes, it is not uncommon that one combines various randomized studies, possibly augmented with observational studies, in order to evaluate safety signals in the data.

Most safety analysis tools involve the application of (parametric) model-based regression methods that rely on unrealistic assumptions. For example, one might employ a Cox-proportional hazards regression model in order to estimate the effect of drug treatment on a certain time to event outcome such as time till a cardiovascular event, in which case one might decide to code a switch in treatment drug exposure as a right-censoring event. The effect measure of interest is now defined as the unknown coefficient in front of treatment in a Cox-proportional hazards regression model that also adjusts for some pretreatment confounders of treatment.

This common approach in safety analysis has some fundamental flaws. Firstly, such models are misspecified, so that the coefficient is not a well-defined effect measure and, even when one properly defines it as a feature of the probability distribution of the data, the rigid model fails to properly adjust for the *measured* confounding. As a consequence, this coefficient cannot be interpreted as a causal effect, even if all confounders are measured. Even when one is convinced that one has specified a correctly specified Cox-proportional hazards regression model, one will have included a number of interaction terms for treatment and baseline covariates, so that there does not exist a single coefficient that represents the desired average treatment effect: instead, one obtains a treatment effect (measured as a relative hazard) within each stratum of the baseline covariates. Putting the issue of misspecification aside, such regression models cannot properly adjust for time-dependent covariates that inform censoring. For example, even when treatment is randomized, and one evaluates the treatment effect by the coefficient in the Cox-proportional hazards regression model that only includes the treatment, this method is biased if the right-censoring event is informed by time-dependent covariates that are also predictive of the time to event outcome. One cannot adjust for the time-dependent covariates through stratification or by model-based stratification, that is, by adjusting for them in these regression models, since that results in biased interpretations of the coefficient in front of treatment, even if the model would be correctly specified.

The real issue behind all these flaws of these approaches is that one has to actually *define* the causal effect of interest and establish a corresponding identifiability result that expresses this causal quantity of interest as a statistical estimand (i.e., as a mapping applied to the data distribution) under a set of assumptions about the data-generating process. This is precisely what causal inference is all about. Causal inference is concerned with posing causal models that code a set of realistic assumptions about the data-generating process and that allow the definition of intervention-specific counterfactual outcomes. Causal effects can now be defined as a difference between the distributions of two intervention-specific counterfactuals. An identifiability result expresses these causal effects as a functional applied to the distribution of the observed data under a set of explicitly stated assumptions. Again, it is important that such identifiability results are established under a minimal set of assumptions so that there is hope that such assumptions are met or could be met by improving the study design. Thus, identifiability results that are based on parametric model assumptions such as parametric structural equation models are not satisfactory, since they only define the desired causal effect under absolutely unrealistic assumptions that are known to be heavily violated, and even the best study design will not make them true. Nonetheless, the assumptions needed in order to express the causal effect as an estimand might be considered unrealistic, and in that case, they should not be viewed as part of the actual causal model.

From the perspective of a statistician, the role of causal inference is to generate interesting estimands that can be interpreted as a desired causal effect under explicitly defined assumptions. Causal inference, if properly conducted, is a field that enriches the field of statistics, by providing a structured approach to define interesting statistical target parameters of interest, which have an enriched interpretation under specified causal assumptions (possibly viewed as unrealistic). The goal is now to construct an estimator of this estimand. From an estimation point of view, only the statistical model, defined as the set of possible distributions of the observed data, is relevant. This statistical model is a by-product of the causal model. The statistical model should be an honest model in the sense that it contains the true distribution of the observed data. If the causal model focuses on only making causal assumptions (i.e., assumptions that put no restrictions on the distribution of the data), then the statistical model is nonparametric, but, of course, the statistical model should incorporate any available statistical knowledge about the data-generating distribution. If the assumption needed in the identifiability result puts unrealistic restrictions on the distribution of the data, then these restrictions should not be included in the statistical model. By making sure that the statistical model is correctly specified, the statistical estimand is now always a well-defined effect measure of interest that properly adjusts for all *measured* baseline and time-dependent covariates, and, under certain possibly unrealistic assumptions, it can be interpreted as a causal effect. For example, one might be able to interpret the estimand as the best possible approximation of the causal effect based on the observed data.

The *pure* statistical estimation problem is now defined. For example, one might observe n independent and identically distributed copies of a random variable O, representing the subject-specific observed longitudinal data structure, with probability distribution P_0 that is known to be an element of a statistical model $\mathcal{M}$. The estimand is now represented by a mapping Ψ that maps any probability distribution $P \in \mathcal{M}$ into a number (or vector) $\Psi(P)$, applied to the true data distribution P_0. Thus, the estimand is defined as $\psi_0 = \Psi(P_0)$ for this target parameter mapping $\Psi : \mathcal{M} \to \mathbb{R}$. The field of statistics is now concerned with constructing an estimator of ψ_0 with excellent statistical properties and whose sampling distribution can be estimated, so that statistical inference in terms of a confidence interval and hypothesis testing can be employed. This happens to be a very challenging estimation problem since the statistical model is extremely large or even nonparametric, so that a (nonparametric) maximum likelihood estimate, (NP)MLE, is ill-defined and we have to incorporate data-adaptive estimation, also called machine learning, while preserving statistical inference for the target parameter ψ_0 based on a central limit theorem.

This challenging field, which integrates the state of the art in machine learning with formal statistical inference, we have named *targeted learning* and our group has developed a fundamental approach/framework called targeted minimum loss-based estimation (TMLE) (van der Laan and Rubin, 2006; van der Laan, 2008; Rose and van der Laan, 2011b) and a variety of corresponding tools for specific estimation problems that involve the innovative application of this general framework: Bembom and van der Laan (2007), van der Laan (2008), Rose and van der Laan (2008, 2009, 2011a), Moore and van der Laan (2009a,b,c), Bembom et al. (2009), Polley and van der Laan (2009), Rosenblum et al. (2009), van der Laan and Gruber (2010), Stitelman and van der Laan (2010), Gruber and van der Laan (2010), Rosenblum and van der Laan (2010), Wang et al. (2010), van der Laan (2010), Stitelman and van der Laan (2011), and van der Laan and Gruber (2011). An original example of a TMLE was presented in Scharfstein et al. (1999) and we also refer to the addendum of Rosenblum and van der Laan (2010). For a detailed bibliography on an alternative approach for construction of double robust efficient estimators, called locally efficient estimating equation methodology (Robins and Rotnitzky, 1992), we refer to Chapter 1 in van der Laan and Robins (2003).

One can return to causal inference at the interpretation stage, at which time a sensitivity analysis that assesses the causal effect under different plausible levels of violations of the identifiability assumptions could be helpful.

The remainder of this chapter is organized as follows. In Section 10.2, we start out with reviewing the roadmap for targeted learning of causal effects (Pearl, 2009; Rose and van der Laan, 2011b; Petersen and van der Laan, 2012), which should also be employed in safety analysis. In order to be reasonably concrete, we focus on a general longitudinal data structure that captures lots of applications of interest. We demonstrate that the causal model allows us to define a large variety of scientific quantities of interest. In Section 10.3, we review a general template for construction of an efficient substitution estimator of the estimand, called TMLE, which also utilizes ensemble machine learning, and discuss its statistical properties. In Section 10.4, we apply it to estimate the causal effect of a multiple time-point intervention or single time-point intervention, allowing for informative right censoring that depends on time-dependent covariates (van der Laan and Gruber, 2011). In Section 10.5, we demonstrate this TMLE in a real safety analysis example in which we estimate the effect of warfarin use on time until death, based on a large observational database of Kaiser Permanente. We conclude with some remarks. This chapter represents a review of previous work, and, due to space limitations, we refer to other articles or books for more extensive presentations of the material.

___

10.2 Roadmap for Safety Analysis

In this section, we describe the roadmap for targeted learning of a causal effect (Pearl, 2009; Petersen and van der Laan, 2012; van der Laan and Rose, 2012).

10.2.1 Observed Data Structure

Suppose that we observe for each subject a time-ordered longitudinal data structure $O = (L(0), A(0), L(1), A(1), ..., L(\tau), A(\tau), Y = L(\tau + 1))$, where $A(0), ..., A(\tau)$ denote the intervention nodes: that is, we wish to understand what the distribution of the data would have been if these intervention nodes would have been assigned by a user-supplied mechanism. For example, in our data analysis example, $A(0)$ is the indicator of being on warfarin or not, while $A(k)$, $k > 0$, denotes the indicator of being right censored, due to either end of study or stopping the treatment as assigned at baseline. The outcome Y could be an indicator of having developed stroke by time $\tau + 1$, while $L(k)$ represents all measurements made on the subject between the two intervention nodes $A(k - 1)$ and $A(k)$. Let P_0 denote the true probability distribution of O. The sample consists of n i.i.d. observations $O_1, ..., O_n$ of O.

10.2.2 Structural Causal Model

We assume that each node $L(k)$ is a deterministic function of its parent nodes $Pa(L(k)) = (\bar{A}(k-1), \bar{L}(k-1))$ and an exogenous error $U_{L(k)}$, $k = 0, ..., \tau + 1$, where $Pa(L(0))$ is the empty set. Here, we use the notation $\bar{X}(t) = (X(0), ..., X(t))$. Similarly, we assume that each intervention node $A(k)$ is a deterministic function of its parent nodes $Pa(A(k)) = (\bar{A}(k-1), \bar{L}(k))$, and an exogenous error $U_{A(k)}$, $k = 0, ..., \tau$. One might assume that a certain subset of these parent nodes have no arrow going into the node. Such an assumption is called an exclusion restriction assumption. For example, if a treatment node is completely randomized, then none of the parent nodes have an arrow going into this treatment node, and its exogenous error has a known distribution and is independent of all other exogenous nodes. Such exclusion restriction assumptions can be visualized with a corresponding causal graph that is obtained by drawing an arrow from one node to another node if that node is a direct parent node. In addition, one might make assumptions about the shape of the deterministic functions, but that is often not warranted and should thus only be done if there is certainly about such a functional relation. This system of structural equations and a model on the distribution of $U = (U_{L(0)}, U_{A(0)}, ..., U_Y)$ describes a collection of distributions of (O, U). This collection of possible distributions of (O, U) is called the structural causal model and we might denote it with $\mathcal{M}^F$. Note that O corresponds with a subset of this complete/full data structure (O, U), and, as a consequence, this structural causal model also implies a statistical model $\mathcal{M}$ of all possible data distributions (Pearl, 1995, 2000).

10.2.3 Causal Quantities

This system of structural equations allows us to define counterfactuals Y_{g^*} corresponding with an intervention g^* on the intervention nodes $\bar{A}$ (Rubin, 1974; Robins, 1987, 1997, 1999a; Neyman, 1990; Pearl, 2000; Gill and Robins, 2001; Yu and van der Laan, 2003). This counterfactual is defined by replacing the equations for the intervention nodes by our user-supplied intervention g^* that defines how each node $A(k)$ is generated in response to its parent nodes. Here, g^* can be coded by a set of conditional distributions g_k^* of $A^*(k)$, given $Pa(A^*(k))$, that specifies how $A^*(k)$ is sequentially generated. Static and dynamic interventions correspond with a degenerate distribution g^*, and nondegenerate choices g^* are often referred to as stochastic interventions (e.g., Didelez et al., 2006; Dawid and Didelez, 2010; Díaz and van der Laan, 2013). For example, g^* might set $A(k) = 1$ for each $k = 0, ..., \tau + 1$. The probability distribution of Y_{g^*} is now indexed by the distribution of (U, O) and the known g^*. One can now define causal quantities that involve contrasts of intervention-specific probability distributions. For example, we might define our quantity of interest as $E_0 Y_{g_1^*} - E_0 Y_{g_2^*}$ for two different interventions. For example, g_1^* might set the treatment $A(0) = 1$ while it sets all censoring nodes equal to *no-censoring*, while g_2^* only differs from g_1^* by setting treatment $A(0) = 0$. This causal quantity represents the causal effect of the single time-point treatment $A(0)$ on the outcome Y in the absence of any censoring. One might also be interested in the effect of a single time-point treatment on the outcome not mediated by an intermediate time-dependent process. In this case, one can include the intermediate time-dependent process in the intervention nodes and define interventions g_1^* and g_2^* that set the intermediate process at a fixed value. Such treatment effects are called the controlled direct effect of treatment. One might also be interested in understanding how the mean of the intervention-specific counterfactual outcome changes as a function of the choice of intervention among a whole class of interventions. This total collection of intervention-specific means can be called a causal dose–response curve, and one could define the target quantity in terms of a marginal structural working model that models this causal dose–response curve according to a user-supplied working model (van der Laan and Petersen, 2007; Robins et al., 2008). This working model defines the target quantity, but it is not an actual assumption and thereby does not affect the causal model and, in particular, the observed data model (Neugebauer and van der Laan, 2007).

Note that a causal quantity is defined as a feature $\psi_0^F = \Psi^F(P_{U,O,0})$ of the true distribution $P_{U,O,0}$ of (U, O), where $\Psi^F : \mathcal{M}^F \to \mathbb{R}^d$ is the underlying target parameter mapping.

In a meta-analysis, one might define a causal effect $\psi_{0,j}^F$ for each study j and define the desired causal effect as a weighted average of these study-specific causal effects, $\psi_0^F = \sum_j w_j \psi_{0,j}^F$. The weights might be a simple function of the sample size for study j, or it might be an unknown function of the variance of the estimator of $\psi_{0,j}^F$. In the latter case, one may take into account that these weights are estimators of true underlying weights, but it is also reasonable to define the target parameter in terms of these data-dependent weights and provide statistical inference for this data-adaptive target parameter.

10.2.4 Identifiability Result

Suppose that $A(k)$ is independent of O_{g^*}, conditional on the observed past $Pa(A(k))$, $k = 0, \ldots, \tau$. This assumption is called the sequential randomization assumption, or no unmeasured confounder assumption. Under this assumption, and the positivity assumption that states that for each possible observed past that follows intervention g^*, there is a positive probability that the next intervention node also follows g^*, one can now prove that the postintervention distribution of O under intervention g^* is given by

$$P_0^F(O_{g^*} = o) = \prod_{k=0}^{\tau+1} P_0(L(k) = l(k) \mid \bar{L}(k-1) = \bar{l}(k-1), \bar{A}(k-1) = \bar{a}(k-1)) \prod_{k=0}^{\tau} g_k^*(o),$$

where g_k^* is the conditional distribution of $A^*(k)$, given its parents. This is called the G-computation formula in causal inference (Robins, 1987). Thus, under the SRA, we can express $E_0 Y_{g^*}^*$ as an estimand $\Psi(P_0)$ of the observed data distribution P_0. In general, such an identifiability result states that for each possible $P_{U,O,0} \in \mathcal{M}^F$, we have $\psi_0^F = \Psi(P_0)$ for some statistical target parameter mapping Ψ. This G-computation formula is just one particularly interesting identifiability result, but many others exist and could be derived, utilizing other known assumptions such as instrumental variable type assumptions (e.g., Pearl, 2000).

10.2.5 Statistical Model, Statistical Target Parameter, and Estimation Problem

The previously mentioned causal inference part of the roadmap has now provided us with a formulation of the statistical estimation problem: We observe n i.i.d. copies of $O \sim P_0 \in \mathcal{M}$ and we want to estimate the target parameter value $\psi_0 = \Psi(P_0)$ corresponding with a target parameter mapping $\Psi : \mathcal{M} \to \mathbb{R} \to$. In addition, we want an estimate of the sampling distribution of the estimator so that we can construct confidence intervals and test one or more null hypotheses. For this purpose, we have proposed TMLE, as described in the next section.

10.2.6 Influence Curve for Inference

We wish to construct an estimator that is asymptotically linear so that $\sqrt{n}(\psi_n - \psi_0) = \frac{1}{\sqrt{n}} \sum_{i=1}^{n} IC(P_0)(O_i) + o_P(1)$ for a so-called influence curve $IC(P_0)$. This assumes that the remainder converges to zero in probability as sample size increases. Asymptotic linearity does not only imply root-n consistency but also establishes that $\sqrt{n}(\psi_n - \psi_0)$ converges in distribution to a normal distribution with mean zero and covariance matrix $\Sigma = E_0 IC(P_0)(O)^2$ equal to the covariance matrix of the influence curve. In addition, it also allows us to estimate this normal limit distribution as $N(0, \Sigma_n)$, where Σ_n is the empirical covariance matrix of the estimated influence curve $IC_n(O_i)$, $i = 1, \ldots, n$. Confidence intervals can now be computed as $\psi_n(j) \pm 1.96 \sqrt{\Sigma_n(j,j)}/\sqrt{n}$, $j = 1, \ldots, d$, and one can also construct simultaneous confidence intervals.

To establish the asymptotic linearity, one might need additional *statistical* assumptions, so that these are now, by necessity, included in the statistical model. These assumptions were not driven by statistical knowledge but they are needed in order to claim valid statistical interpretations of the estimator and confidence interval. Therefore, it is very important to use the most sophisticated estimator (i.e., minimally biased) since a simple estimator might rely on much stronger assumptions, heavily restricting the statistical model under which the conclusions are reliable. In addition, using an efficient estimator will result in smaller confidence intervals and thereby increases the likelihood to have interesting statistical conclusions. Interestingly, efficient estimators are also typically less biased, so that the construction of minimally biased and maximally efficient estimators often corresponds with the construction of efficient estimators. We remind the reader that an estimator is efficient at data distribution P_0 if and only if it is asymptotically linear with influence curve equal to the efficient influence curve $D^*(P_0)$, where the efficient influence curve is defined as the canonical gradient of the pathwise derivative of $\Psi : \mathcal{M} \to \mathbb{R}^d$ along all possible parametric submodels of $\mathcal{M}$ through P_0 (Bickel et al., 1997; van der Vaart, 1998).

10.2.7 Multiple Testing Based on Joint Distribution

If ψ_0 is multivariate, then one might be interested in a multiple testing procedure that tests the collection of null hypotheses $H_0(j)$: $\psi_0(j) \leq 0$, $j = 1, \ldots, d$. For example, the outcome Y might be multivariate and $\psi_0(j)$ might represent the estimand for the causal effect of treatment $A(0)$ on outcome $Y(j)$. Under the asymptotic linearity assumption on ψ_n, we have $\sqrt{n}(\psi_n - \psi_0) \sim N(0, \Sigma_n)$, where Σ_n is the empirical covariance matrix of the estimated influence curve vectors $IC_n(O_i) \in \mathbb{R}^d$. This allows the construction of a simultaneous confidence interval for ψ_0 of the form $\psi_n(j) \pm c_{0.95}\sqrt{\Sigma_n(j,j)}/\sqrt{n}$, where $c_{0.95}$ is the 0.95-quantile of $\max_j |Z(j)|$ with $Z \sim N(0, \rho_n)$, where ρ_n is the correlation matrix of Σ_n. The single-step multiple testing procedure is now defined as follows: reject $H_0(j)$ if $T_n(j) = \sqrt{n}\psi_n(j)/\sqrt{\Sigma_n(j,j)} > c_{0.95}$. This controls the family-wise error rate at level 0.05. One obtains a more powerful multiple testing procedure by using a step-down procedure that iteratively calculates this common cutoff on a shrinking set of null hypotheses ordered from most significant to least significant, stopping the rejections when this common cutoff is larger than the most significant test statistic. In this manner, one fully utilizes the potentially strong dependence between the different test statistics, resulting in multiple testing procedures that are potentially much more powerful than Bonferroni or corresponding Holmes step-down method (Dudoit and van der Laan, 2008).

10.2.8 Sensitivity Analysis

If the causal model did not include the identifiability assumptions, then it is unknown if ψ_0 equals the desired causal effect ψ_0^F. A variety of sensitivity analyses could be employed, which involve posing a new causal model that still allows identification of the desired causal effect but that represents a deviation from the original causal model. Such a causal model is indexed by a sensitivity parameter α that is assumed known, and for each α-specific causal model, one redevelops the identifiability result and estimator with corresponding statistical inference (Robins et al., 1999; Scharfstein et al., 1999; Rotnitzky et al., 2001). In a recent article by Díaz and van der Laan (2013), we develop a much simpler sensitivity analysis that simply defines the sensitivity parameter as the bias $\psi_0 - \psi_0^F$ or an upper bound thereof, and for each plausible value of this bias, it reports the estimator and possibly conservative confidence interval for the causal effect ψ_0^F. The latter method relies on fewer assumptions, and does not involve any extra work.

10.3 TMLE

We have developed a general template for construction of asymptotically efficient substitution estimators of ψ_0 fully respecting the statistical model. It works as follows. Firstly, one determines a representation of ψ_0 as a function of a smaller parameter Q_0 instead of the whole P_0. Let F be the parameter space of Q_0. For notational convenience, we will refer to this mapping as Ψ again. Thus, $\psi_0 = \Psi(Q_0)$. We now assume the existence of a loss function $L(Q)$ (O) for Q_0 so that $Q_0 = \arg\min_{Q \in \mathcal{F}} E_0 L(Q)(O)$, and a corresponding submodel $\{Q(\epsilon, g) : \epsilon\} \subset F$ is chosen so that the linear span of $\left. \frac{d}{d\epsilon} L(Q(\epsilon, g)) \right|_{\epsilon=0}$ includes the efficient influence curve $D^*(Q, g)$, where g is another nuisance parameter so that the canonical gradient $D^*(P) = D^*(Q(P), g(P))$. One can now define a targeted minimum loss-based estimator by starting with an initial estimator Q_n^0

and g_n, computing an update $Q_n^1 = Q_n^0(\epsilon_n^0)$ with $\epsilon_n^0 = \arg\min_\epsilon \sum_i L(Q_n^0(\epsilon, g_n))(O_i)$, and iterate this updating process till convergence at which point $\epsilon_n^k \approx 0$. By construction, it follows that the final update Q_n^* satisfies $\sum_i D^*(Q_n^*, g_n)(O_i) = 0$; that is, it solves the efficient influence curve equation. The TMLE of ψ_0 is now defined as the corresponding plug-in estimator $\psi_n^* = \Psi(Q_n^*)$. This iterative algorithm can be modified by only updating one component of Q_0 at the time and one might also update g_n to make the resulting TMLE (Q_n^*, g_n^*) solve additional equations of interest. Either way, any of such iterative updating algorithms are tailored to increase the fit of Q_0 while solving the efficient influence curve equation $\sum_i D^*(Q_n^*, g_n^*)(O_i) = 0$, thereby not affecting the asymptotic properties of the TMLE ψ_n^*.

The fact that Q_n^* solves the efficient influence curve equation allows one to establish asymptotic linearity and efficiency using the following template. We will use the notation $Pf \equiv \int f(o)dP(0)$. Firstly, one notes that by definition of the canonical gradient and a second-order tailor expansion, $P_0 D^*(Q, g) = \psi_0 - \Psi(Q) + R(Q, Q_0, g, g_0)$ for a second-order term R that involves second-order difference such as integrals involving $(Q - Q_0)(g - g_0)$ or $(Q - Q_0)^2$ or $(g - g_0)^2$. We can apply this general identity to $P_0 D^*(Q_n^*, g_n) = \psi_0 - \psi_n^* + R(Q_n^*, Q_0, g_n, g_0)$. If one now assumes that $R(Q_n^*, Q_0, g_n, g_0) = o_P(1/\sqrt{n})$, then, using $P_n D^*(Q_n^*, g_n) = 0$, it follows that

$$\psi_n^* - \psi_0 = (P_n - P_0)D^*(Q_n^*, g_n) + o_P\left(\frac{1}{\sqrt{n}}\right).$$

If one also assumes that $D^*(Q_n^*, g_n)$ is a random function that falls with probability tending to 1 in a P_0-Donsker class of functions and that $P_0\{D^*(Q_n^*, g_n) - D^*(Q_0, g_0)\}^2 \to 0$ in probability as n converges to infinity, then it follows that

$$\psi_n^* - \psi_0 = (P_n - P_0)D^*(Q_0, g_0) + o_P\left(\frac{1}{\sqrt{n}}\right),$$

which thereby establishes that ψ_n^* is an asymptotically efficient estimator of ψ_0. The asymptotic covariance matrix of ψ_n^* can thus be estimated as $P_n D^*(Q_n^*, g_n)^2/n$ and inference proceeds accordingly.

10.3.1 Super Learning

In order to construct good estimators Q_n and g_n of Q_0 and g_0, so that these asymptotic linearity conditions are most likely to hold, we propose to use super learning due to its theoretical finite sample and asymptotic optimality properties (van der Laan and Dudoit, 2003; van der Vaart et al., 2006;

van der Laan et al., 2006, 2007; Polley et al., 2011; van der Laan and Petersen, 2012). Super learning is an ensemble learning algorithm based on cross validation (also called stacking in the regression/classification literature (LeBlanc and Tibshirani, 1996; Hastie et al., 2001)), which defines a library of candidate estimators of Q_0 (or g_0), a loss function, and a parametric family of combinations of the candidate estimators such as weighted averages, and it uses cross validation to determine the optimal weighted combination of the candidate estimators. As long as the loss function at the weighted candidate estimators is uniformly bounded, one can prove a finite sample oracle inequality for the cross-validation selector that proves that if none of the candidate estimators of Q_0 in the library converges as fast to Q_0 as a parametric model-based estimator for a correctly specified parametric model, then this super learner of Q_0 is asymptotically equivalent to the oracle selector that selects the weighted combination that is closest to the true Q_0. This asymptotic equivalence with the oracle selector remains valid even if the number of estimators in the library converges to infinity as a polynomial power of sample size. By including highly adaptive estimators in the library, one is guaranteed that the super learner will be consistent. In addition, by including optimal estimators for subspaces of the parameter space, one also guarantees to achieve the minimax adaptive optimal rate of convergence that equals the minimax rate of convergence for the unknown smallest subspace that still contains the true Q_0.

For a large class of statistical models and target parameters, one can establish that the efficient influence curve remains unbiased at misspecifications of (Q_0, g_0); for example, the efficient influence curve for the treatment-specific mean and many other causal quantities is so-called double robust so that $P_0 D^*(Q, g) = 0$ if $\Psi(Q) = \psi_0$ and either $Q = Q_0$ or $g = g_0$ (Robins, 2000; Robins et al., 2000; Robins and Rotnitzky, 2001; van der Laan and Robins, 2003). In such cases, one can still establish asymptotic linearity of the TMLE even when (Q_n^*, g_n^*) is inconsistent. We refer to the Appendix in Rose and van der Laan (2011b) for a detailed presentation of an even more general collaborative double robustness of the efficient influence curve, a collaborative TMLE fully utilizing this robustness (Gruber and van der Laan, 2010, 2012; van der Laan and Gruber, 2010), and for generalized asymptotic linearity theorem that allows for corresponding inconsistent estimators (Q_n^*, g_n^*).

10.4 TMLE of Intervention-Specific Mean

Suppose that we wish to estimate the EY_1, where $Y_1 = Y_{g_1^*}$ and g_1^* is a static intervention that assigns treatment $A(0) = 1$ and sets the censoring nodes $A(k) = 0$ for $k = 1, \ldots, \tau$. In this case, we utilize the effective representation of the G-computation formula due to Bang and Robins (2005), which is based on the observation that EY_1 can be evaluated as an iterative conditional expectation starting at the expectation w.r.t. $P(L(\tau+1) \mid Pa(L(\tau+1)))$ moving backward till a final expectation w.r.t. $P(L(0))$. Thus, the first step states that

$$EY_1 = EE(Y \mid \bar{L}(\tau), A(0) = 1, \bar{A}(1:\tau) = 0).$$

We can now define $\bar{Q}_{\tau+1}^1 = E(Y \mid \bar{L}(\tau), A(0) = 1, \bar{A}(1:\tau) = 0)$. Again, applying the rule of iterative conditional expectations, we define $\bar{Q}_\tau^1 = E(\bar{Q}_{\tau+1}^1 \mid \bar{L}(\tau-1), A(0) = 1, \bar{A}(1:\tau-1) = 0)$ and note that $EY_1 = E\bar{Q}_\tau^1$. In general, we define $Q_t^1 = E(\bar{Q}_{t+1}^1 \mid \bar{L}(t-1), A(0) = 1, \bar{A}(1:t-1) = 0)$, $t = \tau+1, ..., 1$, and finally $EY_1 = \bar{Q}_0^1 = E\bar{Q}_1^1$. In this manner, we have expressed EY_1 in terms of a sequence of iteratively defined conditional expectations $\bar{Q} = (\bar{Q}_0^1, \bar{Q}_1^1, ..., \bar{Q}_{\tau+1}^1)$, all directly computable from the observed data distribution. Thus, $EY_1 = \Psi(\bar{Q}) \equiv \bar{Q}_0^1$. Our statistical target parameter Ψ is pathwise differentiable and its efficient influence function is given by

$$D^*(\bar{Q}, g) = \bar{Q}_1^1 - \Psi(\bar{Q}) + \sum_{t=1}^{\tau+1} \frac{I(A(0) = 1, \bar{A}(1:t-1) = 0)}{g_{0:t-1}(O)} (\bar{Q}_{t+1}^1 - \bar{Q}_t^1),$$

where
$$\bar{Q}_{\tau+2}^1 \equiv Y$$
$$g_{0:t}(O) = \prod_{s=0}^{t} P(A(s) \mid Pa(A(s)))$$

A TMLE can now be defined as follows. Consider an estimator g_n of g_0. One starts with estimation of $\bar{Q}_{\tau+1}^1$ by regressing $Y = L(\tau+1)$ onto $\bar{L}(\tau)$, $A(0) = 1$, $\bar{A}(1:\tau) = 0$ using logistic regression. One then uses this fit as an offset in a logistic regression model with clever covariate $\frac{I(A(0)=1, \bar{A}(1:\tau)=0)}{g_{0:\tau,n}(O)}$, and one fits the coefficient in front of this clever covariate with logistic regression. This defines now the targeted fit $\bar{Q}_{\tau+1,n}^*$ of $\bar{Q}_{\tau+1}^1$. This univariate logistic regression submodel generates as score $\frac{I(A(0)=1, \bar{A}(1:\tau)=0)}{g_{0:\tau,n}(O)} (\bar{Q}_{\tau+2}^1 - \bar{Q}_{\tau+1,n}^{1,*})$, which is the last component of the efficient influence curve. One now regresses this targeted fit $\bar{Q}_{\tau+1,n}^{1,*}$ onto $\bar{L}(\tau-1)$, $A(0) = 1$, $\bar{A}(1:\tau-1) = 0$, to obtain an initial fit of $\bar{Q}_\tau^1$. One then uses this initial fit of $\bar{Q}_\tau^1$ as an offset in a logistic regression model with clever covariate $\frac{I(A(0)=1, \bar{A}(1:\tau-1)=0)}{g_{0:\tau-1,n}(O)}$, and one fits the coefficient in front of this clever covariate with logistic regression. This updated fit now defines the targeted fit $\bar{Q}_{\tau,n}^{1,*}$ of $\bar{Q}_\tau^1$. This process is iterated till we end up with the targeted fit $\bar{Q}_{1,n}^{1,*}$. Finally, one takes the empirical mean of the latter to obtain the targeted fit $\bar{Q}_{0,n}^{1,*} = \psi_n^*$. The latter is the TMLE Ψ_n^* of ψ_0.

By construction, this TMLE solves the efficient influence curve estimating equation:

$$0 = \sum_{i=1}^{n} D^* \left(\bar{Q}_n^*, g_n \right)(O_i) = 0.$$

In fact, it solves the empirical mean of each of the $\tau + 1$ components of the efficient influence curve.

In each of these regressions, one respects the fact that if the past tells us that the person already died, then the regression is degenerate and equals 1 (i.e., death) as well, and the update only modifies the regression conditional on not having died yet. Note that the intervention mechanism factorizes in a censoring mechanism and treatment mechanism (i.e., propensity score), and these should be fit separately.

For further details regarding this TMLE, we refer to the original article by van der Laan and Gruber (2011), which also provides R-code. We have released an R-package that implements this TMLE and generalized TMLE of marginal structural working models (Schwab et al., 2013).

If one is willing to assume that the intervention mechanism is estimated consistently, then statistical inference can be based on the working influence curve $D^*(\bar{Q}_n^*, g_n)$, which is asymptotically conservative relative to the true influence curve of the TMLE when $\bar{Q}_n^*$ is misspecified and is asymptotically consistent if $\bar{Q}_n^*$ is consistent for the true $\bar{Q}_0$ (van der Laan and Robins, 2003, Theorem 2.3, Section 2.3.7).

10.5 Safety Analysis of Warfarin: The Causal Effect on Nonstroke Death in Atrial Fibrillation

Atrial fibrillation (Afib) is the most common cardiac arrhythmia in the United States. Persons with Afib are at increased risk of blood clots in the heart that in turn may lead to thromboembolic stroke. The medication warfarin is often prescribed as a preventative measure. While the effectiveness of warfarin in the prevention of thromboembolic stroke is well established, its physiological mechanism of action also puts users at higher risk for other adverse events, for example, bleeding, whose health consequences may be just as devastating. Randomized clinical trials of warfarin have reported favorable results in both effectiveness and safety analyses. However, some clinicians have expressed doubt as to whether these results validly represent the situation in the general Afib patient population. The reason is that typical Afib patients tend to have more comorbidities and may not be as healthy as trial participants.

In practice, the clinical decision to prescribe warfarin is not randomized and involves a delicate balance between reductions in the risk of thrombo-embolic stroke and increases in the risk of adverse side effects, including bleeding events and possibly nonstroke death. Persons at high risk for stroke who might benefit from warfarin may be declined due to a perceived risk of falls, bleeding, or other medical issues that may be aggravated by warfarin. Because treatment is dependent on an individual's medical history, safety analysis of warfarin from observational data must deal with informative treatment assignment and time-dependent right censoring. In this section, we study *nonstroke death* as an adverse outcome and illustrate the TMLE estimation of the intervention-specific marginal probability of this adverse outcome within a 1-year time frame. The first intervention consists of sustained warfarin usage and enforces no censoring, while the second intervention withholds warfarin therapy and enforces no censoring. The causal effect of warfarin is then defined as the difference in these two intervention-specific marginal probabilities.

In this section, we construct the TMLE for this causal effect and compare the resulting estimator to the conventional unadjusted Kaplan–Meier (KM) and the so-called inverse probability of censoring weighted (IPCW) estimating equation-based estimator (Robins, 1999b; Hernan et al., 2000), which attempts to adjust for confounding through the *g*-factors of the likelihood.

10.5.1 ATRIA-1 Cohort Data

One of the largest and most recent observational studies of Afib is the ATRIA-1 cohort study at Kaiser Permanente Northern California (KPNC). This cohort comprised 13,559 individuals who were members of KPNC with a diagnosis of Afib during years 1996–2003. These individuals were followed longitudinally for comorbidities, lab values, warfarin treatment, and stroke and death outcomes. The effectiveness of warfarin in reducing the risk of *stroke or death* was recently addressed in Brooks et al. (2013). The safety analysis in this chapter focuses on the estimation of the causal effect of warfarin on the probability of *nonstroke death* within the first year of follow-up.

At baseline, there were 5289 subjects on warfarin and 8270 who were not. There were 3495 censored individuals because they switched warfarin status; around 1228 were taken off the drug and 2267 were put on the drug within the first year of follow-up. About 168 subjects were censored due to disenrollment from KPNC, and none were administratively censored within the 1-year time frame. Among the noncensored individuals, there were 517 nonstroke deaths within the first year of follow-up. The remaining 9379 were observed to be alive and had not switched treatments at the end of the first year of study. The working data set contained 13,559 persons with 3,926,292 person-days of follow-up until death or censoring. Our analysis regards the $n = 13,599$ subjects as an empirical i.i.d. sample drawn from a probability distribution, $O_1, \ldots, O_{13,559} \sim P_0$.

10.5.2 Causal Effect of Warfarin Parameter

Our parameter of interest is the intervention-specific probability of *nonstroke death* within 1 year. In epidemiology, this probability is called the cumulative incidence. We denote the intervention-specific parameter $\psi^{\bar{a}} = E[Y^{\bar{a}}]$, where Y is the indicator that a person died from a cause other than stroke within the first year of follow-up. We consider two treatment regimens $\bar{a}^1 = (1,0,\ldots,0)$ and $\bar{a}^0 = (0,\ldots,0)$, corresponding with sustained warfarin use and no warfarin use, respectively. The causal effect is then defined as the difference in these two intervention-specific cumulative incidences, that is, $\psi = \psi^{\bar{a}^1} - \psi^{\bar{a}^0}$.

10.5.3 TMLE Implementation

The warfarin treatment assignment mechanism was estimated with a super learner for the negative Bernoulli log-likelihood loss function (log-loss). The time-dependent censoring mechanism was estimated with a super learner for the risk defined as the expectation of the sum of the log-losses for right censoring at each time point. To deal with potential empirical positivity violations, we truncated the estimates of these g-factor estimates to the range [0.01,0.99]. The initial fit for the terminal Q-factor corresponding to the time-dependent event intensity mechanism at day 365 was estimated with a super learner for the risk defined as the expectation of the sum of the log-losses for the non-stroke death event at each time point. The initial fits for the remaining time-point-specific Q-factors were fit with super learners for the log-losses at each time point, respectively. All super learners used fivefold cross validation and a library containing (1) unconditional mean, (2) logistic regression, (3) linear discriminant analysis, (4) three-layer perceptron neural network with two hidden units, and (5) recursive partitioning decision tree candidate estimators. The super learners and the TMLE algorithm were implemented in SAS v9.3.

10.5.4 Results

The super learner for the fixed treatment assignment mechanism was a weighted average of the unconditional mean (9%), logistic regression (37%), linear discriminant analysis (12%), neural network (39%), and decision tree (1%). The super learner for the time-dependent censoring mechanism only involved the neural network (8%) and decision tree (92%). These same super learners were used in the construction of both the TMLE and the IPCW estimators. The range of these fitted g-factors did not suggest practical positivity violations. The super learner for the event intensity, used in the TMLE but not the IPCW, was a weighted average of logistic regression (66%), linear discriminant analysis (4%), neural network (20%), and decision tree (10%). The super learners for remaining Q-factors (364 in total) are omitted here for brevity.

Table 10.1 presents the KM, IPCW, and TMLE estimators and estimates of their standard errors based on their respective influence curves.

TABLE 10.1

Causal Effect of Warfarin on the Probability of Nonstroke Death within 1 Year

Estimator	Warfarin	No Warfarin	Difference	SE	p-Value (2-Sided)
KM	0.0385	0.0558	−0.0173	0.0042	0.0001
IPCW	0.0470	0.0547	−0.0077	0.0070	0.2713
TMLE	0.0510	0.0557	−0.0047	0.0069	0.4907

The KM suggested that warfarin was associated with a 1.7% (SE 0.4%, $p < 0.001$) reduction in the cumulative incidence of nonstroke death. This estimate, however, must be interpreted with caution, as the KM is a stratified estimator that is not adjusted for informative treatment or censoring. Because warfarin treatment decisions are made according to each patient's medical history, that is, are not randomized, the KM is not a consistent estimator of the marginal probability and cannot be interpreted causally.

The TMLE and IPCW estimators are of primary interest, because these may (under identifiability assumptions discussed previously) consistently estimate the marginal causal effect of interest. As discussed previously, the TMLE is robust in the sense that the estimator is consistent for the statistical parameter if either the g-factors, that is, the treatment assignment and right-censoring mechanisms, or the Q-factors of the likelihood are estimated consistently. The IPCW, by contrast, relies solely on consistent estimation of the g-factors.

Neither the TMLE nor IPCW suggested statistically significant causal effects. This null result is more consistent with our a priori knowledge about the physiological mechanism of warfarin, as there is no known reason why warfarin might decrease the risk of nonstroke death. The physiological mechanism of warfarin suggests that, if anything, it should increase the risk of nonstroke death most likely due to bleeding events. The fact that both the TMLE and IPCW point estimates were reductions of 0.8% (SE 0.7%) and 0.5% (SE 0.7%) may indicate that all confounders were not adequately measured. This underscores the need for careful thought in the initial phases of observational study design. Although the TMLE and IPCW estimates were quite similar, it is worth noting that the TMLE point estimate, as compared with that of the IPCW, was closer to the null value. For a richer comparison of TMLE, IPCW, and doubly robust variants of the IPCW in simulation and data analysis, we refer the reader to Brooks et al. (2013).

The TMLE (and IPCW) results indicate that the benefits suggested by the KM analysis are the result of confounding bias. From an epidemiological perspective, the bias is a product of a phenomenon known as confounding by contraindication. The unadjusted estimator suggests that the drug is associated with a lower risk of adverse outcomes because clinicians withhold prescriptions from patients thought to be at high risk of adverse outcomes a priori. The TMLE (and IPCW) adjusts for this to the extent that confounding factors are actually measured and produces a result that is more in line with our true knowledge of the physiological action of the drug.

10.6 Concluding Remark

This chapter provides a broad introduction to the state of the art in general causal inference methods with an eye toward safety analysis. In brief, the estimation roadmap begins with the construction of a formal structural causal model of the data that allows the definition of intervention-specific counterfactual outcomes and causal effects defined as functionals of the distributions of these counterfactuals. The establishment of an identifiability result allows the causal parameter to be recast as an estimand within a statistical model for the observed data, thus translating the causal question of interest into an exercise in statistical estimation and inference. This exercise is nontrivial in (typically nonparametric) statistical models that are large enough to contain the true data-generating distribution.

The fundamental approach presented here, TMLE, allows formal statistical influence curve-based inference (i.e., standard errors and confidence intervals) for estimates built upon extremely flexible nonparametric data-adaptive machine learning algorithms. The TMLE discussed here is directly applicable to the estimation of scalar or vector parameters in randomized controlled trials or observations studies, with general longitudinal or point treatment data structures in nonparametric models. Although our warfarin data analysis focused on the a parameter indexed by a static treatment at one time point, the TMLE may be extended to estimate parameters indexed by more realistic time-dependent dynamic treatment rules, in which a time-dependent treatment for a particular individual may vary according to time-dependent medical characteristics. The TMLE can also estimate an entire counterfactual survival curve along with simultaneous confidence intervals as in Quale and van der Laan (2000), an analysis that may prove useful in many safety analyses.

For a full introductory text on TMLE for causal inference, we refer the reader to the textbook by Rose and van der Laan (2011b).

References

Bang, H. and J.M. Robins. Doubly robust estimation in missing data and causal inference models. *Biometrics*, 61:962–972, 2005. doi: 10.1111/j.1541-0420.2005.00377.x.

Bembom, O., M.L. Petersen, S.-Y. Rhee, W.J. Fessel, S.E. Sinisi, R.W. Shafer, and M.J. van der Laan. Biomarker discovery using targeted maximum likelihood estimation. Application to the treatment of antiretroviral resistant HIV infection. *Stat Med*, 28:152–72, 2009.

Bembom, O. and M.J. van der Laan. A practical illustration of the importance of realistic individualized treatment rules in causal inference. *Electron J Stat*, 1:574–596, 2007.

Bickel, P.J., C.A.J. Klaassen, Y. Ritov, and J. Wellner. *Efficient and Adaptive Estimation for Semiparametric Models*. Springer-Verlag, New York, 1997.

Brooks, J., M.J. van der Laan, D.E. Singer, and A.S. Go. Targeted maximum likelihood estimation of causal effects in right-censored survival data with time-dependent covariates: Warfarin, stroke, and death in atrial fibrillation. *J Causal Infer*, 1:235–254, 2013.

Dawid, A. and V. Didelez. Identifying the consequences of dynamic treatment strategies: A decision theoretic overview. *Stat Surveys*, 4:184–231, 2010.

Díaz, I. and M.J. van der Laan. Sensitivity analysis for causal inference under unmeasured confounding and measurement error problems. *Int J Biostat*, 9:149–160, 2013.

Didelez, V., A.P. Dawid, and S. Geneletti. Direct and indirect effects of sequential treatments. In *Proceedings of the 22nd Annual Conference on Uncertainty in Artificial Intelligence*, Cambridge, MA, 2006, pp. 138–146.

Dudoit, S. and M.J. van der Laan. *Resampling Based Multiple Testing with Applications to Genomics*. Springer Series of Statistics, New York, 2008.

Gill, R. and J.M. Robins. Causal inference in complex longitudinal studies: Continuous case. *Ann Stat*, 29(6), 2001.

Gruber, S. and M.J. van der Laan. An application of collaborative targeted maximum likelihood estimation in causal inference and genomics. *Int J Biostat*, 6(1), Article 18, 2010.

Gruber, S. and M.J. van der Laan. Consistent causal effect estimation under dual misspecification and implications for confounder selection procedure. *Stat Methods Med Res*, February 2012.

Hastie, T., R. Tibshirani, and J.H. Friedman. *The Elements of Statistical Learning: Data Mining, Inference, and Prediction*. Springer Verlag, New York, 2001.

Hernan, M.A., B. Brumback, and J.M. Robins. Marginal structural models to estimate the causal effect of zidovudine on the survival of HIV-positive men. *Epidemiology*, 11(5):561–570, 2000.

LeBlanc, M. and R.J. Tibshirani. Combining estimates in regression and classification. *J Am Stat Assoc*, 91:1641–1650, 1996.

Moore, K.L. and M.J. van der Laan. Application of time-to-event methods in the assessment of safety in clinical trials. In K.E. Peace (ed.), *Design, Summarization, Analysis & Interpretation of Clinical Trials with Time-to-Event Endpoints*. Chapman & Hall, Boca Raton, FL, 2009a.

Moore, K.L. and M.J. van der Laan. Covariate adjustment in randomized trials with binary outcomes: Targeted maximum likelihood estimation. *Stat Med*, 28(1):39–64, 2009b.

Moore, K.L. and M.J. van der Laan. Increasing power in randomized trials with right censored outcomes through covariate adjustment. *J Biopharm Stat*, 19(6):1099–1131, 2009c.

Munoz, I.D. and M.J. van der Laan. Population intervention causal effects based on stochastic interventions. *Biometrics*, 68:541–549, 2012.

Neugebauer, R. and M.J. van der Laan. Nonparametric causal effects based on marginal structural models. *J Stat Plan Infer*, 137:419–434, 2007.

Neyman, J. On the application of probability theory to agricultural experiments. *Stat Sci*, 5:465–480, 1990.

Pearl, J. Causal diagrams for empirical research. *Biometrika*, 82:669–710, 1995.

Pearl, J. *Causality: Models, Reasoning, and Inference*. Cambridge University Press, Cambridge, U.K., 2000.

Pearl, J. *Causality: Models, Reasoning, and Inference*, 2nd edn. Cambridge University Press, New York, 2009.

Petersen, M.L. and M.J. van der Laan. A general roadmap for the estimation of causal effects. Unpublished, Division of Biostatistics, University of California, Berkeley, CA, 2012.

Polley, E.C., S. Rose, and M.J. van der Laan. Super learning. In M.J. van der Laan and S. Rose (eds.), *Targeted Learning: Causal Inference for Observational and Experimental Data*. Springer, New York, 2011, pp. 43–66, Chapter 3.

Polley, E.C. and M.J. van der Laan. Predicting optimal treatment assignment based on prognostic factors in cancer patients. In K.E. Peace (ed.), *Design, Summarization, Analysis & Interpretation of Clinical Trials with Time-to-Event Endpoints*. Chapman & Hall, Boca Raton, FL, 2009, pp. 441–454.

Quale, C.M. and M.J van der Laan. Inference with bivariate truncated data. *Lifetime Data Anal*, 6:391–408, 2000.

Robins, J.M. A graphical approach to the identification and estimation of causal parameters in mortality studies with sustained exposure periods. *J Chron Dis* 40(Suppl. 2):139s–161s, 1987.

Robins, J.M. Causal inference from complex longitudinal data. In M. Berkane (ed.), *Latent Variable Modeling and Applications to Causality*. Springer Verlag, New York, 1997, pp. 69–117.

Robins, J.M. Choice as an alternative to control in observational studies: Comment. *Stat Sci*, 14(3):281–293, 1999a.

Robins, J.M. Marginal structural models versus structural nested models as tools for causal inference. In *Statistical Models in Epidemiology: The Environment and Clinical Trials*. Springer, Berlin, Germany, 1999b.

Robins, J.M. Robust estimation in sequentially ignorable missing data and causal inference models. *Proc. Am. Statist. Assoc. Sect. Bayesian Statist. Sci.* 2000:6–10.

Robins, J.M., L. Orellana, and A. Rotnitzky. Estimation and extrapolation of optimal treatment and testing strategies. *Stat Med*, 27:4678–4721, 2008.

Robins, J.M. and A. Rotnitzky. Recovery of information and adjustment for dependent censoring using surrogate markers. In N.P. Jewell, K. Dietz, and V.T. Farewell (eds.), *AIDS Epidemiology*. Birkhäuser, Basel, Switzerland, 1992.

Robins, J.M. and A. Rotnitzky. Comment on the Bickel and Kwon article, "Inference for semiparametric models: Some questions and an answer". *Stat Sinica*, 11(4):920–936, 2001.

Robins, J.M., A. Rotnitzky, and D.O. Scharfstein. Sensitivity analysis for selection bias and unmeasured confounding in missing data and causal inference models. In M.E. Halloran and D. Berry (eds.), *Statistical Models in Epidemiology, the Environment and Clinical Trials*, IMA Volumes in Mathematics and Its Applications. Springer, New York, 1999.

Robins, J.M., A. Rotnitzky, and M.J. van der Laan. Comment on "On Profile Likelihood" by S.A. Murphy and A.W. van der Vaart. *J Am Stat Assoc—Theory Methods*, 450:431–435, 2000.

Rose, S and M.J. van der Laan. Simple optimal weighting of cases and controls in case–control studies. *Int J Biostat*, 4, Article 19, 2008. http://www.bepress.com/ijb/vol4/iss1/19/.

Rose, S. and M.J. van der Laan. Why match? investigating matched case–control study designs with causal effect estimation. *Int J Biostat*, 5, Article 1, 2009, http://www.bepress.com/ijb/vol5/iss1/1/.

Rose, S. and M.J. van der Laan. A targeted maximum likelihood estimator for two-stage designs. *Int J Biostat*, 7(17), Article 17, 2011a.

Rose, S. and M.J. van der Laan. *Targeted Learning: Causal Inference for Observational and Experimental Data*. Springer, New York, 2011b.

Rosenblum, M., S.G. Deeks, M.J. van der Laan, and D.R. Bangsberg. The risk of virologic failure decreases with duration of HIV suppression, at greater than 50% adherence to antiretroviral therapy. *PLoS One*, 4(9):e7196, 2009. doi: 10.1371/journal.pone.0007196.

Rosenblum, M. and M.J. van der Laan. Targeted maximum likelihood estimation of the parameter of a marginal structural model. *Int J Biostat*, 6(2):Article 19, 2010.

Rotnitzky, A., D. Scharfstein, S. Ting-Li, and J.M. Robins. Methods for conducting sensitivity analysis of trials with potentially nonignorable competing causes of censoring. *Biometrics*, 57(1):103–113, 2001.

Rubin, D.B. Estimating causal effects of treatments in randomized and nonrandomized studies. *J Educ Psychol*, 64:688–701, 1974.

Scharfstein, D.O., A. Rotnitzky, and J.M. Robins. Adjusting for nonignorable drop-out using semiparametric nonresponse models (with discussion and rejoinder). *J Am Stat Assoc*, 94:1096–1146, 1999.

Schwab, J., S. Lendle, M. Petersen, and M. van der Laan. ltmle: Longitudinal targeted maximum likelihood estimation, 2013. http://CRAN.R-project.org/package=ltmle. R package version 0.9.3. Accessed February 1, 2014.

Stitelman, O.M. and M.J. van der Laan. Collaborative targeted maximum likelihood for time to event data. Technical Report 260. Division of Biostatistics, University of California, Berkeley, CA, 2010.

Stitelman, O.M. and M.J. van der Laan. Targeted maximum likelihood estimation of effect modification parameters in survival analysis. *Int J Biostat*, 7(1), 2011.

van der Laan, M.J. The construction and analysis of adaptive group sequential designs. Technical Report 232. Division of Biostatistics, University of California, Berkeley, CA, March 2008.

van der Laan, M.J. Targeted maximum likelihood based causal inference: Part I. *Int J Biostat*, 6(2):Article 2, 2010.

van der Laan, M.J. and S. Dudoit. Unified cross-validation methodology for selection among estimators and a general cross-validated adaptive epsilon-net estimator: Finite sample oracle inequalities and examples. Technical Report. Division of Biostatistics, University of California, Berkeley, CA, November 2003.

van der Laan, M.J., S. Dudoit, and A.W. van der Vaart. The cross-validated adaptive epsilon-net estimator. *Stat Dec*, 24(3):373–395, 2006.

van der Laan, M.J. and S. Gruber. Collaborative double robust penalized targeted maximum likelihood estimation. *Int J Biostat*, 6, Article 17, 2010.

van der Laan, M.J. and S. Gruber. Targeted minimum loss based estimation of an intervention specific mean outcome. Technical Report 290. Division of Biostatistics, University of California, Berkeley, CA, 2011. To appear in IJB, also technical report http://www.bepress.com/ucbbiostat/paper290. Accessed February 1, 2014.

van der Laan, M.J. and M.L. Petersen. Causal effect models for realistic individualized treatment and intention to treat rules. *Int J Biostat*, 3(1), Article 3, 2007.

van der Laan, M.J. and M.L. Petersen. Targeted learning In C. Zhang and Y. Ma (eds.), *Ensemble Machine Learning*. Springer, Boston, MA, 2012, pp. 117–156.

van der Laan, M.J., E. Polley, and A. Hubbard. Super learner. *Stat Appl Gen Mol Biol*, 6(25), Article 25, 2007.

van der Laan, M.J. and J.M. Robins. *Unified Methods for Censored Longitudinal Data and Causality*. Springer, New York, 2003.

van der Laan, M.J. and S. Rose. *Targeted Learning: Causal Inference for Observational and Experimental Data*. Springer, New York, 2012.

van der Laan, M.J. and D. Rubin. Targeted maximum likelihood learning. *Int J Biostat*, 2(1), Article 11, 2006.

van der Vaart, A.W. *Asymptotic Statistics*. Cambridge University Press, Cambridge, U.K., 1998.

van der Vaart, A.W., S. Dudoit, and M.J. van der Laan. Oracle inequalities for multi-fold cross-validation. *Stat Dec*, 24(3):351–371, 2006.

Wang, H., S. Rose, and M.J. van der Laan. Finding quantitative trait loci genes with collaborative targeted maximum likelihood learning. *Stat Prob Lett*, 2010, published online November 11, doi: 10.1016/j.spl.2010.11.001.

Yu, Z. and M.J. van der Laan. Double robust estimation in longitudinal marginal structural models. Technical Report. Division of Biostatistics, University of California, Berkeley, CA, 2003.

11

Safety Graphics

Susan P. Duke, Qi Jiang, Liping Huang, Mary Banach,
and Max Cherny

CONTENTS

11.1 Introduction

The focus of this book is the quantification of safety. One might ask, what is the value of a graph in the quantification of safety, when analytic methods (and tables) for safety more quantitatively characterize the results? It is critical with safety data (and the large amount of data that entail) to effectively present and communicate them.

The value of a graph is experienced by anyone when leafing through a book or browsing the Internet. Graphs and pictures attract the eye far more strongly than table or text. This experience makes sense, knowing that roughly half of the human brain is devoted to vision.

In this chapter, we will look specifically at elements of good graphics; common safety questions and recommended graphics in the areas of adverse events (AEs), ECGs and vital signs, and laboratory parameters (especially liver toxicity); guidance from regulatory agencies; and the future of graphics in benefit–risk and interactive graphics.

Not all graphs are created equal. Just as with statistical methods, mindful attention to the objectives (or messages) the graph intends to convey is a key factor in producing impactful, transparent graphs.[1-3] Utilization of the basics of graphics design principles is important as well.[3]

Well-designed graphs are particularly useful for safety signal detection and safety review. Because new safety signals cannot be anticipated, the power of the human brain via graphical perception can be used to *explore* safety, *identify (or rule out)* safety signals, and *effectively communicate safety findings to decision-makers*. Safety graphs add value throughout the drug life cycle management process: from planning for the creation of the program safety analysis plan (PSAP) to the first internal safety review by the drug development team to regulatory approval, scientific publications, and post approval. Simply said, *use of well-designed, transparent graphics improves a decision-maker's ability to make a well-informed decision*.

Are graphics used effectively in pharmaceutical research? In 2007, Pocock et al. reviewed 77 reports of randomized controlled trials in five medical journals and concluded that there is "considerable scope for authors to improve their use of figures in clinical trial reports."[4] It is the premise of an entire book, *A Picture Is Worth a Thousand Tables: Graphics in Life Sciences*, that "drug development processes are still largely reliant on tables and listings."[5]

It did not start out that way. Over 25 years ago, Cleveland and McGill's[6] experiments described human graphical perception capabilities and their application to the development of graphical methods by assessing individuals' abilities to perceive various shapes and lines (Figure 11.1).

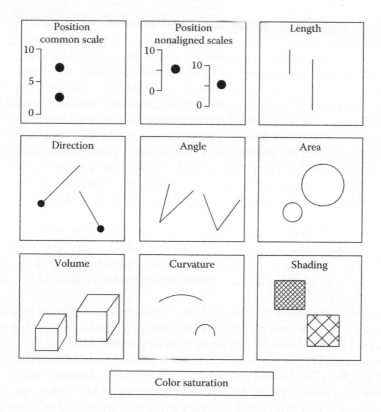

FIGURE 11.1
Elementary perceptual tasks. (From Cleveland, W. and McGill, R., *J. Am. Stat. Assoc.*, 79, 531, 1984.)

They confirmed that humans discriminate in the following order (most to least accurate):

1. Position along a common scale
2. Positions along nonaligned scales
3. Length, direction, angle
4. Area
5. Volume, curvature
6. Shading, color saturation

A number of influential statisticians contributed to the statistical graphics literature of that time: Efron, Kruskal, Tukey, and Tufte are all cited by Cleveland and McGill.[6] In addition to Cleveland and McGill's research, Tufte[1,7–9] has contributed much to the area of graphic design. Cleveland and other statisticians have also written entire books on the topic.[5,10–12]

For some reason, this earlier interest in graphical perception by well-respected statisticians is not emphasized in statistical training or how statisticians typically view their role in analyzing safety data today. Often during the drug development process, the development team relies on a statistician to deliver the data's messages in efficient and effective ways, including graphics. Clearly, it is the statistician's role and responsibility to be as well-versed in statistical graphics design as statistical methods. As an example of good graphics design concepts, consider the graphs in Figure 11.2. Both graphs include population size of the 20 most populated countries, one sorted alphabetically and the other by size. Which one is easier (and quicker) to understand? Use of graphic design features that address Cleveland and McGill's conclusions is arguably just as important as choosing an appropriate statistical test or model.

In the past few years, some progress has been made to create standard, well-designed graphs that answer commonly asked safety questions. With the publication of Amit et al.,[2] there was a heightened interest in the use of graphics for safety analysis, which grew into an initiative across FDA, industry, and academia.

To develop a palette of graphics for the visualization of clinical trial safety data and to promote best practices for statistical graphics, a working group was formed in 2009 consisting of more than 20 statisticians from FDA, industry, and academia. The group has created a publicly available repository of sample graphics, including graphical presentations, datasets, and coding examples.[13] To more effectively answer commonly asked safety questions, the displays are based on good graphics principles (further described in the group's best practices[14]).

With the Safety Graphics Home Page,[14] common graphics types for safety analysis are more accessible, which is encouraging, given that regulatory reviewers have also expressed that graphs could improve the communication of safety signals in submissions,[15] ultimately a benefit to the patients we serve.

Questions regarding patients' safety often arise during or after clinical studies. Some could be simple to interpret, but often they are complicated and might indicate emerging safety signals of an experimental drug. Capturing and confirming emerging safety signals at an earlier stage is highly desirable and a challenge. Conventionally, tables and listings have been used to summarize and report safety-related data. With voluminous pages, it can be challenging for the various parties responsible for safety (Data and Safety Monitoring Boards, safety staff and clinical drug development staff, and regulators) to interpret and act on expeditiously. Regulatory agencies are looking at more efficient and effective ways of communicating safety data.[16–18] Through graphics, we are offering one way to more efficiently and effectively report safety data. The purpose of this chapter is to illustrate how we can use graphical display to assist investigators and reviewers to answer safety-related questions.

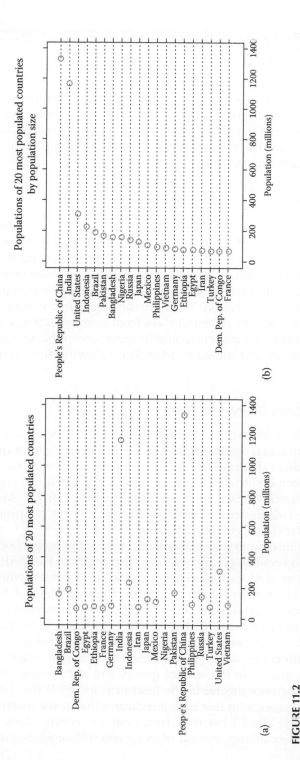

FIGURE 11.2

(See color insert.) Population size of the 20 most populated countries. The graphical perception techniques of table lookup and pattern recognition are displayed in (a) and (b), respectively. Graphs constructed by Susan Duke. (From concepts from Cleveland, W., *The Elements of Graphing Data*, 2nd edn., Hobart Press, Summit, NJ, 1994.)

11.2 Common Safety Questions and Their Graphs

Safety is one of the most important areas to consider during the drug development and approval process. An understanding of safety affects dose level, whether there are any subgroups at more risk (and whether there is a way to mitigate those risks), and the safety labeling in the package insert. Graphics can clarify these and other topics that may come up during the course of developing and marketing the drug.

11.2.1 Adverse Events

Adverse event reporting is usually summarized by percentage of subjects with AEs or by incidence rate, and compared between treatment groups by risk ratio, incident rate ratio, etc., at System Organ Class (SOC) preferred term (PT) level based on Medical Dictionary for Regulatory Activities (MedDRA). With 26 SOCs and nearly 19,000 PTs, there are many ways to summarize and present safety-related information. What are the most effective and efficient ways to allow us to identify potential risks from numerous SOCs and PTs and further to identify patients who could be most susceptible to these risks in an effort to prevent and mitigate risks? The following four graphs are examples.

11.2.1.1 Adverse Event Double Dot Plot

The AE double dot plot is a type of graph designed to compare safety data between an experimental treatment and its control group with a quantitative risk measurement (Figure 11.3). It has also been used in meta-analysis in randomized controlled trials and observational studies. In AE reporting, double dot plots are often used to present the frequency of key AEs along with quantitative risk measurements. Having both types of information side-by-side helps reviewers to identify elevated AEs.

The left panel of this graph shows the relationship of the proportion of AEs in the treatment and control groups. AEs are represented at the MedDRA PT level, categorizing by SOC, high-level term, and so on. Different symbols and colors are used to identify the treatment groups. The right panel of the graph shows the corresponding risk differences and their 95% confidence intervals. The risk differences are ordered so that AEs of highest risk in the treatment group are plotted first. A vertical line through zero representing no difference between treatment groups is also plotted. Left of zero indicates the risk of AE occurrence greater in the control group, and right of zero indicates the risk of AE occurrence greater in the treatment group. If the confidence interval for a PT crosses with this line, it indicates that at the given level of confidence (i.e., 95%), the PT between treatment and control does not differ significantly. Alternatively, instead of using risk difference as a measure,

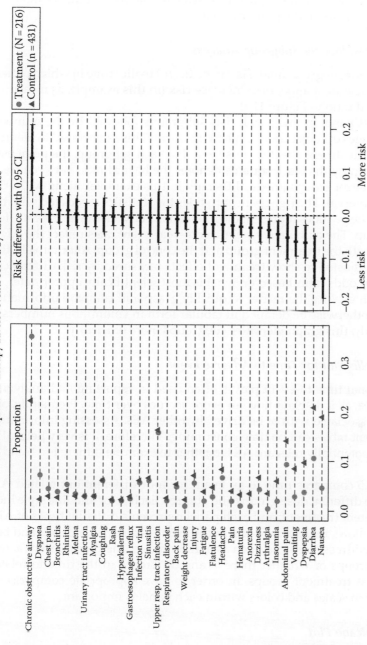

FIGURE 11.3

(See color insert.) Adverse event double dot plot for most frequent adverse events. (From CTSpedia, http://www.ctspedia.org/do/view/CTSpedia/ClinAEGraph000, accessed January 8, 2013. R-code contributed by Frank Harrell, SAS code contributed by Sanjay Matange.)

other quantitative risk measurements such as odds ratio, risk ratios, and hazard ratios can be used. If the odds ratio or the risk ratio is used, the graph may be plotted on a natural logarithmic scale so that the confidence intervals are symmetrical about the point estimates.

11.2.1.2 Forest Plot for Subgroup Analysis

A forest plot for subgroup analysis can be useful to illustrate in which patient groups the treatment appears to increase risk (in this example, as measured by the hazard ratio in Figure 11.4).

The far left panel in this graph lists subgroups followed by the numbers and percentages of subjects associated with them. This can be used to identify whether the treatment effect is consistent. For each subgroup, the point estimate indicated by the black squares denotes the hazard ratio between treatment and control groups (medical therapy in this example), along with 95% confidence interval. The right panel provides the statistics associated with the forest plot. The p-values represent the interaction between the treatment effect of a subgroup. The cumulative estimated 4-year event rates reflect the AE of interest for both the treatment group and the control group. Alternatively, this plot can be adjusted to show treatment effects measured by risk differences, risk ratios, or odds ratio, based on what is most scientifically appropriate.

This graph enables the reader to quickly grasp and compare information that would otherwise be detailed in tabular form in voluminous numbers of pages. Clearly, this graph identifies subgroups more efficiently.

11.2.1.3 Trellis Plot for Time to Adverse Event

The traditional time-to-event plot is often used in reporting and analyzing survival data, as it displays how an event accrues over time. Frequently, one AE can be associated with or be driven by other AEs. Assessing information from different tables can be time consuming. Therefore, it is very helpful to present all potentially relevant AEs together to examine if there is a trend or pattern.

Figure 11.5 compares time to an event for four different but related AEs across three different dose groups and control. Each color and line style indicates a different treatment, and each panel represents different types of AEs. The graph allows reviewers to assess the events side by side or top and bottom. It can address questions such as a difference in the time to event across treatment groups or whether there is any difference in the time to related events across treatment groups. In order to have appropriate comparisons, consistency in scales and colors within each panel is important.

11.2.1.4 Volcano Plot

The volcano plot (Figure 11.6) displays risk differences against their statistical significance. This plot allows us to present many different AEs in one graph,

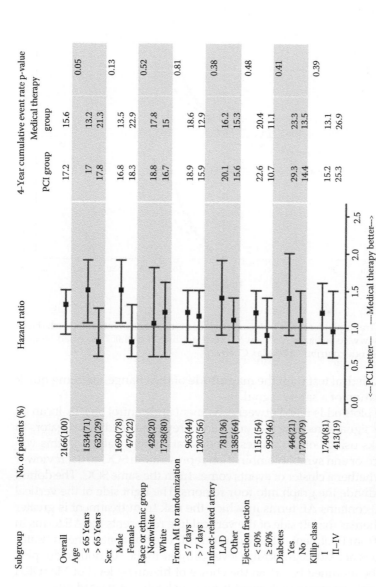

The p-value is from the test statistic for testing the interaction between the treatment and any subgroup variable

FIGURE 11.4

Forest plot for subgroup analysis of percutaneous coronary intervention (PCI, commonly known as angioplasty) versus medical therapy. (From CTSpedia, http://www.ctspedia.org/do/view/CTSpedia/ClinAEGraph001, accessed January 8, 2013. SAS code contributed by Sanjay Matange; Soukup, M., Communicating clinical trial results the statistical graphic way, Joint Statistical Meetings, Boston, MA, 2010.)

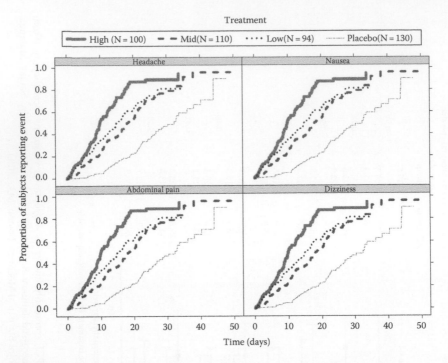

FIGURE 11.5
(See color insert.) Trellis plot for time to adverse event in four adverse events of interest. (From CTSpedia, http://www.ctspedia.org/do/view/CTSpedia/ClinAEGraph002, accessed January 8, 2013. R- and SAS-code contributed by Max Cherny.)

combining a statistical test with the magnitude of the change, enabling quick visual identification of a safety signal.

This example plots odds ratio between treatment and control by PTs. It can be used as a safety signal screening tool and allows reviewers and investigators to evaluate AE risks using both estimates of odds ratio and p-values. In this volcano plot, each color and symbol combination represents an SOC so the reviewer can easily see whether a cluster of events comes from the same SOC. The dotted reference lines divide the graph into four regions. The right side of the vertical line through 1.0 contains AE terms in which the risk from treatment is greater than control, whereas the left side of the vertical line represents the AE terms in which the risk from treatment is less than control. The horizontal line indicates if the odds ratio is at a 0.05 significance level. Similar to the AE double dot plot, this graph can be modified based on the choice of hierarchy level of MedDRA dictionary and risk measures such as odds ratio, risk ratio, or hazard ratio.

11.2.2 ECG Analyses: QTc Prolongation in Clinical Trials

For a subset of clinical trials, sponsors need to assess if there is a potential for the experimental drug to delay cardiac repolarization, an effect that can be

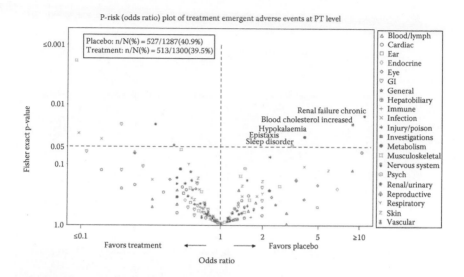

FIGURE 11.6

(See color insert.) Volcano plot of risk differences for treatment-emergent adverse events. (From CTSpedia, http://www.ctspedia.org/do/view/CTSpedia/ClinAEGraph003, accessed January 8, 2013. SAS code contributed by Qi Jiang, Haijun Ma and Jun Wei.)

measured as prolongation of the QT interval measured by electrocardiogram (ECG).[19] However, the QT interval is highly related to heart rate. In order to appropriately evaluate if there is a QT interval prolongation, first a correction to QT interval is made to remove the association of QT with RR interval, and then to examine if the change relative to baseline is a clinical concern. There are many ways to present the corrections as well as to explore the changes at individual and population levels. The following three graphs are specifically designed to elucidate the important clinical aspects of ECG-related safety data.

11.2.2.1 Individual QTc Display

QTc display at an individual level allows us to answer questions such as how an individual's values track over time and if there are any individuals whose values may indicate a safety signal. Figure 11.7 illustrates the individual QTc in time relative to first dose. The red reference line is a proposed threshold QTc value (defined as *prolonged QTc*). It is important to keep the scales consistent across all panels so they are easily compared.

11.2.2.2 QTc Change from Baseline

Figure 11.8 plots the change from baseline at T_{max} (time to maximum drug concentration in the plasma). The x-axis is the baseline value, and the y-axis represents the QTc change from baseline. Horizontal reference lines at 30 and 60 ms allow the reader to quickly identify any changes from baseline above

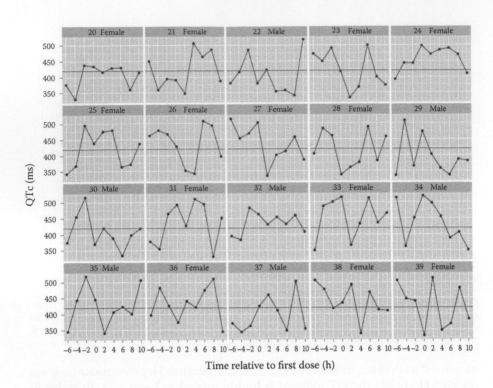

FIGURE 11.7
(See color insert.) Individual QTc values relative to first dose (based on simulated data). The 4 h time point presented here represents T_{max} of the experimental drug. All QTc values corrected by the Fridericia method. (From CTSpedia, http://www.ctspedia.org/do/view/CTSpedia/ClinECGGraph003, accessed January 8, 2013. R-code contributed by Max Cherny.)

these thresholds. Diagonal reference lines at 450, 480, and 500 ms highlight any on-study values above these lines. The overlap area above horizontal and diagonal lines in the upper right corner indicates the region of greatest clinical concern (i.e., on-study values above 450 ms and > 30 ms change from baseline). By simultaneously plotting change from baseline against baseline itself, one can examine the changes from baseline and the baseline raw QTc values.[16]

This graph type could also prove useful for each hour posttreatment administration in a thorough QT design.

This graph design[19] was successfully used at an FDA Advisory Committee meeting to demonstrate that while there were large changes from baseline in QTc, they were not ones of clinical concern. It is an excellent example of how safety graphs can be used to provide greater insight in situations that are multifactorial. The data were graphed in a manner to transparently shed light on the question of simultaneously high changes from baseline and on-study values. The committee was able to better understand the data as it related to their concern, and a potential risk of this medicine that may otherwise have been considered of cardiac concern was discharged.

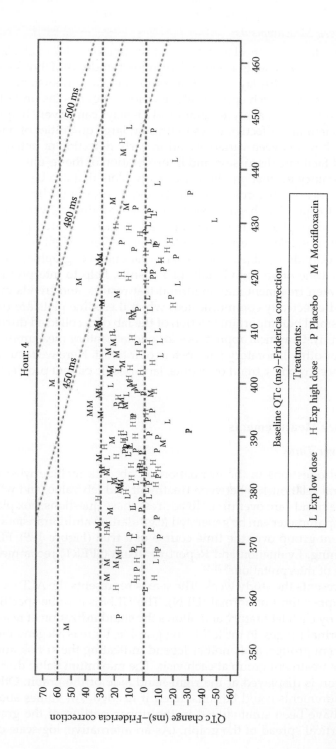

FIGURE 11.8
(See color insert.) QTc change from baseline. In this example, the 4 h time point is shown (representative of T_{max} for the experimental drug). (From CTSpedia, http://www.ctspedia.org/do/view/CTSpedia/ClinECGGraph000, accessed January 8, 2013. R-code contributed by Max Cherry. SAS code contributed by Sanjay Matange.)

11.2.2.2.1 Laboratory Measurements

Clinical laboratory data are critical to the safety evaluation of almost all clinical trials. Laboratory assessments alone may result in hundreds of thousands of observations. How to display so that stakeholders can understand the breadth of the results and reach appropriate decisions? Besides the underlying tabular data and numeric analyses, graphical displays can be very helpful by providing efficient and effective ways to present large quantities of data, examine relationships between variables, aid in early detection of potential safety issues, and facilitate discussion and interpretation of the results.

In clinical laboratory evaluation, there are many plots that can be useful. One of the important areas regards liver toxicity and monitoring of liver function tests (LFTs) for possible hepatotoxicity and drug-induced liver injury (DILI).[19,20] Statistical graphics at the population level, such as scatter plots or box plots, identify possible patterns and effects, and offer insight into possible outliers or individual subjects of concern. Patient-level graphics, such as line plots over time of an individual's LFT results, help to interpret the relationship between treatment and abnormal test results, and trends may be identified (and effectively communicated) when these line plots are trellised. LFTs are a small but important fraction of the lab data collected during the course of a study. Graphical approaches for one laboratory measurement can be adapted for other laboratory parameters of interest. Here we focus on four key laboratory graphics: trend over time, lab-specific patient profile, lab shift, and DILI.

11.2.3 Laboratory Measurements

11.2.3.1 Trend over Time

Common clinical questions in the evaluation of lab data include whether there is a temporal relationship between treatment and lab value, and what the toxicity-grade trends are over time. To address these questions, box plots of the laboratory parameter can be generated (e.g., alanine aminotransferase, ALT), by treatment group over the time course of a trial (Figure 11.9). FDA and Safety Planning, Evaluation and Reporting Team (SPERT) recommend boxplots for ease of interpretation.[18,21,22]

The x-axis represents the study week. The y-axis represents the ALT value divided by the upper limit of normal (ULN). The ULN is a value specified for each variable by each laboratory and allows the standardization of results with different normal ranges. In the following graphic, there is a legend indicating the treatment groups, and another legend indicating the at-risk number of subjects by treatment group at each visit. The maximum value during the study duration is displayed at the right-hand side of the graph. Other alternative measurements could be considered if needed. All values above twice the ULN have been summarized in the upper portion of the graph to reduce the vertical spread of the graph. (As an alternative, log-scale can

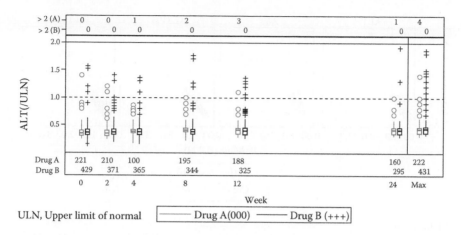

FIGURE 11.9
(See color insert.) Distribution of alanine aminotransferase (ALT) by time and treatment. Maximum value for each subject over the course of the 24 weeks is displayed in the right panel. (From CTSpedia, http://www.ctspedia.org/do/view/CTSpedia/ClinLFTGraph007, accessed January 8, 2013. Contributed by Robert Gordon; original plot cited in Amit, O. et al., *Pharm. Stat.*, 7, 20, 2008.)

be used for the y-axis.) Depending on the lab parameter and level of concern, clinically meaningful values (e.g., > 2 times ULN) and interpretation may differ.

11.2.3.2 Lab Shift

Another common clinical question in the evaluation of lab data is how can we easily identify safety signals relative to baseline.

To address this question, the maximum change in a lab parameter's values over the course of a study for each patient is displayed relative to baseline to examine whether a shift has occurred from normality to levels of clinical concern. Utilizing the original plot from Amit et al.[2] as a basis, the working group re-created the plot with the CTSpedia dataset. Figure 11.10 shows a set of scatter plots displaying the maximum ULN value during the study against the baseline value for the four LFTs. The x-axis represents the baseline values for the parameter graphed. The y-axis represents the maximum on-study value for each patient. Horizontal reference lines are added to show levels of clinical concern. The diagonal reference line is included to note the location on the plot where a patient's lab values have neither increased nor decreased since baseline. An increase in a lab value from baseline would be evidenced by the data point being located above the diagonal line. Data points in this region of the graph would indicate a subject with normal baseline and elevations post-baseline, or below normal baseline values with post-baseline elevation. Outliers in the upper left quadrant are worthy of further investigation for clinical concern.

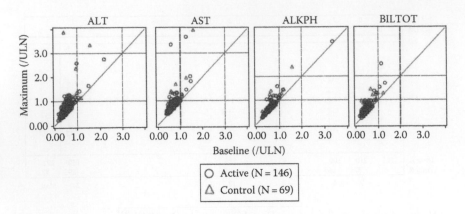

FIGURE 11.10

(See color insert.) Lab shift plots: baseline versus maximum. ALT, alanine aminotransferase; AST, aspartate aminotransferase; ALKPH, alkaline phosphatase; BILTOT, total bilirubin. The clinical concern level is 2× ULN (two times upper limit of normal) for BILTOT and ALKPH. It is usually 3× ULN for AST and ALT, but for some therapeutic areas, 5× ULN, 10× ULN, and 20× ULN are used to assess toxicity grades. (From CTSpedia, http://www.ctspedia.org/do/view/CTSpedia/ClinLFTGraph002, accessed January 8, 2013. Code contributed by Sunil J Mistry and Mark Jones, modified by Sanjay Matange, Robert Gordon, and Max Cherny.)

Although an ALT, AST, or total bilirubin (TBL) increase from baseline is usually of clinical concern, some laboratory parameters, such as white blood cell count, are of greater concern when they decrease from baseline during study.

An important design element of the shift plot is having equal scale on the x- and y-axes to create a square figure for each of the separate panels, which allows for accurate representation of the distance from the 45° reference line. The shift plot shown in Figure 11.10 can easily be altered to a different layout (e.g., trellised 2 × 2 instead of 4 × 1). In the 4 × 1 configuration, several rows can be graphed, each at specific time points. This is consistent with International Conference on Harmonisation (ICH) recommendation of showing baseline values on the x-axis and most extreme on-treatment values on the y-axis, as then outliers can easily be detected.[18] FDA, ICH, and SPERT recommend shift plots for interpreting laboratory data.[22,23]

11.2.3.3 Lab-Specific Patient Profile

One common clinical question is about the patient's profile over time,[19] also recommended by SPERT (22) Figure 11.11 depicts a trellised, multilab line plot to evaluate the time course of selected laboratory values for a set of at-risk patients. Each panel represents the laboratory data for each at-risk patient, displaying the same laboratory parameters for each of them. The x-axis represents the study day that the lab test was measured. It is important to keep this scale consistent across each panel. The y-axis represents the relationship of the values to the ULN for the parameters graphed. Reference

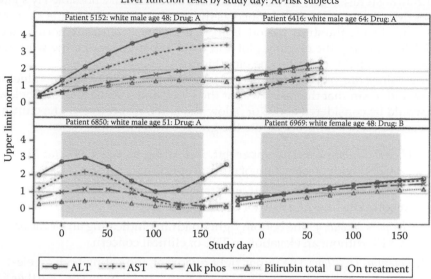

Liver function tests by study day: At-risk subjects

FIGURE 11.11

(See color insert.) Subject liver function tests by trial day: at-risk subjects. For alanine aminotransferase (ALT), aspartate aminotransferase (AST), and alkaline phosphatase (Alk Phos), the clinical concern level (CCL) is two times upper limit of normal (ULN); for total bilirubin, the CCL is 1.5× ULN. Shaded region indicates time on treatment. (From CTSpedia, http://www.ctspedia.org/do/view/CTSpedia/ClinLFTGraph005, accessed January 8, 2013. Code by Susan Schwartz, modified by Robert Gordon.)

lines have been added for the clinical concern levels (1.0× ULN, 1.5× ULN, and 2.0× ULN for this example). Exposure to drug is depicted through the use of the shaded regions. Depending on the lab parameter and the level of concern, reference ranges and interpretation can differ. More subjects can be added and, therefore, more panels would be added to the graphic. However, it is important to maintain clarity, and it may be appropriate to have multiple pages. SPERT[23] recommends use of patient profiles.

11.2.3.4 Drug-Induced Liver Injury

One of the important areas for clinical laboratory evaluation is the monitoring of LFTs for possible hepatotoxicity and DILI. DILI can be nonreversible and fatal. It is the most common adverse effect causing withdrawals and refusals to approve,[24] the second most common reason for termination due to safety during drug development, and the most frequent cause of acute liver failure in patients evaluated for liver transplantation.[19,20]

One way to address the assessment of DILI is with FDA's eDish (evaluation of drug-induced serious hepatotoxicity) system,[20,25] also recommended by SPERT. (22) eDish generates a scatter plot of maximum alanine aminotransferase (ALT) versus maximum TBL to identify possible Hy's law cases. Maximum values are defined as the maximum post-baseline value at any time during the study period. Both x- and y-axes values are normalized to ULN. Utilizing the original plot and the CTSpedia dataset, the working group re-created the eDish-type plot (Figure 11.12).

The values are plotted in log scale for values of ULN, where a value of 1.0 would mean that the result was at the ULN, and a value above or below 1.0 would be considered above or below the ULN, respectively. The clinical concern reference ranges separate the graphic into four quadrants:

1. *Hyperbilirubinemia*: The upper left, indicating elevated TBL (at least 2× ULN) with no elevation of clinical concern in ALT (< 3× ULN).

2. *Normal range*: The lower left quadrant, indicating that the elevations in both TBL and ALT are not of clinical concern.

3. *Temple's corollary*: The bottom right quadrant, indicating an elevation in ALT without an elevation in TBL of clinical concern.

4. *Potential Hy's law*: The upper right quadrant, indicating both an elevation of greater than 3× ULN in ALT and 2× ULN in TBL. Further investigation would be needed in this case.

Clinical judgment is critical to interpret these associations. In addition, it is perhaps more clinically relevant to match the maximum value of ALT with the TBL at the same time point and vice versa. It is certainly

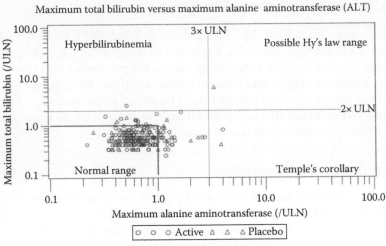

Maximum total bilirubin versus maximum alanine aminotransferase (ALT)

• Maximum values are those maximum values that occur post-baseline
(no time constraints and not necessarily concurrent events)

FIGURE 11.12
(See color insert.) Scatter plot of maximum transaminase versus maximum total bilirubin. (From CTSpedia, http://www.ctspedia.org/do/view/CTSpedia/ClinLFTGraph000, accessed January 8, 2013. SAS code contributed by Robert Gordon.)

important to perform a comprehensive assessment of hepatotoxicity so that possible cases of DILI are identified. To further investigate a possible case of Hy's law, or to examine the liver function parameters of interest for a subject, other plots could be generated such as a lab-specific patient profile (Figure 11.11) on a subject's liver test values over the time course of the study.

Figure 11.12 could also be used for other lab parameters if there is concern about simultaneous elevations (or reductions) in lab parameters.

11.3 Guidance Documents and Safety Graphics

There are several guidance documents advocating the use of graphics to better communicate clinical trial findings. In this section, we will look at these recommendations and use many of the figures given earlier as examples. This discussion focuses on recommended safety graphs, with specific reference to FDA, ICH, Council for International Organizations of Medical Sciences (CIOMS), and SPERT documents.

11.3.1 FDA

We will look at five specific FDA guidance documents: the formatting of the statistical section of an application,[21] the integrated summary of effectiveness,[26] statistical review template,[27] conducting a safety review for a new product application,[28] and guidance for DILI.[20]

The FDA Guidance Document for format and content of an application recommends that measures be displayed graphically showing frequency and timing. Actual visit times are preferred over visit numbers because visit numbers alone are more difficult to interpret (Figure 11.10).[21] Also recommended are graphical patient profiles showing the value of particular parameter(s) over time, drug dose over that same period, and the timing of AEs and other changes of note (Figure 11.13). Graphs must be well-labeled with clear headings and specific study information.

In the integrated summary of effectiveness, FDA suggests that graphical presentations may be better than tables. One example would be to present cases with similar controls (placebo, active) in forest plots (Figure 11.4).[26]

In the statistical review template, FDA discusses using graphical presentations, such as time to event, to show time to dropout with a description of the pattern of treatment and study discontinuation, and to support efficacy claims with a comprehensive evaluation of safety parameters.[27] Figure 11.5 is a related example, where time to AEs of interest is plotted. Graphical results can be presented for subgroups such as gender, race, age, geographic region, or others.

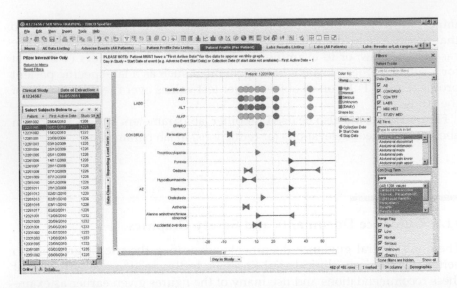

FIGURE 11.13
(See color insert.) Example of patient profile interactive graph, in Spotfire. (This is discussed in detail in Section 11.4.) (Courtesy of Jones, D., Timely safety surveillance of clinical study safety data, Tibco Software, Atlanta, GA; Pfizer, La Jolla, CA, http://www.tibco.com/company/news/releases/2012/tibco-spotfire-to-showcase-graphical-representation-of-safety-data-for-timely-safety-surveillance-during-clinical-trials.)

11.3.2 ICH

ICH Technical Requirements for Registration of Pharmaceuticals for Human Use (representing regulatory bodies of the European Union, Japan, and the United States) has guidelines for a single core clinical study report.[18] ICH recommends that all baseline and demographic characteristics be compared graphically across all treatment groups. This includes relevant individual patient data, "including laboratory values, and all concomitant medication for all individual patients randomized (broken down by treatment and by center for multicenter studies)." (An example can be found in Figure 11.13.)

ICH recommends that three levels of safety-related data are presented graphically. The first is the extent of exposure. Second is common AEs (e.g., Figure 11.3) and laboratory test results for the treatment groups (e.g., Figures 11.9 and 11.10). The third level is serious and significant AEs. Graphics with individual patient data can be helpful in showing parameter(s) over time, drug dose, and the times of particular events (e.g., Figures 11.11 and 11.13). Graphs of laboratory data are helpful in showing group mean or median values, the range of values, and the number of patients with abnormal values or with abnormal values that are of a certain size (Figure 11.9). Outliers can easily be detected when on-treatment values are plotted against their baseline values (Figure 11.10).

11.3.3 CIOMS

CIOMS is an international body concerned with drug safety and the harmonization of pharmacovigilance practices. Its goal is to make recommendations to inform and encourage systematic development of international standards and rules. CIOMS Working Group VI looks at the management of clinical trial information.[17]

CIOMS Working Group VI is concerned with determining that there is a sufficient amount of evidence, noting that the "absence of evidence is not evidence of absence" and encouraging the use of well-designed graphs to aid reviewers in determining such. In this regard, they discuss the use of graphics for laboratory chemistries that are continuous measurements. Figures 11.9 and 11.10 are examples of this type of graphical presentation. Figure 11.10 is an example of their recommendation to use shift plots to determine if there is an overall shift from baseline, as well as large changes.

11.3.4 SPERT

SPERT, a technical group of the Pharmaceutical Research and Manufacturers of America, provides a "pharmaceutical industry standard for safety planning, data collection, evaluation, and reporting" and offers recommendations for the presentation of safety data in graphical displays.[23] Differentiating between exploratory and planned analyses is important for safety reviews. SPERT recommends graphical methods to both review patient data and enhance statistical analysis. In reviewing laboratory data, patient profile graphics (Figure 11.13) that focus on extreme values can be very helpful—particularly when extreme values are linked to severe toxicity. SPERT recommends shift plots as well (e.g., Figure 11.10).

Safety graphics can improve the reporting in each of the three tiers suggested by SPERT for the PSAP. Tier 1 includes all AEs with a prespecified hypothesis. Tier 2 includes all common AEs (usually four or more with the MedDRA PT) that are not prespecified. Finally, tier 3 includes the rest of the AEs—those that do not have a prespecified hypothesis and are infrequent. Of particular note are those that are clinically significant. Figures 11.3 and 11.6 are examples of graphs that are helpful in detecting such outliers. Beyond the three tiers, it may be necessary to look at tier 1 events in multiple ways—for example, liver damage by Hy's law, laboratory data in combination with AE data (Figure 11.12), and subgroup effects (Figure 11.4).

Many concrete examples of graphical displays are given by SPERT in the planning and core analyses for periodic aggregate safety data reviews.[22] Among these examples for the core set of data are the following:

1. Figures that display both incidence and risk estimates (e.g., Figure 11.3)
2. AEs showing risk over time and time to first event (e.g., Figure 11.5)
3. Hazard function and forest plots for subgroups (e.g., Figure 11.4)

4. Patient profiles for those experiencing AEs (e.g., Figure 11.13)
5. QTc and laboratory shift plots (e.g., Figures 11.8 and 11.10, respectively)
6. Boxplots for central tendency of QTc and laboratory data (e.g., Figure 11.9)

11.4 Other Safety Graphics Topics: Benefit–Risk and Interactive Graphics

At the time of this writing, both of these topics are in their infancy. This section is included for completeness. However, it is anticipated that there will be much more to say on both of these important topics in the coming years.

11.4.1 Benefit–Risk

We address benefit–risk graphics in this chapter because the context of both benefit and risk is inherent in drug safety.[30] Conventionally, efficacy and safety have been analyzed separately, with conclusions for benefit–risk of the compound being drawn on purely qualitative and descriptive grounds.

An important question to ask is: Quantitatively, what are the key benefits and risks of this compound to key stakeholders (patients, physicians, regulators, etc.) in the context of their available choices? It has been the authors' experience that presenting benefit and risk together has a significant positive impact on the drug development process. Safety review team leaders report that construction of a value tree[31] (Figure 11.14) focusing on those benefits and risks that could affect the balance gives development teams a strategic one-page view for their own drug development use. Subsequent to the value tree, construction of the benefit–risk graph (Figure 11.15) lends itself to effective communication with internal and external decision-makers. Creation of the compound's benefit–risk value tree and graph is particularly valuable for focusing on the most important aspects of the compound's characteristics.[32]

Graphs are an important aspect of quantifying benefit and risk. Statistical methods for benefit–risk are emerging as well. Worthy of particular note is a benefit–risk graph assessing benefits and risks within each patient over time.[33]

11.4.2 Interactive Graphics

Another emerging area, important for safety signal detection and monitoring, is interactive graphics. With one exception (Figure 11.13), the graphs noted in this chapter are static graphs, for use in reports, publications, and regulatory filings. Interactive graphs are particularly useful in signal detection because the relevant members of the drug development team (including pharmacovigilance specialist, clinician, statistician, and data manager) may

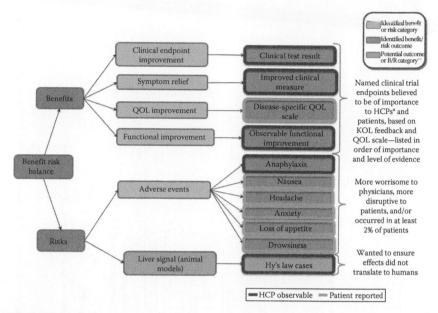

* HCPs = health-care providers

** Relationship to treatment not proven

FIGURE 11.14

(See color insert.) A completed benefit-risk value tree for a hypothetical compound (from PhRMA BRAT framework; EMA has successfully field tested a similar "effects tree").

view relevant safety graphs while the trials are ongoing. With interactive graphs, as data anomalies occur, the user can *drill down* to other views of the data in order to better understand whether the unusual value was a safety signal, data error, etc. One of the most-used interactive graphs is the patient profile (Figure 11.13). Outliers or other patients of interest from summary plots can be investigated in detail by drilling down to a profile such as this. The x-axis is days on study, and the y-axis includes all relevant data. In this example, study medication, medical history, labs, concomitant procedures and medications, and AEs are displayed.

Finding safety signals with interactive graphics depends on (1) well-designed graphs that highlight signals of primary concern and (2) effective and efficient drilldown (for both the user and, from a resource standpoint, the individual configuring the interactive software and loading the data) to the relevant information to confirm or otherwise explain the data anomaly. Interactive graphics is in its infancy. Development of standard, industry-wide graphics for interactive safety signal detection (similar to standards described in Amit et al.[2] and the safety graphics wiki[14] for static graphs) is the next logical step.

Interactive graphics is more complex, however, than static graphs. Due to the nature of the interactivity, they are dependent on software designed specifically for this purpose (e.g., Spotfire,[34] JMP Clinical,[35] and JReview[36]).

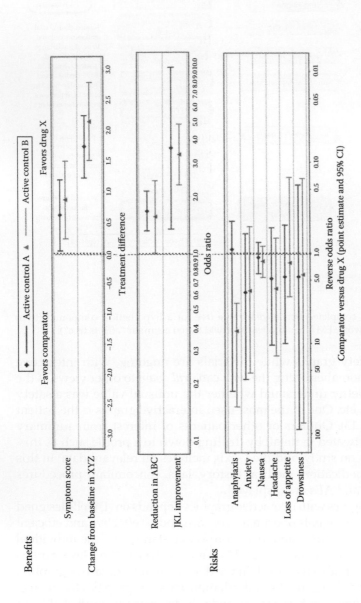

FIGURE 11.15

(See color insert.) Benefit–risk interval plot. Parameters displayed are based on value tree in Figure 11.14. (Graph created by Susan Duke, based on hypothetical data.)

The next challenge with interactive graphics is allowing access to the data and software for all relevant audiences (including sponsor, contract research organization, in-licensing partner, data safety monitoring board, and regulators). This would be analogous to those who would review static graphs in an internal report or regulatory filing. Creation of standards to allow the technology and data security to function effectively for all of these audiences would be a step that encourages the best technology and reduces resource costs that may otherwise be needed to convey safety signal information in a timely manner by other means.

Making the data available (as described at All Trials website, accessed July 5, 2013, http://www.alltrials.net/) would also help when meta-analyses or Cochrane Reviews are performed. "Information on what was done and what was found in these trials could be lost forever to doctors and researchers, leading to bad treatment decisions, missed opportunities for good medicine, and trials being repeated."

11.5 Conclusions

What is the effective use of safety graphics for study reports, integrated safety summaries, and publications of clinical trial findings? These key points are means to transparently and convincingly convey the important safety messages of clinical trial data to the intended audience.

1. Make rich use of the human brain's pattern perception skills.[6]
2. Use standard graph types that answer commonly asked questions.[2,14] For example, is there value in making a novel graph type when standard ones are already familiar to reviewers?
3. Review of the initial draft by the safety review team to ensure the graph clearly conveys the messages of the data for that particular study. The safety review team is advised to make use of graphics best practices[3] as a checklist to ensure that the graph is well designed, and the responsible statistician is advised to consider their safety review team's feedback to ensure that graphs intended to convey key study information are effectively doing just that.
4. Standard displays have been created on a wiki[14] to answer common safety questions. Those highlighted in this chapter are as follows:
   a. What are the frequencies of key AEs, along with a quantitative measure of their risk compared to control? (Figure 11.3)
   b. What are the risk factors for specific subgroups? (Figure 11.4)
   c. What are the respective times to event in AEs of interest? (Figure 11.5)

    d. Which AEs have a significant risk difference (which highlights potential safety signals)? (Figure 11.6)

    e. How do each patient's ECG values track over time? (Figure 11.7)

    f. What is the relationship between change from baseline in QTc levels and their corresponding baseline values? (Figure 11.8)

    g. Is there a temporal relationship between treatment and lab value, and what are the toxicity grade trends over time? (Figure 11.9)

    h. Is there an overall change in laboratory value relative to baseline from this treatment? Are there extreme changes from baseline of clinical concern? (Figure 11.10)

    i. How do a selected set of patient's laboratory parameters of interest vary over time? (Figure 11.11)

    j. Does this compound have any Hy's law cases? (Figure 11.12)

5. Emerging areas in safety graphics include interactive and benefit–risk visualizations.

    a. How do the parameters of interest (dosing, AEs, concomitant medications, and laboratory values) align in patients of interest, and what does this tell us about the medication under study? (Figure 11.13)

    b. What are the key benefits and risks of the compound and indication under study, both those we have measured and those we may consider measuring in future? (Figure 11.14)

    c. What are the quantitative measures of these key benefits and risks? Seeing them simultaneously and understanding the clinical implications of each measure, do the benefits outweigh the risks? (Figure 11.15)

Acknowledgments

The project described was supported by CTSA award No. UL1TR000445 from the National Center for Advancing Translational Sciences. Its contents are solely the responsibility of the authors and do not necessarily represent official views of the National Center for Advancing Translational Sciences or the National Institutes of Health.

The authors are grateful to the fellow members of Safety Graphics Working Group who provided significant input into graphics design of those graphs used in this chapter including Mat Soukup, FDA; Brenda Crowe, Lilly; Rich Anziano, Pfizer; Robert Gordon, J&J; and Frank Harrell, Vanderbilt.

The authors also would like to thank graph contributors: Frank Harrell, Vanderbilt; Sanjay Matange, SAS; Haijun Ma, Amgen; Jun Wei, Amgen;

Robert Gordon, J&J; Sunil J. Mistry, GSK; Mark Jones, Amadeus (formerly GSK); and Susan Schwartz, SAS.

For the use of the interactive graph example, thanks to David Jones, Pfizer, and Michael O'Connell, Tibco.

Also thanks to Jeff Horner and Dale Plummer, Vanderbilt, for maintenance and IT upkeep of CTSpedia.

References

1. Tufte E. *The Visual Display of Quantitative Information*. Cheshire, CT: Graphics Press; 2001.
2. Amit O, Heiberger RM, Lane PW. Graphical approaches to the analysis of safety data from clinical trials. *Pharm Stat* 2008;7:20–35.
3. CTSpedia. Best practices recommendations. NIH CTSA BERD, 2012 (accessed January 8, 2013, at https://www.ctspedia.org/BestPractices).
4. Pocock SJ, Travison TG, Wruck LM. Figures in clinical trial reports: Current practice & scope for improvement. *Trials* 2007;8:36.
5. Krause A, O'Connell M (eds.). *A Picture Is Worth a Thousand Tables: Graphics in Life Sciences*. New York: Springer Science; 2012.
6. Cleveland W, McGill R. Graphical perception: Theory, experimentation, and application to the development of graphical methods. *J Am Stat Assoc* 1984;79:531–554.
7. Tufte E. *Envisioning Information*. Cheshire, CT: Graphics Press; 1990.
8. Tufte E. *Visual Explanations: Images and Quantities, Evidence and Narrative*. Cheshire, CT: Graphics Press; 1997.
9. Tufte E. *Beautiful Evidence*. Cheshire, CT: Graphics Press; 2006.
10. Cleveland, W. *The Elements of Graphing Data*, 2nd edn. Summit, NJ: Hobart Press; 1994.
11. Heiberger R, Holland B. *Statistical Analysis and Data Display: An Intermediate Course with Examples in S-PLUS, R, and SAS*. New York: Springer; 2004.
12. Robbins N. *Creating More Effective Graphs*. Hoboken, NJ: Wiley-Interscience; 2004.
13. CTSpedia. CDISC—FDA test project datasets. NIH CTSI BERD, 2012 (accessed January 8, 2013, at https://www.ctspedia.org/DataSetPageCDISC).
14. CTSpedia. Safety graphics home page. NIH CTSA BERD, 2012 (accessed January 8, 2013, at https://www.ctspedia.org/StatGraphHome).
15. Soukup M. Communicating clinical trial results the statistical graphic way. *Joint Statistical Meetings*, Boston, MA; 2010.
16. Sherman RB, Woodcock J, Norden J, Grandinetti C, Temple RJ. New FDA regulation to improve safety reporting in clinical trials. *N Engl J Med* 2011;365:3–5.
17. CIOMS. Management of safety information from clinical trials. Geneva, Switzerland: Council for International Organizations of Medical Sciences; 2005.
18. ICH. Guideline for industry: Structure and content of clinical study reports. London, U.K.: ICH; 1996.

19. Anziano R, Gordon R. Examples of understanding safety data using graphical methods. *WIREs Comput Stat* 2013;5:78–95.
20. FDA. Guidance for industry drug-induced liver injury: Premarketing clinical evaluation. Washington, DC: FDA; 2009.
21. FDA. Guideline for the format and content of the clinical and statistical sections of an application. Rockville, MD: FDA; 1988.
22. Xia HA, Crowe BJ, Schriver RC, Oster M, Hall DB. Planning and core analyses for periodic aggregate safety data reviews. *Clin Trials* 2011;8:175–182.
23. Crowe BJ, Xia HA, Berlin JA et al. Recommendations for safety planning, data collection, evaluation and reporting during drug, biologic and vaccine development: A report of the safety planning, evaluation, and reporting team. *Clin Trials* 2009;6:430–440.
24. FDA. Hepatotoxicity through the years impact on the FDA, 2001 (accessed February 20, 2013, at http://www.fda.gov/downloads/Drugs/ScienceResearch/ResearchAreas/ucm122149.pdf).
25. Watkins PB, Seligman PJ, Pears JS, Avigan MI, Senior JR. Using controlled clinical trials to learn more about acute drug-induced liver injury. *Hepatology* 2008;48:1680–1689.
26. FDA. Draft guidance for industry: Integrated summary of effectiveness. Rockville, MD: FDA; 2008.
27. FDA. Good review practice: Statistical review template—MAPP 6010.4. Washington, DC: CDER; 2012.
28. FDA. Reviewer guidance: Conducting a clinical safety review on a new product application and preparing a report on the review. Washington, DC: FDA; 2005.
29. Jones D. Timely safety surveillance of clinical study safety data. Atlanta, GA: Tibco Software (accessed August 13, 2014 at http://www.tibco.com/company/news/releases/2012/tibco-spotfire-to-showcase-graphical-representation-of-safety-data-for-timely-safety-surveillance-during-clinical-trials).
30. FDA. E2C(R2) Periodic benefit–risk evaluation report. Washington, DC: FDA; 2012.
31. Levitan BS, Andrews EB, Gilsenan A et al. Application of the BRAT framework to case studies: Observations and insights. *Clin Pharmacol Ther* 2011;89:217–224.
32. Metcalf M, Duke S. Valuing options—A case study. *CIRS Technical Workshop on Benefit–Risk Framework for the Assessment of Medicines: Valuing the Options and Determining the Relative Importance (Weighting) of Benefit and Risk Parameters*, Philadelphia, PA, 2012.
33. Norton J. A longitudinal model and graphic for benefit–risk analysis, with case study. *Drug Inform J* 2011;4:741–747.
34. Spotfire Tibco Software. Tibco (accessed February 15, 2013, at http://spotfire.tibco.com/).
35. JMP Clinical. SAS (accessed February 15, 2013, at http://www.jmp.com/software/clinical/).
36. JReview. Integrated Clinical Systems, Inc. (accessed February 15, 2013, at http://www.i-review.com/).

12

Bayesian Network Meta-Analysis for Safety Evaluation

Bradley P. Carlin and Hwanhee Hong

CONTENTS

12.1 Introduction

Meta-analysis is a statistical technique that compares the effectiveness or safety of two treatments by incorporating the findings from several independent studies (DerSimonian and Laird, 1986). Network meta-analysis (NMA), also sometimes referred to as multiple (or mixed) treatment comparisons (MTCs), is the extension of the traditional meta-analysis of two treatments to simultaneous incorporation of *multiple* treatments, where in most cases none of the studies compared all the treatments at one time. The goal of NMA is to address the comparative effectiveness or safety of interventions while accounting for all sources of data (Hoaglin et al., 2011; Jansen et al., 2011).

In such analysis, there are two types of evidence: *direct* and *indirect* (Gartlehner and Moore, 2008; Lumley, 2002). Suppose that two studies investigate Placebo (P) versus Drug A and three studies investigate Placebo versus Drug B. In this example, every observed comparison (P versus A and P versus B) is a direct evidence; the phrase "direct *head-to-head* comparison" refers to the case where the two treatments are active, such as A versus B within the same trial. To make an inference comparing A and B, conducting a new head-to-head trial is an ideal solution but may be infeasible in many situations, for example, when the active control groups vary for different countries or regions. Thus, we have to rely on the comparative effectiveness of safety between Drugs A and B by using only indirect information (say, the P versus A and P versus B studies). While a traditional meta-analysis depends solely on direct evidence, NMA combines direct and indirect evidence to compare multiple treatments.

Differences between frequentist and Bayesian NMA models are often not large. However, Bayesian methods can handle more complicated models more easily using Markov chain Monte Carlo (MCMC) algorithms. Recently, frequentist methods for complex hierarchical NMA models have been developed by using commercial software such as SAS and Stata, but these methods are not suitable when the data are sparse, a common case (Hong et al., 2013). In the Bayesian approach, we can interpret the findings based on posterior probability summaries such as 95% credible intervals, rather than hard-to-interpret p-values. Also, we can easily calculate the probability of being the best or safest drug using the MCMC samples, summaries of which can estimate both the magnitude and variability of the treatment effects.

A safety outcome could be measured in a number of ways, such as count or severity of side effects or discontinuation of the study drug due to intolerability. Since many trials in NMA investigate approved drugs or treatments, few negative outcomes may be reported, resulting in rare events. In frequentist approaches, studies with no events at all are sometimes dropped, or continuity corrections are applied. However, in a Bayesian framework, we can keep such zero cells in our data retaining this information rather than altering or discarding it. However, the prior distributions for model parameters must be selected carefully, because zero counts could otherwise lead to numerically unstable results (Dias et al., 2011a).

In this chapter, we will provide two different NMA hierarchical models with various prior choices for a single binary outcome and compare them with a corresponding frequentist model. We also illustrate Bayesian tools for choosing the safest treatment. We show our models by applying them to a real NMA example and also offer simulation studies that investigate how well the Bayesian models reveal the safest treatment under various scenarios.

12.2 Methods

12.2.1 Bayesian Homogeneous Random Effects Model

We assume that the event count from each study follows a binomial distribution,

$$r_{ik} \overset{ind}{\sim} Bin(n_{ik}, p_{ik}), \quad i = 1, 2, \ldots, I, k = 1, 2, \ldots, K, \tag{12.1}$$

where
  r_{ik} is the total number of events
  n_{ik} is the total number of subjects
  p_{ik} is the probability of the outcome for the kth treatment in the ith study, with $k = 1$ for placebo

Here, multiarm trials, comparing more than two treatments, can also be included.

We consider a random effects model to allow variability among studies, generalizing the standard fixed effects model. A logistic link can be specified as

$$logit(p_{ik}) = \eta_{iB} + \delta_{iBk}, \tag{12.2}$$

where
  B represents a baseline treatment (usually placebo)
  η_{iB} is the effect of the baseline treatment in the ith study
  δ_{iBk} is a log odds ratio between treatment k and B, with $\delta_{iBk} = 1$ when $k = B$

Defining d_k as the log odds ratio between treatment k and placebo, with $d_1 = 0$, the mean of δ_{iBk} can be replaced by $d_k - d_B$. That is, we can assume an independent normal specification for the δ_{iBk}, such as

$$\delta_{iBk} \sim N(d_k - d_B, \sigma^2). \tag{12.3}$$

In this model, we assume homogeneous variance σ^2 across all k treatments. For a three-arm trial, having two δ_{iBk}, we would further need to consider correlation between them. In this case, we would assume the δ_i vector follows a multivariate normal distribution, typically with common correlation of 0.5 between two log odds ratios (a consequence of the usual assumption of *consistency* between direct and indirect evidence), as suggested by Lu and Ades (2006).

We investigate two sets of prior distributions: one fully noninformative (denoted by Bayes1) and another that shrinks the random baseline effects toward their grand means (Bayes2). For Bayes1, the η_{iB} and d_k are assumed to follow a very vague but still proper $N(0, 100^2)$ distribution. For σ, a vague *Uniform*(0.01, 2) prior is utilized. Turning to Bayes2, we replace the prior on η_{iB} with a $N(m_B, \tau^2)$, where the hyperparameters m_B and τ follow $N(0, 100^2)$ and *Uniform*(0.01, 2) priors, respectively, with other parameters remaining the same as under Bayes1. Here, the upper bound of the *Uniform* distribution 2 is large enough to be slack on the log odds ratio scale. We denote a homogeneous random effects model with the Bayes1 prior as *RE_Bayes1* and with the Bayes2 prior as *RE_Bayes2*.

12.2.2 Arm-Based Bayesian NMA Model

RE_Bayes1 and RE_Bayes2 are examples of *contrast-based* models, which use log odds ratio for inference on *relative* treatment effects. In this section, we introduce an *arm-based* alternative model (denoted by *RE_AB*), which seeks to capture *absolute* effects (say, log odds) in its model parameters (Dias et al., 2011a; Salanti et al., 2008). The arm-based model can be written by respecifying model (12.2) as

$$logit(p_{ik}) = \mu_k + \nu_{ik}, \tag{12.4}$$

where
μ_k is the fixed mean effect of treatment k
the ν_{ik} are study-specific random effects

The distribution of the random effects is assigned as

$$(\nu_{i1}, \ldots, \nu_{iK})^T \sim MVN(0, \Lambda), \tag{12.5}$$

where Λ is a $K \times K$ unstructured covariance matrix containing all possible correlations between treatments. The parameters in arm-based models permit more straightforward interpretation than their contrast-based counterparts, since they offer a direct estimate of a pure treatment effect. Note that arm-based models have a larger number of parameters to be estimated than the homogeneous random effects model in Section 12.2.1, since Λ is unstructured.

For priors, we again assign vague distributions, namely, a $N(0, 100^2)$ to the μ_k, and a *Wishart*(Ω, γ) (having mean $\gamma\Omega^{-1}$) to Λ^{-1}. We select γ to be K, the dimension of Λ, which is the smallest value yielding a proper prior, and Ω to be the identity matrix, giving a broadly noninformative prior that still ensures acceptable MCMC convergence. The *Wishart* prior on Λ^{-1} allows this model to borrow more strength by incorporating structure across both observed and unobserved information and also capture variability more accurately.

Our Bayesian results were obtained in WinBUGS, using two parallel chains of 50,000 MCMC samples after a 50,000-sample burn-in. Jones et al. (2011) recently published SAS code, enabling fitting roughly the same homogeneous random effects model in (12.2) and (12.3) using SAS Proc GLIMMIX. We use this procedure to obtain parameter MLEs with 95% confidence intervals and compare them to our Bayesian results.

12.2.3 Decision Making

The ultimate goal of NMA for safety studies is often to identify the safest treatment with respect to an outcome. In Bayesian analysis, we can generate a probability of being safest under the posterior distribution, containing both the magnitude and variability of the relevant parameter: d_k or μ_k in (12.3) or (12.4), respectively. Suppose P_k is the marginal posterior probability of having a negative event (say, the presence of an adverse event [AE]) under treatment k, perhaps modeled with a logit function. Here, smaller values of P_k indicate safer drugs and we prefer treatments with lower ranks. Denoting the data by $\mathbf{y}$, a natural Bayesian summary to consider is the posterior probability of being ranked safest,

$$Pr\{k \text{ is the safest treatment} \mid y\} = Pr\{rank(P_k) = 1 \mid y\}, \qquad (12.6)$$

a quantity we denote by *Safe1*. (Note that we would simply change 1 to K when the outcome has a positive interpretation, such as the number of subjects remaining in the study at the end of the trial.) Similarly, one can calculate the probability of being the first- or second-safest treatment, say, denoted by *Safe12*, by replacing the right-hand side of Equation 12.6 with $Pr\{rank(P_k) = 1 \text{ or } 2|y\}$, where again lower ranks indicate safer treatments. We use the Safe12 probability for decision making rather than Safe1 when the number of treatment is large (say, > 5) because it could be more reliable through its counting of more than just a single extreme case.

12.3 Example

Urgency urinary incontinence (UI) is defined as involuntary loss of urine associated with the sensation of a sudden, compelling urge to void that is difficult to defer. Our example UI data compare eight pharmacological treatments including placebo for urgency UI in adult women to investigate the absolute and relative safety of the drugs. A total of 47 randomized controlled trials are included, all of which measure the aspect of drug safety, namely, discontinuation of the treatments due to AEs. Each study reports the number of discontinuation events and the sample size for each study arm. To obtain the original data and details, please refer to Appendix in Hong et al. (2013).

Figure 12.1 presents the network among the study drugs for discontinuation due to AE. The size of each node represents the number of studies

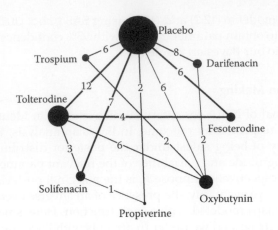

FIGURE 12.1
Network graph of UI data for discontinuation due to AE. The size of each node represents the number of studies investigating the drug, and the thickness of each edge implies the total number of samples for the relation. The number on each line is the number of studies for the relation.

investigating the drug, and the thickness of each edge corresponds to the total number of samples for the relation. The number on each edge is the number of studies for the relation. These data include several head-to-head (not only drug-to-placebo) comparisons, with several studies having nonplacebo baseline treatments.

We fit three Bayesian random effects models (RE_Bayes1, RE_Bayes2, and RE_AB) and one frequentist random effects model (RE_Freq). Figure 12.2 exhibits estimated odds ratios between drugs and placebo, $\exp(d_k)$ (posterior median and MLE for Bayesian and frequentist models, respectively), with associated 95% credible or confidence intervals from the four models. Under RE_AB, we calculate the odds ratios simply by taking differences between the μ_k and μ_1, to compare results with those from contrast-based models. The safest and least safe drugs compared to placebo based on the odds ratios from each model are indicated with closed symbols and thick lines. All models agree that tolterodine is the safest drug and oxybutynin is least safe among seven pharmacological drugs in terms of discontinuation due to AE, and these two odds ratios are significantly different because their credible or confidence intervals do not overlap. Note that the odds ratios for propiverine emerge as having very wide intervals from all models, likely because there are only two studies of the relation between this drug and placebo, and they do not agree regarding its safety.

Although Bayesian and frequentist random effects models yield similar odds ratio estimates, the frequentist model tends to underestimate the random effect variability. The estimate of σ in (12.3) is 0.25 (95% CI is 0.04–0.50) and 0.22 (0.02–0.43) under RE_Bayes1 and RE_Bayes2, respectively, and the square roots of diagonal elements of the estimate of Λ in (12.4) are between 0.47 and 0.66 (note that σ and Λ measure variabilities under different scales, viz., the log

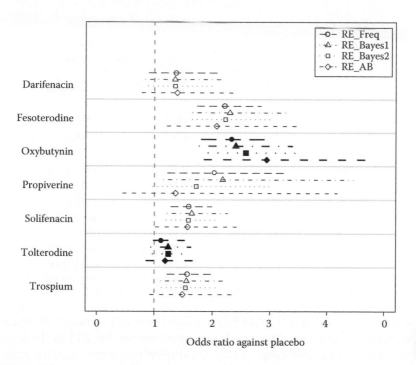

FIGURE 12.2
UI data interval plot of odds ratios between drugs and placebo for discontinuation due to AE under four models. The safest and least safe drugs compared to placebo are indicated with closed symbols and thick lines. RE_Freq, frequentist random effects model; RE_Bayes1, Bayesian random effects model with noninformative prior; RE_Bayes2, Bayesian random effects model with shrinkage prior; RE_AB, Bayesian arm-based random effects model.

odds ratio and log odds, respectively), while the frequentist model estimates it to be 0. That is, the frequentist model cannot capture all possible variabilities, leading to unrealistically narrow 95% confidence intervals in Figure 12.2. On the other hand, the arm-based model gives the widest 95% credible sets because it incorporates all uncertainty even from unobserved data.

Table 12.1 compares the three Bayesian models using the deviance information criterion (DIC) and presents the Safe12 probabilities for decision making. DIC is the sum of $\overline{D}$, a measure of goodness of fit, and p_D, an effective number of model parameters. Smaller DIC values indicate better models, with a DIC difference of 5 or more considered to be practically meaningful (Spiegelhalter et al., 2002). RE_AB fits the data best due to the smallest $\overline{D}$. However, RE_Bayes2 gives the smallest pD (effective number of parameters), resulting in the smallest DIC. That is, RE_Bayes2 is the best overall in consideration of both goodness of fit and complexity. Note that RE_Bayes2's shrinkage prior ($\eta_{iB} \sim N(m_B, \tau^2)$) permits the model to be less complex, whereas RE_AB has relatively more parameters to estimate than RE_Bayes1 and RE_Bayes2 due to the unstructured covariance matrix, leading to the highest p_D value.

TABLE 12.1

Model Selection with DIC and Safe12 Values Arising from the UI Data, with Associated Standard Deviations in Parentheses

	RE_Bayes1	RE_Bayes2	RE_AB
p_D	67.0	56.1	73.0
$\overline{D}$	116.7	121.7	109.1
DIC	183.7	177.8	182.1
Safe12 probability			
Placebo	0.989 (0.10)	0.989 (0.10)	0.881 (0.32)
Darifenacin	0.342 (0.47)	0.298 (0.46)	0.213 (0.41)
Fesoterodine	0.000 (0.01)	0.000 (0.01)	0.011 (0.10)
Oxybutynin	0.000 (0.00)	0.000 (0.00)	0.000 (0.01)
Propiverine	0.032 (0.18)	0.101 (0.30)	0.360 (0.48)
Solifenacin	0.023 (0.15)	0.026 (0.16)	0.054 (0.23)
Tolterodine	0.539 (0.50)	0.526 (0.50)	0.378 (0.48)
Trospium	0.074 (0.26)	0.060 (0.24)	0.103 (0.30)

For the Safe12 probability, placebo is always the safest across all three models, with tolterodine and darifenacin (propiverine for AB_RE) as the second and third place finishers, respectively. Oxybutynin and fesoterodine emerge as the least safe drugs. These findings correspond to those in Figure 12.2, though these Safe12 probabilities do not significantly differ due to somewhat large standard deviations. In general, RE_Bayes1 and RE_Bayes2 give similar Safe12 probabilities, while RE_AB provides noticeably smaller probabilities for tolterodine and placebo but larger for propiverine.

12.4 Simulation Study

12.4.1 Setting

In this simulation, we investigate how Bayesian NMA random effects models perform in finding the safest treatment and compare them to the frequentist approach under various settings. We create 1000 datasets for a single binary safety outcome investigated under four treatments including placebo and fit three models: RE_Freq, RE_Bayes1, and RE_AB. Figure 12.3 illustrates our two true underlying simulated data structures, which we term star network and closed loop. The upper part of the figure shows the network of comparisons between drugs, while the table at the bottom shows details of the data framework; here, each row denotes a particular study design with *o* indicating the treatments involved and the columns giving the numbers of studies for each design over four different cases.

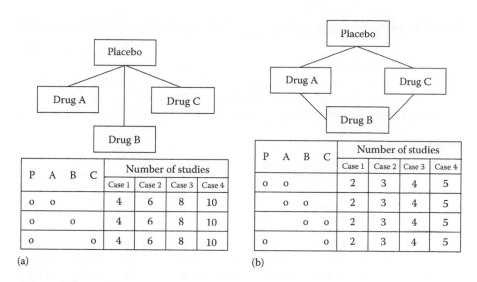

FIGURE 12.3
Simulation scenarios: (a) star network and (b) closed loop.

Under the star network scenario, all trials have only pairwise comparisons between placebo (P) and active drug (A, B, or C), and placebo is always the baseline treatment. We first assign 4 studies to each design (P versus A, P versus B, and P versus C), for a total 12 studies (denoted by Case 1). Then, we gradually increase the number of studies to 6, 8, and 10, denoted by Case 2, Case 3, and Case 4, respectively. Panel (b) displays the closed-loop setting where placebo is not always the baseline treatment across trials: only half of the studies involve the placebo. We again assign four different sets of numbers of studies for Cases 1–4 to let the number of studies for each active drug be the same as that in star network. That is, 4, 6, 8, and 10 studies examine each drug (A, B, or C) for Cases 1, 2, 3, and 4 under both the star network and the closed loop.

We first generate a dataset under model RE_AB, Equation 12.4, pretending that every study observes all four treatments. Then, we randomly select studies for each design and drop arms that we suppose to be unobserved. For instance, in a trial designed to compare P versus A, arms B and C are dropped from the generated full dataset. For the true treatment effect parameters, we set all treatments to have equivalent safety except Drug C (Drug C is less safe than the others), where the effect size for Drug C varies. Specifically, μ_P, μ_A, μ_B are always zero while μ_C is set to 0.2, 0.5, or 0.8, which corresponds to $d_C = 0.2$, 0.5, or 0.8 in the contrast-based models, RE_Freq and RE_Bayes1. We set Λ in (12.5) to have 1 for its diagonal elements and 0.5 for the off-diagonals. Every arm in each trial has sample size $n_i = 25$ or 100. Note that Case 1 with $n_i = 25$ is a relatively small (yet probable) NMA dataset, whereas Case 4 with $n_i = 100$ would be an unusually large NMA dataset.

12.4.2 Results

We calculate the empirical frequency that Drug C is selected as least safe based on Safe1 probabilities or log odds ratios for a Bayesian or frequentist model, respectively, over 1000 simulations. We call this the empirical probability of correct decision (i.e., that Drug C is the worst drug). Tables 12.2 and 12.3 present the probability of correct decision with various parameter

TABLE 12.2

Probability of Correct Decision under Star Network

	$n_i = 25$				$n_i = 100$			
	Case 1	Case 2	Case 3	Case 4	Case 1	Case 2	Case 3	Case 4
$d_C = 0.2$								
RE_Freq	0.403	0.403	0.436	0.467	0.413	0.438	0.454	0.496
RE_Bayes1	0.404	0.413	0.476	0.481	0.423	0.466	0.468	0.508
RE_AB	0.391	0.440	0.461	0.504	0.404	0.474	0.470	0.506
$d_C = 0.5$								
RE_Freq	0.582	0.623	0.676	0.713	0.615	0.640	0.713	0.759
RE_Bayes1	0.567	0.642	0.692	0.714	0.607	0.686	0.728	0.778
RE_AB	0.578	0.659	0.710	0.762	0.603	0.699	0.754	0.807
$d_C = 0.8$								
RE_Freq	0.738	0.779	0.848	0.886	0.787	0.830	0.887	0.918
RE_Bayes1	0.720	0.812	0.846	0.879	0.766	0.856	0.883	0.913
RE_AB	0.753	0.845	0.901	0.921	0.770	0.875	0.922	0.951

TABLE 12.3

Probability of Correct Decision under Closed Loop

	$n_i = 25$				$n_i = 100$			
	Case 1	Case 2	Case 3	Case 4	Case 1	Case 2	Case 3	Case 4
$d_C = 0.2$								
RE_Freq	0.351	0.377	0.419	0.435	0.384	0.375	0.438	0.476
RE_Bayes1	0.358	0.381	0.408	0.435	0.374	0.420	0.407	0.462
RE_AB	0.364	0.383	0.432	0.441	0.374	0.406	0.437	0.483
$d_C = 0.5$								
RE_Freq	0.544	0.582	0.672	0.727	0.580	0.634	0.689	0.751
RE_Bayes1	0.556	0.592	0.654	0.686	0.573	0.640	0.676	0.755
RE_AB	0.575	0.625	0.696	0.742	0.597	0.657	0.716	0.771
$d_C = 0.8$								
RE_Freq	0.722	0.769	0.841	0.898	0.749	0.824	0.882	0.931
RE_Bayes1	0.717	0.779	0.841	0.888	0.753	0.828	0.886	0.931
RE_AB	0.743	0.814	0.873	0.909	0.765	0.848	0.908	0.937

settings under the star network and closed loop, respectively. Overall, the probability of correct decision increases as the effect size, the number of studies, and the size of each arm get bigger. All three models do not perform well when $d_C = 0.2$ because the effect size is apparently too small to be detected in our simulation settings. When the effect size is moderate ($d_C = 0.5$ or 0.8), RE_AB gives the largest probability of correct decision except in Case 1 with $n_i = 100$ under star network. This suggests that the arm-based Bayesian model with its flexible Wishart prior on the unstructured covariance matrix Λ may be more accurate in many balanced data settings.

Figure 12.4a exhibits bias and MSE of $\hat{d}_C$ from all three models when $d_C = 0.5$ and $n_i = 100$ under the closed-loop setting (results under the star network are similar). RE_AB always gives the smallest bias and MSE while RE_Freq provides relatively larger values, although the differences are not that large. We notice that all models produce more reliable estimates as the number of studies increases. Figure 12.4b shows the mean of $\hat{\sigma}$ for both RE_Freq and RE_Bayes1. Note that the arm-based (RE_AB) and contrast-based (RE_Freq and RE_Bayes1) models have different parameterizations so we cannot compare their variability on a common scale. Also, since our simulated data are generated under the RE_AB model, we cannot calculate the exact *true* σ for the RE_Freq or RE_Bayes1 models. However, we can observe that the mean of $\hat{\sigma}$ under RE_Freq is much smaller than that under RE_Bayes1. This in turn leads to narrower 95% intervals for $\hat{d}_C$ in the frequentist approach; the mean of the width of 95% intervals under RE_Freq is 15%–20% narrower than that under RE_Bayes1. This supports the findings from our UI data analysis and Jones et al. (2011) that a frequentist NMA random effects model leads to smaller variability estimates than Bayesian modeling.

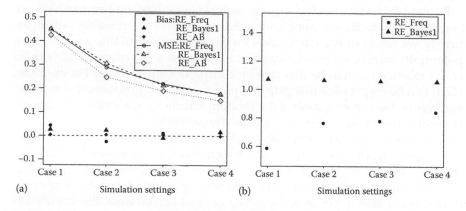

FIGURE 12.4
Simulation results for parameters under closed loop when $d_C = 0.5$ and $n_i = 100$: (a) bias and MSE of d_C and (b) posterior mean for σ.

12.5 Discussion

In this chapter, we have introduced a popular current Bayesian NMA random effects model and a slightly more advanced arm-based model for a single binary safety outcome. These models can be easily applied to nonbinary settings (e.g., continuous or count) with an appropriately modified likelihood and link function. Through simulations and the analysis of a real dataset, we have shown that Bayesian methods appear more suitable for complex models than frequentist approaches. Although here we have only considered homogeneous random effect models, we can certainly fit a heterogeneous random effects model in both approaches. However, frequentist methods do require sufficient information (i.e., we need at least one study for every combination of pairwise comparisons), whereas Bayesian methods, aided by their prior distribution, do not need this condition (Hong et al., 2013). Bayesian posterior probability-based inference gives a more straightforward interpretation of findings (thanks to its ability to directly estimate the probability that a certain treatment is safest or in the top two), and our simulation studies confirm that Bayesian models perform slightly better than frequentist in terms of making the correct decision and accurately and precisely estimating model parameters.

So far, we have run models only under the *consistency* assumption, where we assume that the effect sizes from direct and indirect comparisons are exchangeable. While appealing theoretically and computationally, some NMA data could contain *inconsistency*, possibly arising during the data collection process due to heterogeneous effect modifiers (e.g., dissimilar dose levels or severity of disease across study subjects). Several methods have been developed to detect and address inconsistency; see Dias et al. (2011b) for a summary.

Recent NMA data tend to include *multiple* outcomes, and researchers are beginning to focus on combining findings across all possible outcomes. Briefly, we can fit univariate models for each outcome, but this would ignore potentially significant correlations among outcomes. Bayesian models can be easily extended to handle this case by adding one more index. For example, (12.4) can be respecified to $logit(p_{ik\ell}) = \mu_{k\ell} + \nu_{ik\ell}$, where ℓ indexes the multiple endpoints. We can also control the random effects $\nu_{ik\ell}$ with various assumptions of correlations under the Bayesian framework.

Finally, when we generated datasets for simulation studies in Section 12.4, we assigned trials randomly to each design, leading to a *missing completely at random* mechanism. However, we can consider other, more complex mechanisms, such as *missing at random* or even *missing not at random*. Since nonrandom missingness depends on observed or unobserved data, the imputation method in our arm-based model could outperform traditional Bayesian NMA models by integrating information from observed and unobserved. This view may lead to alternative concepts of inconsistency, a subject of ongoing research.

References

DerSimonian, R. and N. Laird. 1986. Meta-analysis in clinical trials. *Controlled Clinical Trials* 7:177–188.

Dias, S., N.J. Welton, A.J. Sutton et al. 2011a. NICE DSU technical support document 2: A generalised linear modelling framework for pairwise and network meta-analysis of randomised controlled trials. Last updated April 2012; available from http://www.nicedsu.org.uk.

Dias, S., N.J. Welton., A.J. Sutton et al. 2011b. NICE DSU technical support document 4: Inconsistency in networks of evidence based on randomised controlled trials. Last updated April 2012; available from http://www.nicedsu.org.uk.

Gartlehner, G. and C.G. Moore. 2008. Direct versus indirect comparisons: A summary of the evidence. *International Journal of Technology Assessment in Health Care* 24:170–177.

Hoaglin, D.C., N. Hawkins, J.P. Jansen et al. 2011. Conducting indirect-treatment-comparison and network-meta-analysis studies: report of the ISPOR task force on indirect treatment comparisons good research practices: Part 2. *Value in Health* 14:429–437.

Hong, H., B.P. Carlin, T. Shamliyan et al. 2013. Comparing Bayesian and frequentist approaches for multiple outcome mixed treatment comparisons. *Medical Decision Making* 33:704–714.

Jansen, J.P., R. Fleurence, B. Devine et al. 2011. Interpreting indirect treatment comparisons and network meta-analysis for health-care decision making: Report of the ISPOR task force on indirect treatment comparisons good research practices: Part 1. *Value in Health* 14:417–428.

Jones, B., J. Roger, P.W. Lane et al. 2011. Statistical approaches for conducting network meta-analysis in drug development. *Pharmaceutical Statistics* 10:523–531.

Lu, G. and A.E. Ades. 2006. Assessing evidence inconsistency in mixed treatment comparisons. *Journal of the American Statistical Association* 101:447–459.

Lumley, T. 2002. Network meta-analysis for indirect treatment comparisons. *Statistics in Medicine* 21:2313–2324.

Salanti, G., J.P. Higgins, A.E. Ades et al. 2008. Evaluation of networks of randomized trials. *Statistical Methods in Medical Research* 17:279–301.

Spiegelhalter, D.J., N.G. Best, B.P. Carlin et al. 2002. Bayesian measures of model complexity and fit. *Journal of the Royal Statistical Society. Series B (Statistical Methodology)* 64:583–639.

13

Regulatory Issues in Meta-Analysis of Safety Data*

Aloka G. Chakravarty and Mark Levenson

CONTENTS

13.1 Introduction

Meta-analysis is the application of statistical techniques to quantitatively combine the results of independent studies, deemed similar and appropriate to be combined, to allow inferences to be made to the population of interest. It has been used in regulatory decision making, especially in safety evaluations, quite extensively. This framework can be motivated for several reasons:

- To provide a more precise estimate of a measure of risk (and its uncertainty) associated with the use of a regulated product than a single trial can provide
- To rule out a certain threshold risk value for a potential class effect (e.g., cardiovascular risk in type 2 diabetes mellitus products)
- To search for adverse events not previously encountered

---

* This chapter reflects the views of the authors and should not be construed to represent the Federal Drug Administration's views or policies.

Meta-analyses conducted in the regulatory context and for safety evaluation have unique issues compared to traditional meta-analyses. In particular, use in the regulatory context requires high levels of rigor, robustness, and transparency. Concepts well accepted in the demonstration of efficacy for registration purposes are also relevant in this safety evaluation. These concepts include well-defined objectives, prespecification, blinding, clear exposure definitions, good outcome ascertainment, appropriate statistical methodology, data quality, and clear and thorough reporting. The evaluation of safety outcome presents its own challenges compared to meta-analyses of other outcomes such as efficacy endpoints. Safety outcomes, while may be very important in the risk-benefit considerations of a drug, may occur with frequencies of the order 1 per 100 or less. The sparseness of the outcomes creates both procedural and methodological challenges.

In this chapter, we discuss the use of meta-analysis of randomized trials conducted in the regulatory context for the evaluation of safety. The chapter discusses key design, methodological, and reporting issues. We also present two real-life examples that had regulatory consequences and display many of these issues.

The emphasis in this chapter is on meta-analyses designed to explore and refine a hypothesized safety signal. Meta-analyses have roles in other aspects of the drug development cycle. As part of the NDA submission, an integrated summary of safety is most often required. Meta-analysis for this purpose may be an exploratory activity primarily undertaken for possible signal detection, not signal refinement. However, often, there are specific adverse outcomes that may be associated with a drug. Such information may be based on drug class behavior or may have been identified in nonclinical and early clinical phases of the development. The examination of cardiovascular outcomes associated with antihyperglycemic agents is an example of a drug class concern. The U.S. Food and Drug Administration (FDA) guidance *Diabetes Mellitus—Evaluating Cardiovascular Risk in New Anti-Diabetic Therapies to Treat Type 2 Diabetes* discusses some meta-analysis issues for this case. In particular, the guidance discusses ruling out a given relative risk before marketing and a more stringent criterion on the relative risk after marketing. Although a meta-analysis is not required for this purpose, it is an important approach discussed in the guidance.

Before presenting our discussion, we present a brief and select overview of some relevant literature. Peto (1987) discusses some foundational issues in systematic reviews of randomized trials for rare outcomes, in particular the need for such reviews and considerations of heterogeneity of the treatment effect. Hammad et al. (2011) present a thorough overview of the issues associated with the secondary use of randomized trials to evaluate safety. Bradburn et al. (2007) evaluate statistical meta-analysis methods for rare events. Kaizer et al. (2006) present an interesting example of a hierarchical Bayesian method for meta-analysis of safety. Crowe et al. (2009) provide recommendation for a premarket safety program.

13.2 Design Issues

Perhaps the most important concept of meta-analysis in the regulatory context is prespecification. This concept for meta-analysis is more multifaceted than for the case of a single trial. For example, a meta-analysis can be conducted based on completed trials. In this case, it is likely that the results from the completed trials influence the objectives and design of the meta-analysis. If a trial has identified a safety issue for a drug, including such a trial in a meta-analysis to confirm the safety signal is problematic, since in this case the meta-analysis will not provide independent findings.

There are three broad levels of prespecification in meta-analysis. The highest level of rigor is when the meta-analysis is designed prior to conducting the trials to be included in the meta-analysis. In this case, the findings of the trials cannot be expected to influence the design of the meta-analysis. The meta-analysis findings can be interpreted directly without the multiplicity issues associated with reusing data in the hypothesis generation and refinement stages. In addition, the objectives and design of the meta-analysis can influence the design of the trials so that the trials provide useful information to evaluate the safety issue. In particular, the meta-analysis can positively impact the design of the safety outcome ascertainment in the trials. This feature is discussed further below (see Section 13.3) in the outcome ascertainment discussion. The example of evaluating cardiovascular risk for a diabetic therapy can fall into this category of prespecification, since the need to evaluate this risk is known prior to conducting the premarket trials (FDA 2008).

The next level of prespecification rigor is the prospective design of a meta-analysis based on completed trials. Although this situation is not ideal, it is sometimes necessary. Often, a safety concern arises and needs to be evaluated to understand the safe and effective use of a drug. Perhaps, the issue arose based on spontaneous reporting, the findings of a pharmacoepidemiological study, or a new trial. In this case, a detailed protocol and analysis plan can be prospectively developed for the meta-analysis. If this process is performed without reviewing outcome information from the existing trials, then multiplicity concerns may not be great. However, in this situation, the meta-analysis is limited by the use of trials not specifically designed to address the safety issue.

The lowest level of prespecification rigor is a meta-analysis designed with the knowledge of the results of the trials with regard to the safety outcome. In this case, the meta-analysis can be thought of as a summarization exercise with limited signal refinement potential.

When designing a meta-analysis, the inclusion criteria of trials and patients should be carefully considered. Trials should provide information that is relevant to the safety outcome. For example, if the safety issue is a long-latency outcome such as a malignancy, short-term trials may not be relevant and including them may even produce misleading results. Casting a wide net in terms of trial inclusion may provide increased precision. However, broad inclusion criteria may introduce heterogeneity, which may make inference more difficult.

When designing a meta-analysis, there may exists large trials in terms of the number of patients or follow-up time with good safety outcome ascertainment. If these trials are included in the meta-analysis, they may dominate the overall meta-analysis results. It may be worthwhile to design the meta-analysis so as not to include these trials. The resulting meta-analysis would summarize the information from the smaller trials, and the larger trials can be evaluated as additional, independent sources of information. The rosiglitazone meta-analysis example is this chapter takes this approach.

An important aspect that contributes substantially to the quality and strength of evidence is the availability of patient-level data and individual study protocols for each study in the meta-analysis. Such availability allows evaluation of each study's quality and eligibility for inclusion in the meta-analysis. It allows for more precise outcome definition and ascertainment. The patient-level data permit time-to-event and subgroups analyses, including dose–response analyses. These issues are discussed further in the following section below (see Section 13.6).

13.3 Outcome Ascertainment

Good outcome ascertainment is an important component to a meta-analysis in the regulatory context. In the best-case scenario, the meta-analysis is conceived and designed prior to or in conjunction to the design of the trials to be included in the meta-analysis. In this case, the safety outcome can be collected prospectively and actively and if necessary can be adjudicated. By prospectively and actively collecting the outcome, ascertainment procedures can be designed to reduce misclassification and bias between the treatment arms of the trials.

Short of prospectively collecting the safety outcome, the outcome can be retrospectively obtained and adjudicated with a well-defined procedure. This approach was taken in the antidepressant meta-analysis example in this chapter. This approach requires detailed extensive patient-level data. It is also helpful to have the individual study protocol to understand the information collected in the trials. With this version of a retrospective outcome ascertainment, attention should be given to potential differential outcome ascertainment between the treatment arms. If the outcome was not actively solicited, there may be differences in reporting between the arms because of unrelated side effects of the treatment. For example, if a drug has more side effects, the patient may be more likely to have encounters with the study investigators and report the safety outcome.

A lower level of outcome ascertainment is the use of outcomes based on adverse event reporting from the trials. Here, the use of standardized dictionaries and queries such as MedDRA (2013) can be used to create a single outcome definition. However, differences among the trial designs limit the uniformity of the outcome defined in this manner. Using reported trial summaries of outcomes from the literature or other sources has limitations because often, the definitions are not explicit or uniform. In this case, it is often not clear when the outcome occurred and if it occurred on randomized treatment.

In order to avoid potential biases, the choice of the specific outcome should be considered. The previous section emphasized the advantages of prospective collection and possibly adjudication of outcome events. However, even in the presence of randomization, there may be differences in ascertainment between the treatment arms. For example, if an unrelated side effect of a treatment results in additional health care utilization, there may be more opportunity to detect the outcome of interest. The use of *hard outcomes* may offer some protections against this phenomenon. For example, if vital status can be obtained for the whole analysis population, then a death outcome would not exhibit this phenomenon. Stroke and myocardial infarction may be more consistently and unbiasedly ascertained than other cardiovascular events such as arrhythmia and transient ischemic attacks.

13.4 Statistical Methods

There are important statistical considerations in meta-analyses for the evaluation of safety. In particular, statistical methods must be valid in the presence of sparse data. As discussed previously, safety outcomes may be infrequent. Some trials may not have any relevant events. We refer to these trials as *zero-event* trials. The statistical methods should provide estimates with good bias properties and with valid standard errors and confidence intervals in the presence of low event counts and zero-event trials. In any meta-analysis, the overall estimator and associated standard errors and confidence intervals should be stratified by the trials. Simple pooling of data across the trials can result in misleading results because of Simspon's paradox. With stratification, the randomize comparisons within trials are maintained.

Several methods have been found adequate for low event count situations. For odds ratio summary measures, the Mantel–Haenszel, Peto, and exact methods appear to work well. The commonly used method in meta-analysis based on inverse variance weights does not perform well in low event count situations because the weights are not stable with low event counts. The Mantel–Haenszel risk difference appears to work well as well. This method has the added benefit in that unlike the methods for the odds ratio, this method for the risk difference incorporates the information from the zero-event trials into the estimate and its standard error. The use of continuity corrections should be avoided. This approach adds a small fraction of an event to both arms in the zero-event trials. Because of the sparseness of safety outcomes, even this small fraction can have undue influence on the overall effect measure. Careful attention should be given to meta-analysis software to ensure they do not incorporate a continuity correction either in the estimate or the standard error.

In discussion of meta-analysis, there is often much attention given to the random effects model versus the fixed effects model. Random effects models assume that the true treatment effects of the individual trials represent a random sample from some population. The random effects model estimates the population mean of the treatment effects and accounts for the variation in the observed effects. It is sometimes stated that the fixed effects model assumes that the individual trial effects are constant. However, this is not a necessary assumption. An alternative view is that the fixed effects model estimates the mean of the true treatment effects of the trials in the meta-analysis. Senn (2000) discussed the analogy with center effects in multicenter trials. In safety, random effects models may be problematic because of the need to estimate between-trial effects with sparse data. Additionally, the random effects model is less statistically powerful than the fixed effects model, albeit the hypotheses are different. In the fixed effects model, the variance estimate should account for trial effect differences either through stratification, conditioning, or modeling of fixed effects.

It is important to evaluate differences in mean therapy duration and mean follow-up duration between the treatment assignments (Hammad et al. 2011). Unequal duration between the treatments in either of these quantities might imply there is unequal opportunity to experience or observe an event. Additionally, unequal durations may result from informative censoring, in which patients may discontinue therapy or follow-up because of some condition associated with the outcome of interest.

Sensitivity analyses are vital to support and explore the meta-analysis findings. The sensitivity analyses should explore the meta-analyses design choices and methods. These include the trial and patient inclusion criteria, the statistical methods, the effect of large trials, and the effect of zero-event trials. An important sensitivity analysis is to explore the effect of large or otherwise unique trials on the overall meta-analysis summaries. If there are one or a few trials that may have great influence because of their size or their follow-up length, repeating the meta-analysis without these trials is informative. Large trials can often serve as independent sources of information and can be summarized individually.

Subgroups in a meta-analysis can be of direct interest or useful in exploring the consistency of findings. Adjustments for multiplicity from examining subgroups in meta-analysis are usually not feasible because of the limited information from the available data. Patterns and consistency of results may

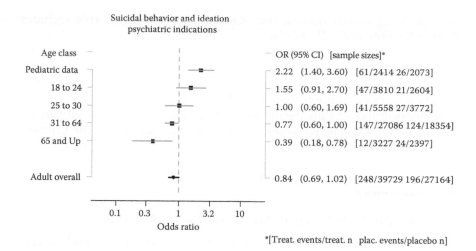

FIGURE 13.1
FDA meta-analysis of antidepressants and suicidal behavior and ideation.

be informative. For example, in antidepressant example in this chapter, there was a clear pattern in the risk for different age groups (see Figure 13.1). Care must be taken when comparing subgroups in a meta-analysis. Different subgroups may come from different trials or not be equally distributed within trials. Although the meta-analysis should maintain the randomized comparisons for the trials, differences in trial characteristics may confound comparisons between subgroups.

13.5 Reporting

Completeness and transparency are key principles in the reporting of a meta-analysis for the evaluation of safety in the regulatory framework. The protocol and the statistical analysis plan should be completed and finalized prior to the conduct of the meta-analysis. Any deviations or additional investigations after the finalization of the protocol should be clearly identified as such.

The protocol should discuss what information was available prior to designing the meta-analysis and what information motivated the research objectives of the meta-analysis. The reporting should clearly state which trials were conducted and which trial results were known by study investigators prior to the design of the meta-analysis. The protocol should state the trial and patient inclusion, including discussion on any possible publication bias. The protocol should state the sources of the trial and patient data.

The protocol should discuss the outcome ascertainment. This includes whether the outcome collection was part of the design of the individual trials or retrospectively collected, whether the outcome was actively collected or passively collected, whether the outcome was adjudicated, and if so how. A clear outcome definition and the outcome window should be stated. The primary outcome definition should be identified.

The protocol should state the primary analysis methods, sensitivity analyses, and subgroups. The software and the specific routines for the primary analysis should be stated.

13.6 Example 1: Antidepressants and Suicidal Events

In order to study the possible association antidepressant drugs and adverse suicidal events in adult patients, FDA planned and conducted a meta-analysis study of randomized trials of antidepressants. The meta-analysis had several key features that supported its quality and utility for regulatory actions; (1) hypotheses generated from previous and independent evidence provided the meta-analysis objectives; (2) the meta-analysis was based on well-defined inclusion criteria and an exhaustive set of trials with patient-level data available; (3) the meta-analysis employed rigorous and consistent outcome definitions across trials and patients; and (4) the meta-analysis was based on prespecified statistical analysis plan.

FDA initiated the adult antidepressant meta-analysis based on findings from a previous FDA meta-analysis of pediatric patients. In 2004, FDA completed a meta-analysis of pediatric patients that showed an association of antidepressant drugs and suicidal behavior and ideation (Hammad et al. 2006). Unsolicited information provided from drug sponsors and published articles motivated this meta-analysis. Based on the meta-analysis and deliberations from a meeting of FDA advisory committees, FDA added a boxed warning to the label of all antidepressants with the warning concerning pediatrics patients in 2005. The FDA advisory committees requested that FDA explore the association in adult patients.

The FDA meta-analysis of adult patients was based on well-defined inclusion criteria and an exhaustive set of trials. FDA requested from all manufacturers of antidepressants all available patient-level data from randomized placebo-controlled trials of antidepressants. Basing the meta-analysis on trials data available to drug makers, while not exhaustive of all potentially available data, has some important advantages. The set trial does not suffer from publication bias, because publication has no bearing on the inclusion of the trials. Because of regulatory requirements, trials from drug manufacturers often contain detailed patient-level data including medical history, baseline characteristics, patient dispositions, patient outcomes, and adverse events. Focusing on the relatively small group of nine drug manufacturers allowed for the timely acquisition of the large amounts of pertinent data. Overall, the meta-analysis obtained data on 372 trials with usable data.

Because detailed patient-level data were available, the meta-analysis employed rigorous and consistent outcome definitions across trials and patients. This meta-analysis was an example of a prospectively planned analysis of preexisting trials. Because the specific outcomes of interest were not systematically collected and adjudicated during the conduct of the trials, FDA directed a retrospective identification and adjudication of events. The outcome definition required that the suicidal behavior and ideation events occurred on randomized treatment or within 1 day of stopping the randomized treatment. Based on adverse event reporting, potential events were identified with a specified algorithm. Based on blinded narratives of the events, qualified personal classified the events into specific outcomes including completed suicide, attempted suicide, preparatory actions toward imminent suicidal behaviors, and suicidal ideation based on the Columbia Classification Algorithm for Suicide Assessment (Posner et al. 2007). The overall process resulted in outcome measures that were consistently and rigorously defined across trials and patients.

The meta-analysis employed a prespecified analysis plan, which specified trial inclusion criteria, hypotheses, outcome definitions, analysis methods, sensitivity analyses, and subgroups. The trial inclusion criteria and the outcome definitions have already been discussed. The primary analysis method stratified by trial and accounted for the sparse nature of the outcome events by using exact statistical methods. The sensitivity analysis examined the possibility and consequences of the following: differential exposure time between the randomized treatment arms, heterogeneity of the effect measure across the trials, and trials with no events (FDA 2006). The patient-level data allowed for the examination of important subgroups, including patient age, and for the examination of changing risk over time.

FDA sought advice from a meeting of FDA advisory committees on the interpretation and possible regulatory actions based on the findings of the meta-analysis. The meta-analysis found that the overall association of antidepressant drugs and suicidal behavior and ideation was not statistically significant in contrast to the FDA meta-analysis of pediatric patients. However, the association was nearly statistically significant for young adults and the meta-analysis showed a clear pattern in the association as related to the patient age (see Figure 13.1). The inclusion of the result from the pediatric meta-analysis supported this trend.

FDA chose to update the boxed warning on all antidepressants to warn about the risk suicidal behavior and ideation associated with antidepressants for young adult patients in addition to pediatric patients. The warning states that the effect was not seen in patients over the age of 24, and for patients aged 65 and older, the risk may be reduced. It should be noted it was the clear pattern in the effect as related to patient age and not solely statistical significance that led to the warning.

13.7 Example 2: Rosiglitazone and Cardiovascular Outcomes

Cardiovascular safety concerns have been raised for the U.S.-approved thiazolidinedione (TZD) class of antidiabetic drugs, which include pioglitazone and rosiglitazone. In 2003, the American Heart Association and American Diabetes Association issued a consensus statement on TZDs and congestive heart failure (CHF) providing recommendations on the use of TZDs with patients with CHF (Nesto et al. 2003). In June 2007, Nissen and Wolski (2007) published an article in the *New England Journal of Medicine* on a meta-analysis that showed a statistically significant risk of myocardial infarction and a nearly statistically significant risk of cardiovascular death associated with rosiglitazone.

In July 2007, FDA convened a joint advisory committee meeting to discuss the cardiovascular risks of the TZDs, with a focus on rosiglitazone. The meeting included presentations and discussions on an FDA patient-level meta-analysis of rosiglitazone and several large, long-term randomized trials of the TZDs. A majority of the committee members voted that the data suggested that rosiglitazone increases the cardiac ischemic risk in patients with type 2 diabetes. In another vote, a majority of the committee members voted that the overall risk–benefit profile of rosiglitazone supported its continued marketing in the United States. In August 2007, FDA placed a boxed warning on the risk of CHF for both TZDs. In November 2007, FDA placed an additional boxed warning on rosiglitazone warning of a potential increased risk of myocardial ischemia.

In July 2010, FDA again convened a joint advisory committee meeting to discuss the cardiovascular risks of the TZDs. The meeting was intended to discuss the available data on the safety of the TZDs, including randomized controlled trials, observational studies, and reanalysis of randomized trials. In preparation for this meeting, FDA reviewers conducted up-to-date meta-analyses of the association of cardiovascular safety outcomes separately for pioglitazone and rosiglitazone.

The 2010 FDA meta-analyses had two goals: (1) to update the 2007 FDA meta-analysis of rosiglitazone of 42 trials with 10 newly available trials and (2) to conduct a pioglitazone meta-analysis in order to compare indirectly the cardiovascular profile of the two drugs. FDA reviewers set the inclusion criteria for trials in the meta-analysis to intentionally exclude the large, long-term trials of the drugs. FDA viewed the large trials as independent sources of information to be summarized individually. FDA intended the meta-analysis to summarize the information provided from the collection of smaller trials. With the exclusions of the large trials, the meta-analyses would not be dominated by these trials. Additionally, the exclusion improved the comparability of the meta-analyses between the two drugs, because the large trials differed in important aspects, such as the control groups.

The 2010 FDA meta-analyses were designed to promote the comparability between the meta-analyses of the two drugs. Both meta-analyses employed a common plan. In particular, the trial and patient inclusion criteria, the outcome definitions, and the analysis methods were the same for both meta-analyses. Importantly, FDA reviewers recognized that the trials from the two drugs differed in important ways. To overcome this situation, the meta-analysis plan called for the use of predefined subgroups of trials that had comparable designs.

Three major categories of trial-level groups were considered: randomized comparator, add-on therapy, and therapy duration. Randomized comparator trial-level groups included two categories: placebo-controlled trials and active-controlled trials. The add-on therapy trial-level groups included two categories: monotherapy trials and add-on therapy trials. Monotherapy trials consisted of trials in which the randomized therapy was the sole assigned agent, whereas in add-on trials, the randomized therapy was given in addition to some predefine therapy given to all patients in the trial. Finally, the last major trial-level groups were based the protocol-specified therapy duration of the trial and consisted three categories: > 2 months to ≤ 6 months, > 6 months to ≤ 1 year, and > 1 year to ≤ 2 years.

The 29 pioglitazone trials and the 52 rosiglitazone trials that met the common inclusion criteria of the meta-analyses differ in the distribution of all three of these trial-level groups. Table 13.1 shows the distribution of the randomized comparator groups of the trials for the two drugs. Overall, rosiglitazone had higher percentages of trials and patients in placebo-controlled trials than pioglitazone. This difference demonstrates the need to control for this factor when comparing the results between the meta-analyses of the two drugs. Similarly, the two drugs differed in their distributions of the add-on therapy and therapy duration categories. Overall, 49% of the pioglitazone patients were in monotherapy trials compared to 32% of the rosiglitazone patients. Overall, 47% of the pioglitazone patients were in trials with ≤6 month of duration compared to 69% of the rosiglitazone patients.

The primary endpoint was major adverse cardiovascular event (MACE), defined as cardiovascular death, stroke, or myocardial infarction. Secondary safety outcomes were cardiovascular death, stroke, myocardial infarction, all-cause death, serious myocardial ischemia, total myocardial ischemia,

TABLE 13.1

Distribution of Trial and Patients by Randomized Comparator for Pioglitazone and Rosiglitazone

	Pioglitazone Meta-Analysis		Rosiglitazone Meta-Analysis	
Randomized Comparator	Trials $N = 29$ n	Sample Size $N = 11,774$ n (%)	Trials $N = 52$ n	Sample Size $N = 16,995$ n (%)
Placebo	18	4,574 (39)	46	13,760 (81)
Active	12	7,350 (62)	13	4,037 (24)
Sulfonylurea	8	4,383 (37)	8	3,106 (18)
Metformin	3	2,232 (19)	4	613 (4)

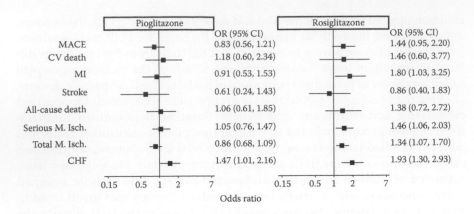

FIGURE 13.2
Overall meta-analysis results by outcome for pioglitazone and rosiglitazone. MACE, major adverse cardiovascular event; CV death, cardiovascular death; MI, myocardial infarction; Serious M. Isch., serious myocardial ischemia; Total M. Isch., total myocardial ischemia; and CHF, congestive heart failure.

and CHF. The primary analysis method was the exact method for an odds ratio and associated 95% confidence interval, stratified by trial. Sensitivity analyses explored among other things: the effect of trials with no outcome events, potential heterogeneity of the trials, and potential imbalances in therapy duration between comparator arms within trials.

Figure 13.2 gives side-by-side forest plots of the overall results of the two meta-analyses for the eight outcomes. Odds ratios greater than 1 indicate that the drug has an adverse effect on the outcome. In addition to the odds ratios, the plots display 95% confidence intervals. Because of the differences in the characteristics of the trials and patients between the two drugs, this comparison may be confounded. It is noteworthy that both drugs had a statistically significant adverse effect on CHF. This adverse effect is well accepted and it included as a boxed warning for both drugs.

The subsets of trials that are placebo controlled were much more similar between than the drugs that the full sets of trials. There were 18 placebo-controlled trials for pioglitazone and 46 for rosiglitazone. Among these trials, 24% of the pioglitazone patients were in monotherapy trials compared to 20% of the rosiglitazone patients. As previously stated, these percentages for the full sets of trials were 49 and 32, respectively. Similarly, among placebo-controlled trials, 83% of the pioglitazone patients were in trials with ≤6 month of duration compared to 82% of the rosiglitazone patients. As previously stated, these percentages for the full sets of trials were 47 and 69, respectively.

Figure 13.3 gives side-by-side forest plots of the results of the two meta-analyses for the placebo-controlled trials for the eight outcomes. For rosiglitazone, the odds ratio estimates were consistently above 1. The 95% confidence interval for myocardial infarction was > 1. For pioglitazone, with

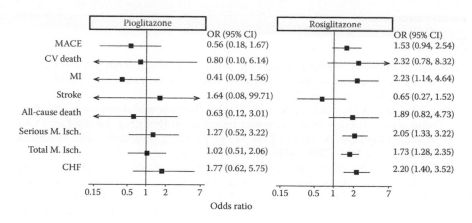

FIGURE 13.3
Placebo-controlled trials' meta-analysis results by outcome for pioglitazone and rosiglitazone. MACE, major adverse cardiovascular event; CV death, cardiovascular death; MI, myocardial infarction; Serious M. Isch., serious myocardial ischemia; Total M. Isch., total myocardial ischemia; and CHF, congestive heart failure.

the possible exception of CHF, there was not clear adverse association between pioglitazone and the outcomes.

FDA asked the members from the July 2010 advisory committee meeting to discuss the strengths and weaknesses of the various sources of data available to address the question of the risk of ischemic cardiovascular in patients treated with rosiglitazone compared to non-TZDS and compared to pioglitazone. There were few studies designed specifically to compare rosiglitazone to pioglitazone. The FDA meta-analyses provided one such source of data designed to make this comparison. The meta-analyses played a large role in the deliberation of the committee members (FDA 2010).

FDA used the deliberations of the committee members and its own considerations of the available data to implement regulatory actions related to rosiglitazone. FDA chose to revise the drug label to restrict the use of the drug to patients already being treated with rosiglitazone and to patients whose blood sugar cannot be controlled with other antidiabetic medicines and who, after consulting with their health care professional, do not wish to use pioglitazone-containing medicines.

13.8 Conclusion

Meta-analysis, when applied appropriately, provides a powerful tool to evaluate safety. However, in the regulatory context, careful attention to rigor, robustness, and transparency is essential when performing meta-analysis.

References

Bradburn M., Deeks J., Berlin J., Localio R. Much ado about nothing: A comparison of the performance of meta-analytical methods with rare events. *Statistics in Medicine*, 2007; 26(1):53–77.

Crowe B.J., Xia H.A., Berlin J.A., Watson D.J. et al. Recommendations for safety planning, data collection, evaluation and reporting during drug, biologic and vaccine development: A report of the safety planning, evaluation and reporting team. *Clinical Trials*, 2009; 6:430–440.

FDA. FDA briefing document for the Psychopharmacologic Drugs Advisory Committee (meeting). http://www.fda.gov/ohrms/dockets/ac/06/briefing/2006-4272b1-01-FDA.pdf, December 13, 2006. Accessed August 7, 2014.

FDA. Guidance for industry: Diabetes mellitus—Evaluating cardiovascular risk in new antidiabetic therapies to treat Type 2 diabetes. U.S. Department of Health and Human Services, Food and Drug Administration, Center for Drug Evaluation and Research (CDER). http://www.fda.gov/downloads/Drugs/GuidanceComplianceRegulatoryInformation/Guidances/UCM071627.pdf, December 2008. Accessed August 7, 2014.

FDA. Joint meeting of the endocrinologic and metabolic drugs advisory committee and the drug safety and risk management advisory committee meeting transcript. http://www.fda.gov/AdvisoryCommittees/CommitteesMeetingMaterials/Drugs/DrugSafetyandRiskManagementAdvisoryCommittee/ucm199874.htm, July 13 and 14, 2010. Accessed August 7, 2014.

Hammad T.A., Laughren T., Racoosin J. Suicidality in pediatric patients treated with antidepressant drugs. *Archives of General Psychiatry*, 2006; 63(3):332–339.

Hammad T.A., Pinheiro S.P., Neyarapalli G.A. Secondary use of randomized controlled trials to evaluate drug safety: A review of methodological considerations. *Clinical Trials*, 2011; 8(5):559–570.

Kaizer E., Greenhouse J., Seltman H., Kelleher K. Do antidepressants cause suicidality in children? A Bayesian meta-analysis. *Clinical Trials*, 2006; 3:73–98.

MedDRA. Medical dictionary for regulatory activities. http://www.meddramsso.com/, 2013. Accessed August 7, 2014.

Nesto R.W., Bell D., Bonow R.O., Fonseca V. et al. Thiazolidinedione use, fluid retention, and congestive heart failure: A consensus statement from the American Heart Association and American Diabetes Association. *Circulation*, 2003; 108:2941–2948.

Nissen S.E., Wolski K. Effect of rosiglitazone on the risk of myocardial infarction and death from cardiovascular causes. *The New England Journal of Medicine*, 2007; 356(24):2457–2471.

Peto R., Why do we need systematic overviews of randomized trials? *Statistics in Medicine*, 1987; 6:233–240.

Posner K., Oquendo M., Gould M., Stanley B., Davies M., Columbia classification algorithm of suicide assessment (C-CASA): Classification of suicidal events in the FDA's pediatric suicidal risk analysis of antidepressants. *The American Journal of Psychiatry*, 2007; 164:1035–1043.

Senn S., The many modes of meta. *DIJ*, 2000; 34:535–549.

14

Bayesian Applications for Drug Safety Evaluation

H. Amy Xia and Karen L. Price

CONTENTS

14.1 Introduction

Safety assessment is a critical part of drug development. In analyzing the safety data from clinical trials and observational studies, a number of statistical difficulties arise. Although demonstrating both safety and efficacy is required for the drug regulatory approval, the use of advanced statistical methods in evaluating safety data has not received as much attention as evaluating efficacy.

There are several important questions in drug safety evaluation: How to detect unexpected adverse drug reactions while handling the multiplicity issue properly? How to synthesize data from different trials, or even different sources? How to deal with rare events? How to evaluate multidimensional, complex safety information as a whole? Can we monitor a potential safety issue in a continuous manner during a trial so that patients can be better protected? These questions lead to some unique statistical challenges in quantitative safety analysis including low power due to rare events, multiplicity, event medical classification, multidimensionality and complexity of safety data, and time dependency of adverse events (AEs) [1,2].

In recognizing that many of these challenges can be difficult to tackle with traditional frequentist approaches, some key opinion leaders in the field commented "Safety assessment is one area where frequentist strategies have been less applicable. Perhaps Bayesian approaches in this area have more promise." [3]. Bayesian methods have many advantages and are suitable for safety assessment primarily in the following aspects:

1. Ability to incorporate prior information from different sources
2. Natural for evidence synthesis or meta-analysis
3. Handling multiplicity through borrowing strength and hierarchical modeling
4. Appealing in dealing with rare events as the model modulates the extremes
5. Ability to handle complex problems via unified modeling, taking all the uncertainty into appropriate account
6. Allowing direct probability inferences on different scales

In this chapter, we present a few case examples for drug safety assessment where Bayesian approaches were utilized. In each example, we describe the Bayesian methods or models, the rationale for Bayesian methods, and the challenges that we encountered. Section 14.2.1 describes a Bayesian hierarchical modeling approach for clinical trial signal detection. Section 14.2.2 discusses how Bayesian methods can be used in meta-analysis for analyzing AE data, especially for rare AEs with nonconstant hazards. Section 14.2.3 presents an example where the Bayesian approach is used to continuously monitor an event of interest (EOI) for an ongoing trial while prior information can be incorporated. Section 14.2.4 describes Bayesian joint modeling for evaluation of safety and/or efficacy outcomes. Section 14.2.5 discusses a Bayesian application in estimating the dose–response relationship. Section 14.2.6 discusses how the Bayesian approach can be useful in the context of mixed treatment comparisons or network meta-analysis. Finally, in Section 14.3, we offer some concluding remarks and directions for future work.

14.2 Specific Applications

14.2.1 Application 1: Clinical Trial Signal Detection

Signal detection is an important task in drug safety evaluation. While Bayesian approaches have been widely used for safety signal detection in the spontaneous AE reporting databases [4,5] and electronic medical records and administrative claims databases [6,7], this section focuses on signal detection in clinical trials using Bayesian hierarchical modeling. Additionally, there has also been other work published in the area of clinical trial safety signal detection using Bayesian methods [8–10].

In a typical late-stage clinical trial, hundreds or even thousands of different types of AEs are collected in the clinical trial database. This results in a challenging statistical problem pertaining to multiplicity in analyzing these AEs. If multiplicity is not considered, excessive false-positive signals may be flagged, potentially causing undue concerns for drug approval, unnecessary safety studies, or postmarketing commitments. On the other hand, adjusting multiplicity with traditional approaches may fail to detect important signals that could have a great public health impact. Thus, statistically, we need to build a screening device that can strike a fair balance between *no adjustment* and *too much adjustment*. Bayesian methods can deal with this issue in a straightforward manner. Multiplicity is handled through hierarchical shrinkage. Clinical trial AE data are typically coded in the Medical Dictionary for Regulatory Activities (MedDRA). Each AE collected in the clinical trial database is then coded in a hierarchical manner with low-level term (LLT), preferred term (PT), high-level term (HLT), high-level group term (HLGT), and system organ class (SOC) from the most to least granular level. The coding system largely (although not perfectly) reflects the biological relationship among different types of AEs. As such, Bayesian hierarchical modeling allows for explicitly modeling AEs within the existing MedDRA coding structure, so that strength can be borrowed within or even across (although less so) SOCs since the AEs within an SOC are more similar than those in different SOCs. It provides *partial correction* that accounts for multiplicity when it is crucial to avoid detecting excessive false-positive signals but on the other hand does not overdo it in a sense to let data tell how much borrowing should incur [11].

Berry and Berry [12] first applied the Bayesian method in analyzing AE data by constructing a three-level Bayesian mixture model for binary outcomes. In that approach, they let $Y_{bj} \sim Bin(N_t, t_{bj})$ and $X_{bj} \sim Bin(N_c, c_{bj})$, where t_{bj} and c_{bj} are event rates for PT $j = 1, ..., J_b$ and SOC $b = 1, ..., B$ in the treatment and control groups, respectively. A logistic regression model was considered as $\text{logit}(c_{bj}) = \gamma_{bj}$, $\text{logit}(t_{bj}) = \gamma_{bj} + \theta_{bj}$ so that γ_{bj} is the logit PT rate in the control group and θ_{bj} is the logarithm of odds ratio between the two treatment groups. A three-level hierarchical mixture model was constructed as follows:

- Stage 1 priors:
  $\gamma_{bj} \sim N(\mu_{\gamma b}, \sigma_{\gamma b}^2)$ and $\theta_{bj} \sim \pi_b \delta(0) + (1 - \pi_b)N(\mu_{\theta b}, \sigma_{\theta b}^2)$, where $0 \leq \pi_b \leq 1$ and $\delta(0)$ is a point mass at 0 (allows a prior probability of π_b that the treatment and control rates are exactly the same, which is appropriate as we expect many AEs are not drug induced).

- Stage 2 priors:

$$\mu_{\gamma b} \sim N(\mu_{\gamma 0}, \tau_{\gamma 0}^2), \sigma_{\gamma b}^2 \sim IG(\alpha_\gamma, \beta_\gamma),$$

$$\pi_b \sim Beta(\alpha_\pi, \beta_\pi), \mu_{\theta b} \sim N(\mu_{\theta 0}, \tau_{\theta 0}^2), \sigma_{\theta b}^2 \sim IG(\alpha_\theta, \beta_\theta).$$

- Stage 3 priors:

$$\mu_{\gamma 0} \sim N(\mu_{\gamma 00}, \tau_{\gamma 00}^2), \tau_{\gamma 0}^2 \sim IG(\alpha_{\gamma 00}, \beta_{\gamma 00}),$$

$$\alpha_\pi \sim exp(\lambda_\alpha)I(\alpha_\pi > 1), \beta_\pi \sim exp(\lambda_\beta)I(\beta_\pi > 1),$$

$$\mu_{\theta 0} \sim N(\mu_{\theta 00}, \tau_{\theta 00}^2), \tau_{\theta 0}^2 \sim IG(\alpha_{\theta 00}, \beta_{\theta 00}).$$

Hyperparameters $\mu_{\gamma 00}$, $\tau_{\gamma 00}^2$, $\alpha_{\gamma 00}$, $\beta_{\gamma 00}$, λ_α, λ_β, $\mu_{\theta 00}$, $\tau_{\theta 00}^2$, $\alpha_{\theta 00}$, $\beta_{\theta 00}$, α_γ, β_γ, α_θ, β_θ are fixed constants.

Given the posterior distribution of this Bayesian hierarchical model, the posterior *exceedance probability* can be used to flag a signal. For example, PT_{bj} is flagged if $Pr(\theta_{bj} > d \mid Data) > p$, where d and p are prespecified constants. While we often set $d = 0$ (i.e., OR = 1) to identify an elevated risk of AE in the treatment group, we may also choose a larger value of d (e.g., log(2)) to better capture a *clinically meaningful* effect to avoid detecting medically unimportant signals because an AE could have a high $Pr(OR > 1 \mid Data)$ but medically unimportant. In addition, in the Bayesian framework, the posterior exceedance probability for other scale measures (e.g., risk difference) can be easily obtained with $Pr(t_{bj} - c_{bj} > d^* \mid Data) > p$ for a prespecified d^*.

Xia et al. [13] extended the work by accounting for various exposure or follow-up times among different subjects under the Poisson likelihood. They did a simulation study to compare the operating characteristics (in terms of false discovery rate [FDR] and power) among the unadjusted method (Fisher exact test), the Benjamini–Hochberg approach [14], the double FDR method [15], and the Bayesian hierarchical modeling approach. In the scenarios that they examined (data coming from a phase 2/3 drug development program), they concluded that the Bayesian hierarchical modeling approach outperforms the other methods included in the simulation, in that it has a combination of low FDR and high power. In addition, they provided the WinBUGS and S-Plus programs for implementing the Bayesian models and the volcano plots, respectively.

One practical question in applying these approaches is whether modeling SOC is too broad and nonspecific as the PTs within the same SOC might not necessarily be medically related. The Bayesian hierarchical modeling allows flexibility in modeling any level of the coding structure, for instance, at the HLGT/HLT/PT or HLT/PT level. However, in certain patient populations in which there are too few AEs in an HLT, there might not be much borrowing allowed if we model HLT/PT. Modeling standard MedDRA queries

(SMQs) might be an interesting future direction as the SMQs represent more well-established medical concepts. Because there is flexibility in modeling different levels of AEs, as a practical advice, it is important to prespecify the multiplicity adjustment method that will be used. Grouping and definitions should be described in the study protocol or the statistical analysis plan, rather than on a post hoc basis.

14.2.2 Application 2: Bayesian Survival Meta-Analysis Using Individual Patient Data

Meta-analysis has been increasingly used to address drug safety issues [16,17]. In particular, meta-analysis using individual patient data (IPD) has been advocated because of its many advantages [2,18].

Bayesian piecewise exponential survival model can be utilized for Bayesian meta-analysis using IPD data [19,20].

The model can be specified as follows:

Let $i = 1, ..., n$ index studies and $j = 1, ..., m_i$ index subjects within studies. Suppose time axis can be divided into K prespecified intervals $I_k = (s_{k-1}, s_k)$ for $k = 1, 2, ..., K$ where $0 = s_0 < s_1 < \cdots < s_K < \infty$, with s_K being the last survival or censoring time, and we assume the baseline hazard λ_k is constant within intervals such that at time t it is $h_0(t) = \lambda_k$ for $t \in I_k$. Let y_{ij} indicate the random survival time of the jth patient in the ith study. Furthermore, let $\mathbf{w} = (w_1, w_2, ..., w_n)$ represent random effects for studies, which captures unobserved heterogeneity, and let $\lambda = (\lambda_1, \lambda_2, ..., \lambda_K)$ be the hazard rates on the time intervals, let σ_w^2 represent the variation of the study-specific random effects, and let $\boldsymbol{\beta} = (\beta_1, \beta_2, ..., \beta_L)$ represent the coefficients for fixed effects. The hazard for an event y_{ij} occurring at time $t_{ij} \in I_k$ is $h_{ij} = w_i \lambda_k \theta_{ij}$, where $x_1, x_2, ..., x_L$ are covariates and $\theta_{ij} = \exp(\beta_1 x_{ijk1} + \beta_2 x_{ijk2} + \cdots + \beta_L x_{ijkL})$ is the contribution of the regression variables to the hazard function. Note that this model can also accommodate time-dependent covariates that may vary with each person–time interval. Inference is based on the posterior distribution of $\boldsymbol{\beta}$, $\mathbf{w}$, and λ, which is proportional to the product of the prior for $(\boldsymbol{\beta}, \mathbf{w}, \lambda)$ and the likelihood.

Per [20], the complete likelihood is expressed as the following:

$$\prod_{i=1}^{n} \prod_{j=1}^{m_i} \left[\left\{ \prod_{k=1}^{g_{ij}} \exp(-\lambda_k \Delta_k \theta_{ij} w_i) \right\} (w_i \lambda_{g_{ij}+1} \theta_{ij})^{\delta_{ij}} \times \exp\{ -\lambda_{g_{ij}+1} (y_{ij} - s_{g_{ij}}) \theta_{ij} w_i \} \right]$$

where we define an indicator for censoring, $\delta_{ij} = 0$ if subject j in study i is censored and 1 if subject j has an event and g_{ij} is such that $y_{ij} \in (s_{g_{ij}}, s_{g_{ij}+1})$.

For the prior specification, noninformative priors in the form of normal distributions with mean 0 and very large variances can be used for β_j. For the piecewise hazards, we assume normal hazards on λ_k so that $\pi(\eta_1) \sim N(0, \kappa)$ for

$\eta_k = \log(\lambda_k)$ and $\pi(\eta_k \mid \eta_{k-1}) \sim N(\eta_{k-1}, \kappa)$ for $k = 2, \ldots, K$. The hyperprior for κ is a finite uniform distribution defined in the positive real line. For the random study terms, we assume $\pi(w_i) \sim N(\mu_w, \sigma_w^2)$ with noninformative hyperpriors on μ_w and σ_w^2.

Such Bayesian hierarchical piecewise exponential survival models were used in a cross-company meta-analysis to investigate the short-term cancer risk in three tumor necrosis factor (TNF) inhibitors [21], in response to an EMA request. In that meta-analysis, 74 randomized controlled trials ($\geq$ 4 week duration) of TNF inhibitors across multiple indications ($n = 22{,}904$) were included. That meta-analysis showed that the relative risk (RR) for all cancers excluding nonmelanoma skin cancer (NMSC) is 0.99 (95% Bayesian credible interval [BCI] = 0.61–1.68), and RR = 2.02 (95% BCI 1.11–3.95) for NMSC. In that Bayesian IPD meta-analysis, time was divided into 0–3 and > 3 months, with constant hazard in each interval, allowing for relaxing the proportional hazards assumption. Class effects and drug-specific effects among three anti-TNF agents were assessed, and differences in *sponsor-specific control-group effect* were investigated. In addition, the model took into account patient-level covariates, between study heterogeneity, and time-dependent covariates. The Bayesian models in that example demonstrated great capability and flexibility in dealing with an extremely complex problem.

A practical challenge of Bayesian meta-analysis for rare AE data is that noninformative priors may lead to convergence failure due to very sparse data. Weakly informative priors may be used to solve this issue. In the example of the previous Bayesian meta-analysis with piecewise exponential survival models, the following priors for log hazard ratio (HR) (see Table 14.1) were considered. Prior #1 assumes a nonzero treatment effect with a mean log(HR) of 0.7 and a standard deviation of 2. This roughly translates to that the 95% confidence interval (CI) of HR is between 0.04 and 110, with an estimate of HR to be 2.0. Prior #2 assumes a 0 treatment effect, with a mean log(HR) of 0 and a standard deviation of 2. This roughly translates to the assumption that we are 95% sure that the HR for treatment effect is between 0.02 and 55, with an estimate of the mean hazard of 1.0. Prior #3 assumes a nonzero treatment effect that is more informative than that of Prior #1, with a mean log(HR) of 0.7 and a standard deviation of 0.7. This roughly translates to the assumption that we are 95% sure that the HR

TABLE 14.1

Three Weakly Informative Prior Specifications for Bayesian Meta-Analysis

Prior	Mean Log(HR)	Standard Deviation	Translated Estimated Mean HR (95% CI)
1	0.7	2	2 (0.04, 110)
2	0	2	1 (0.02, 55)
3	0.7	0.7	2 (0.5, 8.2)

is between 0.5 and 8.2. All of these priors impose weak restrictions on the size of the treatment effect. Priors #1 and #3 are considered skeptical priors, in that they assume an effect of twofold risk and force the data to overcome it. Because this is a safety assessment, the weakly informative Priors #1 and #3 that assume an increased risk are more conservative but are still reasonably weak in the sense that they specify a wide range of possible values for HR. By using these three weakly informative priors as sensitivity analyses, we can assess the robustness of the results against different choices of priors.

14.2.3 Application 3: Continuously Monitoring an Adverse Event of Interest in an Ongoing Clinical Trial

Statistical issues posed by monitoring safety in clinical trials are considerably different from monitoring efficacy. In the presence of a safety concern, timely assessment of new data from an ongoing trial and establishment of a fast response system are important to protect patients participating in the trial. In this context, it is preferred to monitor an EOI continuously, that is, to assess the risk whenever a new event occurs. In contrast, when focusing on efficacy assessment, the number of interim analyses and the number of subjects or events per interim analysis are often predetermined under the framework of traditional group sequential methods. In this example, we propose a novel Bayesian approach for the establishment of statistical guidelines in order to facilitate informed and objective decisions made by the data monitoring committee (DMC).

Here is a motivating scenario. A particular EOI occurs in patients with a particular disease. The background EOI rate is about 13%–14% in this population. EOIs occurred in two out of seven subjects (29%) in a single-arm Phase 1b portion of a study for a novel testing agent X, which presents a new safety signal. Given the small number of patients tested, there is a significant false-positive risk of declaring the new treatment as unsafe.

The Phase 2 portion of the trial is a randomized, double-blind three-arm study to compare Drug X to placebo with a 2:1 randomization ratio and a total sample size of 120. Given the observation in Phase 1b, the DMC of the trial felt that they should continuously monitor the EOI in the Phase 2 portion of the study, so an excessive risk in treatment arm as compared to placebo arm may result in the early termination of study or other actions.

In this example, the Bayesian approach provides an appropriate way to conduct the continuous monitoring of this EOI. Data from literature (for control) and the Phase 1b data (for Drug X) could be used to formulate priors. Each new event on study changes the posterior probability of the treatment effect and a threshold for stopping the study can be tested on a continuous basis.

Beta-binomial models are used as follows:

$$\text{Prior: } p_i \sim \text{Beta}(a_i, b_i),$$

$$\text{Likelihood: } y_i \mid p_i \sim \text{Binomial}(m_i, p_i)$$

$$\text{Posterior: } p_i \mid y_i \sim \text{Beta}(a_i + y_i, b_i + m_i - y_i)$$

where
  i is the treatment arm (1, control; 2, treatment)
  p_i is the EOI rate in arm i
  a_i, b_i are the parameters of beta distribution for prior EOI rate in arm i,
    which can be understood as the prior belief of number of subjects with
    events (ai) among total number of subjects in arm i ($ai + bi$)
  y_i is the number of subjects who experienced EOI events in arm i
  m_i is the total number of subjects in arm i

We formulate the priors by using Phase 1b data for Drug X arm and literature
data with down-weighting for placebo arm. Thus, for the drug arm, Beta(2, 5)
is used based on the Phase 1b data. As for the placebo arm, literature shows
that 32 out of 241 (13%) subjects had the same EOI in a similar population.
We use Beta(5, 35) as the prior for the placebo arm by down-weighting the
literature data to approximately 1/6.

The decision rule is to consider to stop or take other actions if the posterior
probability Pr(risk difference > cutoff) > bound, where cutoff is the minimum risk difference that is clinically meaningful and bound is the minimum
unacceptable posterior probability of risk difference exceeding the cutoff.
For example, if cutoff = 10% and bound = 80%, we say that the combination
of prior information and the observed data indicates that the posterior probability that the EOI rate in the drug X arm is at least 10% higher than the
placebo arm exceeds the minimum unacceptable bound of 80%.

Simulation study can be conducted to guide selection of the decision rule
in achieving good operating characteristics in terms of power and type I
error. Table 14.2 shows the operating characteristics of this monitoring plan

TABLE 14.2

Operating Characteristics of the Monitoring Plan in Terms of Type I Error and Power

	Rules: Cutoff; Bound	Stopping Probability (Overall and By Look)							Average Sample Size
		Overall	1	2	3	4	5	...	
Type I error	0.1; 0.7	0.007	0.002	0.001	0	0.001	0.001	...	119
	0.1; 0.8	0.004	0.001	0.001	0	0	0.001	...	120
	0.1; 0.9	0.001	0.001	0	0	0	0	...	120
Power	0.1; 0.7	0.786	0.028	0.018	0.025	0.022	0.037	...	90
	0.1; 0.8	0.649	0.009	0.007	0.008	0.007	0.012	...	102
	0.1; 0.9	0.432	0.001	0.001	0.001	0.003	0.004	...	112

by assuming that the rates in placebo and drug X are 13% versus 33% for power and 13% versus 13% for type I error, respectively. For example, when evaluating type I error, assuming cutoff = 0.1 and bound = 0.7, the stopping probability is 0.007 and it leads to an average sample size of 119. On the other hand, when evaluating power, assuming the same cutoff and bound, the stopping probability is 0.786 and it results in an average sample size of 90.

In this example, we propose a novel Bayesian approach for the establishment of statistical guidelines in order to facilitate informed and objective decisions made by the DMC. The study is designed and powered with demonstrating efficacy as the primary objective so tight type I and II error control is not feasible for the safety endpoint of interest given the predetermined sample size. However, optimal operating characteristics can be explored before the study start in determining the safety monitoring rules (in terms of cutoff and bound) through simulations. Note that Bayesian posterior exceeding probabilities serve as safety monitoring guidelines, rather than *stopping rules,* because ultimate decision on whether to stop a trial is a more complicated decision-making process by taking risk–benefit tradeoff of the drug into consideration. The approach allows for incorporation of both internal (i.e., within a drug development program) and external information (i.e., outside a program) and offers flexibility and efficiency. In the context of safety monitoring, it is desirable to conservatively incorporate prior information from Phase 1b data. There are some challenges from both statistical and operational perspectives that need to be considered. For example, some consideration should be given to how much to down-weight the prior information for the control arm. Simulation can be used to study the impact of different down-weights of the priors and sensitivity analyses need to be conducted to determine the optimal design. From the operational standpoint, timely acquisition of clinical trial data without delays in entering/cleaning data is important to ensure successful monitoring according to the plan and timely decision making.

14.2.4 Application 4: Joint Modeling of Safety and Efficacy

Throughout a drug's development, it is important to evaluate the relationship between efficacy and safety to better understand the drug's benefit–risk balance. The use of Bayesian methods for joint modeling has been discussed recently in the literature. Gould et al. [22] provided an overview of joint modeling of longitudinal and survival outcomes. The authors mentioned that in the Bayesian paradigm, asymptotic approximations are not necessary, model assessment is more straightforward, computational implementation is typically much easier, and historical data can be incorporated easily into the inference procedure. The paper reviewed currently available methods for carrying out joint analyses and reviewed applications of the methodology. While the focus of the paper was on efficacy outcomes, concepts are readily applicable to efficacy and safety outcomes. The authors commented that the models could be used for safety endpoint modeling during early stages of development, such as time to occurrence of toxicity in phase I oncology trials where late-onset (LO) toxicities become a serious concern for the development of targeted therapies.

Others have discussed the benefits of joint modeling of survival and patient-reported outcomes (PROs), including Hatfield et al. [23] and Guo and Carlin [24]. While PROs tend to be considered as focused on efficacy, clearly if safety events happen, then PROs would also be impacted. In [22], the authors note that approval of new oncology therapies often depends on demonstrating prolonged survival. Particularly when these survival benefits are modest, consideration of therapeutic benefits to PROs may add value to the traditional biomedical clinical trial endpoints. The authors extended a popular class of joint models for longitudinal and survival data to accommodate the excessive zeros common in PROs by building hierarchical Bayesian models that combine information from longitudinal PRO measurements and survival outcomes. In the oncology setting, joint models provided more efficient estimates of the treatment effects on the time to event and the longitudinal marker and reduced bias in the estimates of the overall treatment effect [22]. In addition, while these examples tend to focus on a combination of safety and efficacy, joint modeling of a longitudinal and time-to-event outcome can be useful in the setting of two safety endpoints. For example, within a clinical trial (or across several clinical trials), it is of interest to understand how the safety events (AEs) relate to relevant longitudinal laboratory data, such as joint modeling of time to thrombotic events and the associated laboratory information for hemoglobin. This type of analysis could provide insights into whether hemoglobin level predicts the thrombotic event.

Another relevant and recent example of the use of Bayesian methods for joint modeling of a safety and efficacy outcome was from Zhao et al. [25], who discussed three different methods to model the joint distribution of HbA1c and hypoglycemic episodes. They demonstrated the value of the joint modeling and the Bayesian approach, particularly in this setting when efficacy and safety are correlated. The authors conducted simulations and applied the models to a dataset from a diabetes study.

In accordance with the paper, let $\{(X_1, Y_1), \ldots, (X_n, Y_n)\}$ denote the observations for a sample of n patients, where X indicates the efficacy outcome and Y denotes the safety outcome. The three models provided explore different mechanisms to model the correlation between X_i and Y_i, $i = 1, \ldots, n$. These are called the normal induced copula estimation (NICE) model, the conditional model, and the hierarchical model. The NICE model directly models the joint distribution (X_i, Y_i) by a copula, the conditional model establishes $Y_i \mid X_i$ and X_i, and the hierarchical model introduces the correlation between $E(X_i)$ and $E(Y_i)$.

We now describe the NICE model to demonstrate the methods. More information regarding the other two models is available in [25].

Suppose there are n patients randomized to two treatments in a randomized clinical trial, with n_1 patients randomized to treatment 1 and n_2 patients randomized to treatment 2. In [18], the authors assume that for patient i ($i = 1,...,n$)

$$X_i \sim \text{Normal}(\mu_{xi}, \sigma^2), \text{ and } \mu_{xi} = a_1 + b_1{}^*\text{trt}_i;$$

$$Y_i \sim \text{Negative Binomial}(\mu_{xi}, \text{size}), \text{ and } \log(\mu_{xi}) = a_2 + b_2{}^*\text{trt}_i,$$

where trt_i is the treatment assigned to patient i.

The authors transform X_i and Y_i into uniform random variables U_i and V_i as

$$U_i = F_X(X_i) \sim \text{Unif}(0,1); \ V_i = F_Y(Y_i) \sim \text{Unif}(0,1),$$

where F_X and F_Y are the cdf's of X_i and Y_i.

Next, U_i and V_i are transformed into standard normal random variables Z_i and W_i and then the joint distribution of Z_i and W_i introduces the correlation through a bivariate normal distribution.

The other two models were described in [25] and demonstrate additional ways that correlation can be introduced for X_i and Y_i.

The models were applied to a dataset from a randomized controlled diabetes study to compare safety and efficacy in patients with type 2 diabetes. There were a total of 1939 patients, and the efficacy measurement of interest was changed from baseline in HbA1c after a 6-month treatment period and the safety measurement was total nocturnal hypoglycemic events over the duration of 6 months. The three models were fit to the data and Bayesian predictive checks were implemented to evaluate goodness of fit. The hierarchical model tended to overestimate the proportion of patients with zero hypoglycemic events, whereas the conditional and NICE models seemed to adequately fit the data. The NICE model yielded a smaller DIC than the conditional model. Note that the correlation parameter estimates for both of these models was negative, which was consistent with the hypothesis that a patient with better efficacy will have more hypoglycemic events. The parameter b_1 (see the previous description of NICE model) represented the treatment effect of HbA1c. The posterior summaries indicated that the treatment arm has a statistically significantly stronger efficacy effect than the control. The AE, hypoglycemic effect, was not statistically significant.

14.2.5 Application 5: Dose-Finding Studies

Knowledge regarding whether an AE of interest exhibits a potential dose–response relationship is important in the overall assessment of a compound. In order to thoroughly evaluate whether an AE rate increases with increasing dose level, evidence from multiple clinical trials needs to be combined and analyzed, often with differing dose levels

Since each clinical trial is designed to answer specific questions (typically about efficacy) for a specific indication, separate trials often include different dose levels. As a consequence, there is likely no single trial that includes all studied doses. Furthermore, even if there are studies with common dose levels, restricting to only those clinical trials results excludes important and relevant information, which is particularly problematic as often there is limited statistical power for evaluating safety data. These issues create challenges for evaluating the AE dose–response (AEDR) relationship.

From an analysis standpoint, often only the trials with all common doses are included, which can exclude important information. One may also crudely pool patients at the same dose level across studies, but the drawback is that it breaks down the randomization. Bayesian methodology has been shown to be useful in the context of indirect and mixed treatment comparison methods, to combine information from different therapies in different studies in order to make treatment effect inferences, but instead of modeling different dose arms in different studies, we extend the methodology to allow for assessment of the dose–response relationship across multiple clinical trials.

In [26], three Bayesian indirect/mixed treatment comparison models were proposed to assess AEDR. These three models were designed to handle binary responses and time to event responses. To illustrate as an example, one of the models is a linear constrained model, provided as

$$\text{logit}(p_{ij}) = \alpha_i D_j + \varphi_i,$$

$$\alpha_i \sim N\left(\alpha, \sigma_\alpha^2\right),$$

$$\varphi_i \sim N\left(\mu_\varphi, \tau_\varphi^2\right), \quad i = 1, 2, \ldots, S,$$

where
  D_j is the dose for the jth treatment
  i is the ith study for $i = 1, \ldots, S$
  α is the parameter of primary interest

This model can be extended to the log linear model to account for study duration.

Note the interpretation of α in these models is very useful in decision making. Specifically, when the posterior probability that α is significantly > 0 is high, it indicates an AE that may indicate a dose–response relationship. Further evaluation of the potential dose–response relationship would then be warranted. On the other hand, if the posterior probability that α is significantly < 0 is high, it indicates that the particular AE decreases with decreasing dose levels. If the 95% credible interval covers 0, there is no indication of a potential dose-related AE rate. Thus, the Bayesian model allows direct evaluation of potential dose–response relationship via parameter α.

The model was demonstrated through a case study utilizing a safety database that contains 30 randomized clinical trials with one primary drug of interest [27]. Posterior probabilities were obtained to assess the potential dose–response relationship, including evaluation of whether or not any potential relationship is clinically meaningful.

The case example provided in the paper indicates that for the AE of interest, we have

$$\Pr(\alpha > 0 \mid \text{Data}) = 98.2\%$$

Therefore, there is a 98.2% probability that there is an increasing AE rate with increasing dose levels. Furthermore, within the Bayesian approach, it is straightforward to assess clinical relevance, as given large sample sizes one may detect a statistically meaningful but clinically irrelevant difference. In the case example, the authors highlighted that the difference between the event rates for the lower and higher dose levels is 4.6%. The information can then be used to evaluate clinical relevance.

14.2.6 Application 6: Network Meta-Analysis

The use of Bayesian methods to conduct network meta-analysis is rapidly growing. For example, recent publications include the *Evidence Synthesis Technical Support Documents* series found on the website of the Decision Support Unit (DSU) of the National Institute for Health and Care Excellence [27]. The statistical methods used in these documents were primarily Bayesian, and WinBUGS was used as the main software platform for data analysis. The documents can be downloaded from the site http://www.nicedsu.org.uk/Evidence-Synthesis-TSD-series%282391675%29.htm.

Recently, the DIA Bayesian Scientific Working Group published a paper that explores the use of Bayesian methods when applied to drug safety meta-analysis and network meta-analysis [18]. Guidance was presented on the conduct and reporting of such analyses and the work was illustrated through a case study. While this paper focused on a safety example, aspects discussed in the paper are readily applicable to efficacy outcomes. Specifically, the case example focused on analyzing available evidence on cardiovascular safety of nonsteroidal anti-inflammatory drugs. This particular example was also published with analysis results from Trelle et al. [28]. The conclusions of the analyses from [28] indicate that although uncertainty remains, little evidence exists to suggest that any of the investigated drugs are safe in cardiovascular terms. Naproxen seemed least harmful. In [18], the authors focused on comparison of a variety of network meta-analysis models to evaluate these data. The authors explored the impact of different model choices, the interplay between prior and parameterization, and extensions based on the analysis of multiple outcomes. As provided in the discussion, the analysis showed that the outcomes can be particularly sensitive to the choice of model and outcome weighting scheme, highlighting the importance of prespecifying literature search and analysis plans, and the need for model sensitivity analysis along with transparency regarding assumptions and potential limitations when reporting results. This is an important example of utilizing Bayesian network meta-analysis in the context of safety evaluation. In particular, prescribers further understand the importance of accounting for cardiovascular risk when prescribing any nonsteroidal anti-inflammatory drug.

In general, the use of Bayesian network meta-analysis has broad applicability to evaluate AEs between related drugs. These methods can provide insight to prescribers and also assess cost-effectiveness. The direct probability statements that result from the Bayesian approach are helpful to decision makers evaluating a variety of medical products for a given therapeutic area/indication. Furthermore, the information obtained from the network meta-analysis can be utilized throughout the medical product development life cycle in simulations to design future clinical trials.

14.3 Concluding Remarks

In this chapter, we have discussed many of the challenges associated with the assessment of safety data, both in premarketing and postmarketing settings. We have highlighted some of the advantages of the use of Bayesian methods in this context, in particular, the ability to incorporate prior information from a variety of sources, flexibility in evidence synthesis, dealing with multiplicity and rare events, handling complex problems, and obtaining direct probability statements. Furthermore, externally, we have seen that the use of Bayesian methods in the safety setting have been encouraged throughout drug development [29].

We have also provided several advanced examples of the use of Bayesian methods in safety signal assessment. These examples indicate that, while progress has been made, more work is needed to continue to expand their appropriate use. We recommend that whenever these methods are employed, then expertise should be on board to ensure appropriate implementation. In all of these cases, the use of Bayesian methods enhanced the evaluation of potential safety signals and enabled improved decision making.

References

1. Council for International Organizations of Medical Sciences (CIOMS) Working Group VI. Management of safety information from clinical trials, Geneva, Switzerland, 2005.
2. Xia HA and Jiang Q. 2014. Statistical evaluation of drug safety data. *Therapeutic Innovation & Regulatory Science* 48(1): 109–120.

3. Chi G, Hung HMJ, and O'Neill R. 2002. Some comments on "Adaptive Trials and Bayesian Statistics in Drug Development" by Don Berry. Pharmaceutical report 9, pp. 1–11.
4. DuMouchel W. 1999. Bayesian data mining in large frequency tables, with an application to the FDA spontaneous reporting system (Disc: p190–202). *The American Statistician* 53: 177–190.
5. Bate A, Lindquist M, Edwards IR et al. 1998. A Bayesian neural network method for adverse drug reaction signal generation. *European Journal of Clinical Pharmacology* 54: 315–321.
6. Schuemie MJ. 2011. Methods for drug safety signal detection in longitudinal observational databases: LGPS and LEOPARD. *Drug Safety* 20: 292–299.
7. Norén G, Hopstadius J, Bate A, Star K, and Edwards I. 2010. Temporal pattern discovery in longitudinal electronic patient records. *Data Mining and Knowledge Discover* 20(3): 361–387.
8. Gould A, Lyslig T, Lu Y, Fu H, Ma H, and Madigan D. Methods and issues to consider for detection of safety signals from spontaneous reporting databases: A report of the DIA Bayesian safety signal detection working group. *Therapeutic Innovation and Regulatory Science*, Published online May 8, 2014. DOI: 10.1177/2168479014533114. Available from: http://dij.sagepub.com/content/early/2014/05/07/2168479014533114. Accessed August 10, 2014.
9. Gould AL. 2008. Detecting potential safety issues in clinical trials by Bayesian screening. *Biometrical Journal* 50: 837–851.
10. DuMouchel W. 2011. Multivariate Bayesian logistic regression for analysis of clinical study safety issues. *Statistical Science* 27(3): 319–339.
11. Berry SM, Carlin BP, Lee JJ, and Muller P. 2010. *Bayesian Adaptive Methods for Clinical Trials*. Boca Raton, FL: Chapman & Hall/CRC Press.
12. Berry S and Berry D. 2004. Accounting for multiplicities in assessing drug safety: A three-level hierarchical mixture model. *Biometrics* 60(2): 418–426.
13. Xia HA, Ma H, and Carlin BP. 2011. Bayesian hierarchical modeling for detecting safety signals in clinical trials. *Journal of Biopharmaceutical Statistics* 21(5): 1006–1029.
14. Benjamini Y and Hochberg, Y. 1995. Controlling the false discovery rate: A practical and powerful approach to multiple testing. *Journal of the Royal Statistical Society, Series B* 57(1): 289–300.
15. Mehrotra DV and Heyse JF. 2004. Use of the false discovery rate for evaluating clinical safety data. *Statistical Methods in Medical Research* 13: 227–238.
16. Nissen SE and Wolski K. 2007. Effect of rosiglitazone on the risk of myocardial infarction and death from cardiovascular causes. *The New England Journal of Medicine* 356(24): 2457–2471.
17. Bridge JA, Iyengar S, Salary CB et al. 2007. Clinical response and risk for reported suicidal ideation and suicide attempts in pediatric antidepressant treatment: A meta-analysis of randomized controlled trials. *JAMA* 297(15): 1683–1696.
18. Ohlssen D, Price KL, Xia HA, Hong H, Kerman J, Fu H, Quartey G, Heilmann CR, Ma H, and Carlin BP. 2014. Guidance on the implementation and reporting of a drug safety Bayesian network meta-analysis. *Pharmaceutical Statistics* 13(1): 55–70.
19. Berry SM, Berry DA, Natarajan K, Hennekens CH, and Belder R. 2004. Bayesian survival analysis with nonproportional hazards: Meta-analysis of combination pravastatin-aspirin. *JASA* 99: 36–44.

20. Ibrahim JG, Chen MH, and Sinha D. 2001. *Bayesian Survival Analysis*. New York: Springer Verlag.
21. Askling J, Fahrbach K, Nordstrom B, Ross S, Schmid CH, and Symmons D. 2011. Cancer risk with tumor necrosis factor alpha (TNF) inhibitors: meta-analysis of randomized controlled trials of adalimumab, etanercept, and infliximab using patient level data. *Pharmacoepidemiology and Drug Safety* 20: 119–130.
22. Gould L, Boye M, Crowther M, Ibrahim J, Quartey G, Micallef S, and Bois F. 2014. Joint modeling of survival and longitudinal non-survival data: Current methods and issues. Report of the DIA Bayesian joint modeling working group. *Statistics in Medicine*, Published online; March 14, 2014 . DOI: 10.1002/sim.6141. Available from: http://onlinelibrary.wiley.com/doi/10.1002/sim.6141/pdf. Accessed August 10, 2014.
23. Hatfield L, Boye M, Hackshaw M, and Carlin B. 2012. Multilevel Bayesian models for survival times and longitudinal patient-reported outcomes with many zeros. *Journal of the American Statistical Association* 107: 875–885.
24. Guo X and Carlin B. 2004. Separate and Joint modeling of longitudinal and event time data using standard computer packages. *The American Statistician* 58: 16–24.
25. Zhao Y, Shen W, and Fu H. 2013. Joint modeling of clinical efficacy and safety with an application to diabetes studies. *Journal of Biopharmaceutical Statistics* 23: 1155–1171.
26. Fu H, Price K, Nilsson M, and Ruberg S. 2013. Identifying potential adverse events dose–response relationships via Bayesian indirect and mixed treatment comparison models. *Journal of Biopharmaceutical Statistics* 23: 26–42.
27. Dias S, Welton N, Sutton A, and Ades A. 2011. NICE DSU Technical support document 1: Introduction to evidence synthesis for decision making; last updated April 2012; http://www.nicedsu.org.uk/Evidence-Synthesis-TSD-series%282391675%29.htm. Accessed August 10, 2014.
28. Trelle S, Reichenback S, Wandel S, Hildebrand P, Tschannen B, Villiger P, Egger M, and Juni P. 2011. Cardiovascular safety of non-steroidal anti-inflammatory drugs: network meta-analysis. *BMJ (Clinical Research Edition)* 342: c7086.
29. Institute of Medicine of the National Academies. 2012. *Ethical and Scientific Issues in Studying the Safety of Approved Drugs*. Washington, DC: National Academies Press.

15

Risk-Benefit Assessment Approaches

Chunlei Ke, Qi Jiang, and Steven Snapinn

CONTENTS

15.1 Introduction

Every medicinal product is targeted to have some benefits for patients but inevitably comes with certain risks, also known as side effects. Benefit–risk assessment (BRA) refers to a judgment made based on available evidence regarding whether or not the benefits outweigh the risks for a particular product in the context of treatment options available to patients. Therefore, BRA is important and common for drug development and life-cycle management. Sponsors perform a BRA of a candidate drug to determine whether further clinical development is warranted and, if so, create the development plan to accumulate evidence. For regulators, BRA in the context of a new drug application is a central element of the scientific evaluation of a marketing authorization application and related variations (EMA 2007). Health-care providers and patients evaluate benefits versus risks for the available treatment options to decide on the best treatment course for individual patients.

The challenges with BRA have been well recognized (e.g., EMA 2007; Guo et al. 2010). It involves multifactoral sources of evidence and is informed by science, medicine, and policy (U.S. FDA 2013). Available clinical data are usually quite limited. Human judgment, which is unavoidably subjective, plays an essential role in BRA as data and models do not make a decision. In the past, BRA has been done informally, and the methodology used has been mostly descriptive, thereof lacking consistency or transparency. There is no defined and agreed methodology to combine benefits and risks to allow direct comparisons. Vast scattered pieces of evidence are available for evaluation, which requires a structured or systematic approach.

In light of these challenges, regulatory agencies and pharmaceutical industry have launched a series of projects or initiatives to review the current practice and to develop and streamline the BRA methodology and practice. European Medicines Agency (EMA) established a working group in 2006 with the goal of providing recommendations on ways to improve the methodology, consistency, transparency, and communication of the BRAs conducted by the Committee for Medicinal Products for Human Use (CHMP). This working group produced a reflection paper on BRA methods in 2007. The EMA began the Benefit–Risk Methodology Project in 2009, with the aim to explore methodologies that can increase the consistency and transparency of the BRA for medicinal products. This work continued through 2012 and provided recommendations with respect to benefit–risk methodologies (EMA 2010). In 2012, the European pharmacovigilance legislation was released. As part of this legislation, the periodic safety update report (PSUR) will include new subsections for benefit evaluation and BRA, also known as the Periodic Benefit–Risk Evaluation Report (PBRER) (EMA/CHMP/ICH/544553/1998). In addition, EMA coordinated a project of Pharmacoepidemiological Research on Outcomes of Therapeutics by a European Consortium (PROTECT) to strengthen monitoring of the benefit–risk of medicines in Europe. The PROTECT team reviewed and developed a set of tools and methods for BRA and presented some real-life case studies (IMI-PROTECT 2013).

The U.S. Food and Drug Administration (FDA) internally piloted a benefit–risk framework with the intention to provide a standard structure. The Institute of Medicine (IOM) also recommended that FDA create a publicly available BRA management plan. In 2013, FDA released its benefit–risk framework and implementation plan and committed to a series of meetings and workshops during 2013–2018 to develop a benefit–risk framework (U.S. FDA 2013). The Pharmaceutical Research and Manufacturers of America (PhRMA) established a working team for BRA in 2005, developing their BRAT framework and providing several case studies using the framework. PhRMA later transferred the benefit–risk work to the Centre for Innovation in Regulatory Science (CIRS) who further refined and developed an unformed framework. In 2013, a benefit–risk working group has been formed mainly among U.S. statisticians under the sponsorship of Quantitative Sciences in the Pharmaceutical Industry (QSPI), a committee affiliated with the Society for Clinical Trials.

In this chapter, we will provide an overview of the key methodologies for the BRA, focusing on the benefit–risk frameworks and several useful quantitative methods. In Section 15.2, an overview of the BRA approaches is provided. A simulated example is introduced in Section 15.3 and will be used throughout this chapter for illustration. In Section 15.4, benefit–risk frameworks are described. Section 15.5 discusses several selected quantitative benefit–risk approaches. The chapter concludes with discussions of some challenging issues with the BRA.

15.2 Overview of BRA Approaches

A benefit is any known favorable effect for the target population caused by the product. Benefits are clinically meaningful improvements to a patient that result in an improvement in health state or quality of life (QoL). A risk or harm is any known or unknown detrimental effect that can be attributed to the product.

BRA approaches refer to any qualitative or quantitative methods or processes that can be used to integrate or balance, explicitly or implicitly, information on benefits and risks to assist or facilitate a comprehensive BRA. Many approaches have been proposed or adapted in the literature for BRA for different considerations with various scopes. For example, Guo et al. (2010) reviewed 18 methods, and the IMI-PROTECT working group reviewed and evaluated 47 methods (IMI-PROTECT 2013). Several efforts were attempted to categorize and appraise various methods. For example, IMI-PROTECT (2013) categorized the 47 methods into BRA framework, metric indices, estimation equation, and utility survey technique and provided recommendations in each category of the methods. We consider that general BRA approaches fall into two categories: process or frameworks, on one hand, and quantitative methods, on the other. In this chapter, we will provide an overview of selected benefit–risk approaches and provide references for more detailed and comprehensive review (e.g., Guo et al. 2010; IMI-PROTECT 2013).

Because significant judgment is necessary in the BRA, there are limitations of and concerns with using quantitative methods. It is agreed that the BRA relies on expert judgment, but quantitative methods are helpful to quantify components in the BRA as an important part of the process to facilitate decision making.

15.3 Hypothetical Case Study

A simulated hypothetical example, motivated by the experiences from a phase III clinical trial of an experimental oncology drug, is used throughout this chapter to illustrate the methods discussed. The simulated dataset includes two efficacy endpoints, progression-free survival (PFS) and over-all survival (OS) time, and incidences of two adverse events (AEs) of inter-est (AE1 and AE2) on 500 patients for each of two treatment arms, placebo or treatment. The simulated data assumes a treatment effect on the PFS but no effect on OS. The treatment is assumed to increase the incidence of both AEs: for AE1, the rate is higher with active treatment than with placebo, and for AE2, the rate is zero with placebo and greater than zero with active treatment.

Figure 15.1 shows the Kaplan–Meier (KM) estimate of the survival prob-ability for the endpoints of PFS and OS. Median PFS time was estimated at 23.9 months and 30.0 months for placebo and treatment, respectively, and HR = 0.82 for treatment relative to placebo with 95% CI (0.69, 0.97) and p-value = 0.02, indicating an 18% reduction of the risk for disease progression or death. OS time was similar between two arms (HR = 0.99 [0.79, 1.24], p = 0.93). The rates of AE1 were 10% with placebo and 14.2% with active treatment, and the rates of AE2 were 0% with placebo and 6.4% with active treatment. The risk for AE2 increased with increasing exposure to treatment. We also assume that AE2 is a more serious event than AE1. It is of interest to conduct the BRA of the treatment based on these results.

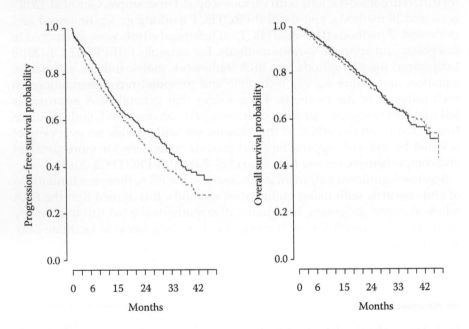

FIGURE 15.1
(See color insert.) KM estimate of the survival probability for progression or death and overall death (red solid lines, treatment; blue dotted lines, placebo).

15.4 BRA Frameworks

A framework is a set of principles, guidelines, and tools to guide the decision makers in selecting, organizing, understanding, and summarizing evidences relevant to decision making. Due to the complex nature of the BRA, it is necessary to have a systematic framework to structure and guide the process of BRA. Several frameworks are described in this section.

Firstly, the PhRMA established a working team to address the need for improved BRA in 2005. The PhRMA working group developed the Benefit–Risk Action Team (BRAT) framework (Coplan et al. 2010), a general platform that seeks to incorporate all relevant aspects of the BRA. The BRAT framework consists of six steps as summarized in Table 15.1. The framework focuses on describing the BRA process that structures and assists decision making, but does not replace clinical judgment.

The first step, defining the decision context, involves specifying the therapeutic context, specifying the comparators, defining the time horizon (e.g., the duration of exposure and the follow-up period), and specifying the stakeholder perspective. The next step is to identify benefits and risks outcomes or criteria. Usually, multiple benefits or risks are relevant and the value tree

TABLE 15.1

Six Steps in the BRAT Framework

Step 1: Define the decision context.	Define drug, dose, formulation, indication, population, comparator(s), time horizon for outcomes, and perspective of the decision makers.
Step 2: Identify and select benefit and risk outcomes.	Select all important outcomes. Create the value tree. Define preliminary set of measures for each outcome. Document rationale for outcomes to be included or excluded.
Step 3: Identify and extract source data	Determine and document all data sources (e.g., clinical trials, observational studies). Populate data source with relevant data.
Step 4: Customize the framework.	Update the value tree based on further review of the data and clinical expertise. Refine the outcomes and measures.
Step 5: Assess outcome importance.	Apply or assess ranking or weighting of outcome importance to decision makers or other stakeholders.
Step 6: Display and interpret key benefit–risk measures.	Summarize data in tabular or graphical displays to aid interpretation. Review summary measures and source data, and identify and fill information gaps. Interpret summary information. Conduct sensitivity analysis to assess impact of uncertainty in data sources on displays or summary measures.

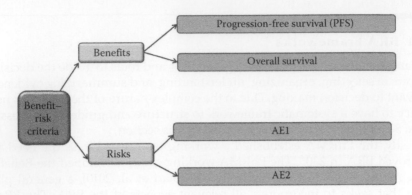

FIGURE 15.2
(See color insert.) Example of a value tree for the case study.

is an effective graphic tool to visually display and summarize the criteria identified. Figure 15.2 shows a value tree of the hypothetical case study. See Coplan et al. (2010) for additional examples.

At Step 3, data sources for the BRA are identified. Step 4 customizes the framework based on the quality and limitation of the data. BRA requires a comparison of benefits and risks among treatment options, for which Step 5 assesses relative importance or weighting of all benefits and risks. Finally, the last step of the framework displays and interprets the key BR metrics. The BRAT framework advocates graphical presentation of the data and metrics by use of such tools as forest plots for the case study shown in Figure 15.3.

EMA benefit–risk methodology project recommends the framework of PrOACT-URL, a generic framework that provides a generic problem structure to be considered when facing a decision problem (Hammond et al. 2002). The acronym PrOACT-URL represents the eight steps of this framework: problems, objectives, alternatives, consequences, trade-offs, uncertainty, risk attitudes, and linked decisions. The first five elements (PrOACT) are very general and pertinent to any decision problem. The last three elements (URL) are more specific and address the uncertainties of the decisions. See IMI-PROTECT (2013) for details on this framework.

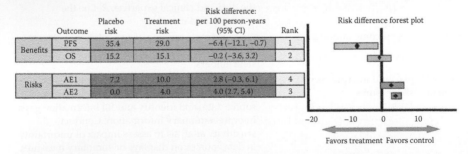

FIGURE 15.3
(See color insert.) Example of benefit–risk summary table for the case study.

TABLE 15.2

FDA Benefit–Risk Framework

Decision Factor	Evidence and Uncertainties	Conclusions and Reasons
Analysis of condition Unmet medical need Benefit Risk Risk management	Identify and present facts, uncertainties, and assumptions that need to be made to address what is not known for each factor.	Conclusion that must be made about each decision factor. Capture implications of facts, uncertainties, and assumptions with respect to regulatory decision making, draw conclusions from the evidence and uncertainties, and explain the basis for these conclusions.
	Benefit–risk summary and assessment	

To address the concern regarding improvement in the clarity and transparency of FDA's BRA in drug review, FDA started an initiative in 2009 to develop a structured approach for BRA. FDA considered that the BRA needs to be qualitative and developed the basic structure of its qualitative benefit–risk framework (U.S. FDA 2013). The FDA framework (Table 15.2) includes the following categorization of key decision factors: analysis of condition, current treatment options, benefit, risk, and risk management. Within each of these categories, there are two considerations that inform the regulatory decision: identifying facts as well as uncertainties and any assumptions that need to be made to deal with what is not known and conclusions that must be made about each decision factor. The final component of the framework is the benefit–risk summary assessment, a succinct well-reasoned summary that clearly explains FDA's rationale for the regulatory action including important clinical judgments that contributed to the decision.

In addition, the UMBRA initiative by CIRS incorporates many benefit–risk activities and proposed a unified benefit–risk framework with the following eight steps embedded within a core of decision-making elements (CIRS).

In summary, all these frameworks proposed are similar in that they depict the steps and aspects of the BRA process. Given the complexity of the BRA process and large amount of evidence from various sources, a benefit–risk framework should be used to structure and organize the evaluation, to provide clarity and transparency, and to help to identify potential gaps for the assessment. At the early stage of drug development, the framework also informs study design and evidence accumulation.

15.5 Quantitative Approaches

Quantitative benefit–risk approaches refer to methods that can combine or synthesize benefit and risk outcomes to allow for an overall assessment or balancing of benefits and risks. While the BRA still relies on expert judgment, quantitative approaches can provide valuable and instrumental information to aid the benefit–risk decision, in particular for cases where the benefit–risk decision is not obvious or involves large number of benefits and risks with various sources of evidence. In this section, several useful quantitative methods are discussed.

15.5.1 Number Needed to Treat

The number needed to treat (NNT) refers to the number of patients who need to be treated with a given therapy to prevent one additional event compared with a control therapy or no treatment (Laupacis et al. 1988). The NNT based on crude incidences is calculated as

$$NNT = \frac{1}{p_1 - p_2}$$

where p_1 and p_2 are the proportion of subjects with an event in the control and treatment group, respectively. NNT is a useful measure of clinical benefit, intuitive, interpretable, and very popular in practice (Holden et al. 2003).

Several extensions have been made to incorporate the risks into NNT to allow for an overall BRA. AE-adjusted NNT (AE-NNT) estimates the number of patients needed to treat to prevent one additional event without inducing treatment-related AEs (unqualified success) compared with the control (Schulzer et al. 1996):

$$AE\text{-}NNT = \frac{1}{(p_1 - p_2) \times [1 - (q_2 - q_1)]}$$

where
  q_1 and q_2 are the proportion of subjects with the AE of interest in the control and treatment group, respectively
  AE-NNT is useful for a situation where there is only one AE of interest and the benefit event and AE are independent

Alternatively, for an AE of interest, number needed to harm (NNH) can be calculated similarly as

$$NNH = \frac{1}{q_2 - q_1}$$

NNH is interpreted as the number of patients who need to be treated to cause one more AE compared with a control therapy or no treatment. The overall BRA can be made through a comparison of NNH and NNT, for example, calculating the ratio $r = NNH/NNT$. The NNH/NNT ratio refers to the number of benefit events prevented for each AE caused. Favorable balancing of the benefit and risk may be warranted if the ratio is large enough. The weakness of this method is that the importance of the AE in relation to the benefit is not incorporated, which may imply that the benefit and risk are of the same severity. A threshold L can be set for the ratio to claim that benefit outweighs the risk, that is, $r > L$, where the value of L depends on the relative clinical importance of the beneficial and harmful events.

One can also include the relative importance or relative weight of the benefit and risk into the calculation of NNH and NNT as an adjusted ratio. For example, the relative importance is represented by the utility value (V) (Guyatt et al. 1999), and the adjusted ratio is defined as

$$r = \frac{1/[(q_2 - q_1) \times v_R]}{1/[(p_1 - p_2) \times v_B]}$$

A favorable benefit–risk profile is concluded when $r > 1$.

In addition, the adjusted NNH/NNT ratio can be extended to multiple benefits and risks:

$$r = \frac{1/\left[\sum_{j=1}^{J} (q_{j2} - q_{j1}) \times v_{jR}\right]}{1/\left[\sum_{k=1}^{K} (p_{k1} - p_{k2}) \times v_{kB}\right]}$$

However, the intuitive interpretation of the ratio for single benefit and risk is lost when using this approach. We also notice that the ratio $r > 1$ is equivalent to

$$\sum_{j=1}^{J} [(p_{j1} - p_{j2}) \times v_{jB}] + \sum_{k=1}^{K} [(q_{k1} - q_{k2}) \times v_{kR}]$$

$$= \left[\sum_{j=1}^{J} p_{j1} \times v_{jB} + \sum_{k=1}^{K} q_{k1} \times v_{kR}\right] - \left[\sum_{j=1}^{J} p_{j2} \times v_{jB} + \sum_{k=1}^{K} q_{k2} \times v_{kR}\right] > 0 \quad (15.1)$$

The left-hand side (15.1) is of a special form of the benefit–risk score in the MCDA method to be discussed next. MCDA is more flexible to handle multiple benefits and risks, and the NNH/NNT ratio is commonly used for the situation where a single benefit and a single risk are considered for BRA.

The concept of NNT has been extended to the time-to-event endpoint or to consider the benefit–risk profile over time. One can select a time point and calculate NNT based on the estimate (e.g., KM product limit estimate) of the cumulative incidence rate at the selected time point (Altman and Andersen 1999). However, the NNT depends upon the time point selected, and sometimes it is not sufficient to evaluate benefit and risk based on a fixed time point. Alternatively, one can calculate NNT using the inverse of the difference of the exposure-adjusted incidence rates in the two treatment groups in place of the difference in the crude incidence rates (Lubsen et al. 2000; Mayne et al. 2006). If the hazard rate deviates greatly from being constant over time or from an additive hazard model for both treatment arms, the value of NNT would still depend upon the follow-up time. It is prudent to check the hazards when using the exposure-adjusted NNT. NNT has also been extended to recurrent event (Halpin 2005; Aaron and Fergusson 2008).

In order to address the sampling variation, a 95% CI for NNT and NNH can be calculated based on inverse of the 95% CI for the absolute risk difference. However, if the CI for the absolute risk difference contains zero, the 95% CI is seen as ill-defined and the interpretation is challenging (Altman 1998). In fact, the 95% CI is not commonly reported in practice. The inference of NNH/NNT ratio can be based on the delta method or by use of resampling-based methods.

In the example in Section 15.3, because the OS was balanced between the two groups and AE1 is of less importance than AE2, whether or not benefits outweigh risks will be mainly determined by PFS and AE2. The crude NNT = 15.2 for PFS and NNH = 15.6 for AE2, which gives a ratio of about 1. Therefore, the treatment would not be favored unless a disease progression or death is considered of greater importance than AE2. As the endpoint is also concerned with the event time, we calculated the subject-year-adjusted NNT = 15.6 and NNH = 23.8 years, leading to a ratio of 1.5. The larger subject-year adjusted NNH was due to the fact that the risk for AE2 was increased overtime and the events occurred late. Because the risk for AE2 is not the same across time, it would be informative to evaluate the benefit–risk across the time as well. Figure 15.4 shows the NNH/NNT ratio for a series of time

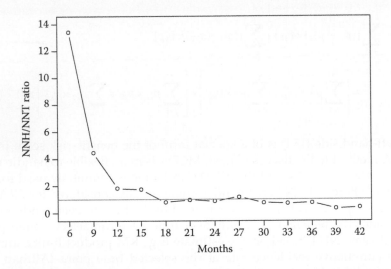

FIGURE 15.4
(See color insert.) Ratio of NNH and NNT calculated based on the KM estimate over time (reference line of NNH/NNT ratio = 1 in red).

points calculated based on the KM estimate, and it is clear that a more favorable BR balancing was observed for the first 15 months than with continued exposure to the treatment.

15.5.2 Multicriteria Decision Analysis

Conducting a BRA requires applying evaluative judgments on the benefit and risk outcomes. A pure mathematical method is not appropriate, but the use of a formal decision-making process involving decision analysis is suitable for and has been applied to the BRA. As the BRA needs to account for multiple benefit and risk criteria, multicriteria decision analysis (MCDA), using a decision-tree model, is appropriate and has been adapted for the BRA to allow for systematic decision making in complex situations.

According to Mussen et al. (2007), there are three key phases in the MCDA process: problem identification and structuring, model building and use, and development of action plan. These phases are specified through the seven-step approach (Mussen et al. 2007). Steps 1–5 of MCDA are similar to those in the BRA frameworks discussed in Section 15.4. Therefore, MCDA is sometimes called a quantitative benefit–risk framework (IMI-PROTECT 2013). However, the MCDA process further stipulates the calculation of the weighted scores at each level and calculation of the overall weighted BR scores to facilitate a benefit–risk decision at step 6. MCDA also emphasizes sensitivity analysis to assess the impact of weights and other assumptions in the data at step 7.

Denote $C_i, i = 1, \ldots, m$ as estimates of the identified benefit and risk outcomes. Usually, C_i's are of different scales or data types, for example, a proportion for a binary outcome and a change from baseline for a continuous parameter. To be able to apply the MCDA approach, the C_i's need to be transformed into the same scale to be combined together, for example, between 0 and 1, through some monotone value functions $v_i(x)$. A simple example of a value function is a linear interpolation when the minimum and maximum levels of a criterion can be identified and their values are set as 0 and 1, respectively. The minimum and maximum levels could be specified based on expert judgment or based on the percentile of the distribution of the criterion.

Let $V_i = v_i(C_i)$. Denote W_i as the corresponding weight for V_i with $\Sigma W_i = 1$. Weight selection is among the most challenging components in BRA and will be discussed further in Section 15.7. Then the overall weighted benefit–risk score is calculated for each treatment arm as

$$S = \sum_{i=1}^{M} W_i V_i \qquad (15.2)$$

The comparison among the treatment options may be undertaken based on the score S. The difference or ratio of the benefit–risk scores between treatment and control can be evaluated to measure the magnitude of the benefit–risk balancing:

$$\Delta S = S_1 - S_2 \quad \text{or} \quad r_S = \frac{S_1}{S_2}$$

The difference of the benefit–risk scores can be rewritten as

$$\Delta S = \sum_{i=1}^{M} W_i^1 V_i^1 - \sum_{i=1}^{M} W_i^2 V_i^2 = \sum_{i=1}^{M} W_i^1 (V_i^1 - V_i^2)$$

and $W_i^1(V_i^1 - V_i^2)$ represents the contribution from the ith criterion to the overall benefit–risk balancing.

The original MCDA approach is focused on the point estimate of the BR score. Conditional on the value functions and weights selected, an interval estimate can be constructed for S, ΔS, and r_S to account for the sampling variation of the data. The correlation matrix needs to be estimated or a resampling-based method can be employed to construct the interval estimate.

Linear combination in (15.2) is a natural way of combining multiple components, and it may be generalized to some nonlinear functional of V_i and W_i:

$$S = \varphi(V_1, \ldots, V_M; W_1, \ldots, W_M)$$

where $\varphi(.)$ is a multivariate monotone increasing function. For example, $\varphi = \prod_{i=1}^{M} V_i^{W_i}$ when V_i is a relative measure. In the case where the benefits and risks are time-to-event measurements, one could use the subject-year-adjusted rate or estimate of the cumulative incidence rate at a prespecified time point for C_i. Using a series of time points allows for investigating the benefit–risk profile over time. In general, C_i can be the outcome of any statistical analysis, including adjusted analysis for covariates. It should be noted that the BR score (15.2) may be evaluated at the patient level and statistical inference and analysis can be based on the patient-level score data.

It is also worthwhile to note a few other benefit–risk approaches similar to MCDA. The approach of a clinical utility index (CUI) has been used in the early stage of drug development (Ouellet 2010). CUI helps to make the trade-off decision between greater magnitude of response and higher risk of adverse effects and thereby define the therapeutic window over the entire dose/concentration range. The index is a weighted linear combination of utilities of efficacy and risk attributes, similar to the weighted BR score of MCDA except that V_i is defined as the utility of each benefit and risk criterion in CUI.

In addition, the benefit–less-risk (BLR) approach (Chuang-Stein 1994) is concerned with BRA based on one defined efficacy endpoint and more uncertain drug-induced AEs. In it, the safety data are organized into classes and the risk is quantified by considering the frequency and severity of the side effects in the classes. Then a risk-adjusted benefit measure is constructed as the difference between benefit and risks using a proportionality constant that corresponds to weighting and is determined based on the relative importance of the side effects. Therefore, the BLR approach is a simple MCDA analysis.

MCDA recognizes important assumptions made in the approach and thus emphasizes sensitivity analyses to demonstrate the impact any assumption has on the final weighted BR score and the BR conclusion. The sensitivity analysis should take into account, among other things, endpoint selection, value functions, and weights.

In the hypothetical case study, we take C_i as the KM estimate of event-free survival probability at month 12 as an example. Based on some clinical judgment, the weight vector was chosen as (0.48, 0.29, 0.04, 0.19) with the highest weight for PFS and lowest weight for AE1. As all endpoints are on the same scale, no further transformation was applied (i.e., $v_i(x) = x$).

Table 15.3 shows the details of the calculations. Slightly favorable benefit–risk balancing to the treatment was observed and the PFS endpoint contributed the most to the benefit–risk score. A much higher weight was given AE2 than AE1, but AE2 contributed to the overall BR score almost the same as AE1 due to its lower incidence. Figure 15.5 shows the sensitivity analysis by varying the ratio of the total weight for benefits to risks while maintaining the relative weight within benefits and within risks. When the total weight for benefits is less than or about the same as that for the risks, the overall benefit–risk score was seen negative. The more weight given to benefits, the more favorable the benefit–risk score is for the treatment. It should be noted that, in this example, OS was balanced between placebo and treatment and so did not directly contribute to the BRA. However, it contributed indirectly by impacting the weight distribution.

The stepwise MCDA approach is easy to understand, allowing more explicit and transparent BRA to enhance consistency and objectivity. It is also flexible and can be fine-tuned by adding or changing criteria. MCDA can handle missing data and uncertainty using appropriate modeling and weights. However, the scoring and weighting still requires judgments beyond the available data. It may be an oversimplification to place any

TABLE 15.3

MCDA Analysis of the Case Study at Month 12

	Placebo	Treatment	Value	Weight	Benefit–Risk Score
PFS	0.705	0.741	0.036	0.48	0.017
OS	0.877	0.887	0.009	0.29	0.003
AE1	0.927	0.863	−0.064	0.04	−0.003
AE2	1.000	0.980	−0.020	0.19	−0.004
Overall (95%CI)	—				0.014 (−0.021,0.049)

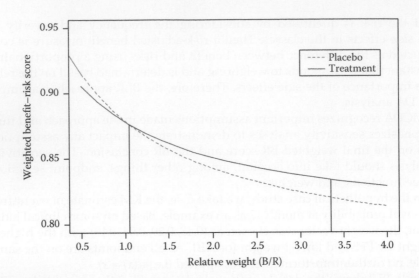

FIGURE 15.5
(See color insert.) Weighted benefit–risk scores versus ratio of total weights for benefits and risks.

emphasis on such an overall BR score for regulators and industry. However, MCDA assessments should be used as a tool by the decision makers and cannot substitute for actual decision making.

15.5.3 Stochastic Multicriteria Acceptability Analysis

The MCDA approach mainly relies on the estimate of the overall BR score in BR assessment, which does not account for the uncertainties associated with sampling variation in the data for the benefit and risk endpoints. In addition, MCDA requires an explicit supply of weights for each criterion. Such weights are not easy to derive and are likely to differ among different decision makers. Stochastic multicriteria acceptability analysis (SMAA) was extended for BRA to deal with these limitations in MCDA (Tervonen et al. 2011).

Consider a set of K alternatives under evaluation. Denote the vector of endpoint values as $V = (V_1, ..., V_M)$, which could be the original endpoint measurement transformed via the value functions. Assume that V is a random vector with density function $f_V(v)$ in the evaluation space $X \times R^n$. Let W be a vector of weights for the criteria. Instead of soliciting fixed values for W, assume that W is a vector of random variables with a joint distribution $f_w(w)$ in the feasible weight space:

$$\Omega = \left\{ w \in R^n : w \geq 0, \sum_{j=1}^{n} w_j = 1 \right\} \quad (15.3)$$

When there is no preference made or available for the criteria, W is a uniformly distributed on Ω. Instead, when there is some information available on W, the prior information can be incorporated in the model by properly constraining the weight space (15.3).

Given the value functions, the weighted BR score $S = S\,(V,\,W)$ is a random variable. Based on the observed values for V, $\hat{S} = S(\hat{V}, W)$. The benefit–risk comparison between treatment options will be made based on $\hat{S}$. When there are no data available on W, several statistical indices were proposed to facilitate such comparison (Tervonen et al. 2011): rank acceptability index, center weight vectors, and confidence factors.

Rank acceptability index, denoted by b_i^r, refers to the probability of treatment alternative i being ranked at place r among the K options as

$$b_i^r = \int\limits_{\xi \in X^M} f_{X^M}(\xi) \int\limits_{w \in \Omega_i^r(\xi)} f_w(w)dw\,d\xi$$

where $\xi = (V_1,\,...,\,V_M)$, $f_X^M(\xi)$ usually is the product of $f_v(v)$, and Ω_i^r refers to the subset of Ω, where alternative i is rank at place r given ξ. For example, b_i^1 is the probability of alternative i ranked the first. The preferred alternatives would be those with high index for the best rank.

Weight is implicitly used in the calculation of the rank acceptability index. As weighting is the key information for the decision makers to judge the balance of benefits and risks, the central weight vector in SMAA is used to show what kind of typical weights would lead to favoring one alternative over the others. The central weight vector for alternative i, w_i^c, is the expected center of gravity of Ω_i^1 defined as

$$w_i^c = \frac{\int_{\xi \in X^m} f_{X^m}(\xi) \int_{w \in \Omega_i^r(\xi)} w f_w(w)dw\,d\xi}{b_i^1}$$

The central weight vector is useful to complement the rank acceptability index.

Finally, the confidence factor, p_i^c, is defined as the probability that alternative i is ranked first when the weight vector is the same as w_i^c. It represents how the data (criteria values) can discern each treatment option. Treatment options with low first-rank acceptability indexes and small confidence factors are unlikely to be considered as the preferred alternative.

In practice, if there exists information on the weights, either having an expected weight or having a preference order of benefits and risks, the rank acceptability index can be used to compare the alternatives. Otherwise, if no preference information is available, the central weight vectors and confidence factors can be used to explore comparisons of the alternatives to assist the BRA.

In the case study, we considered the weight vector with a uniform distribution, that is, no prior preference information available, and with a Dirichlet distribution with parameter vector $w_0 = (0.48, 0.29, 0.04, 0.19)$ so that the average weights are the same as their relative importance. Evaluation was based

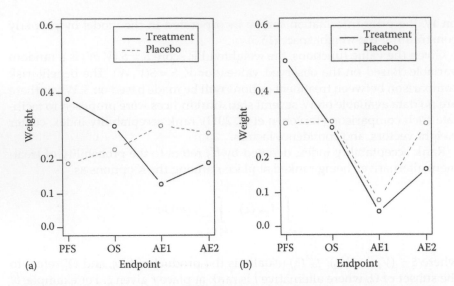

FIGURE 15.6
(See color insert.) Central weight vectors for uniform (a) and Dirichlet (b) distributions.

on 5000 samples of the data and weight vector draws from the distributions. The treatment rank acceptability indices were $b_T^1 = 0.31$ for uniform distribution and 0.66 for the Dirichlet distribution. The central weight vectors are shown in Figure 15.6, indicating that large weight needs to be given to PFS to prefer the active treatment.

SMAA provides a general approach to addressing the sampling variability and uncertainty of the weights with interpretable indices to summarize the outcome. However, understanding and interpreting the results from SMAA involve mathematical knowledge of stochastic phenomena and uncertainty. The computation is relatively complicated. Consideration of the weight as a random variable and averaging over the weight space may not be always meaningful for a decision maker. Still, decisions have to be made for some important components such as the choice of the alternatives and value function.

In addition, Zeng et al. (2013) also extended the MCDA approach to incorporating the variations in the data and weighting. They introduced a counterfactual p-value to account for the sampling variations and proposed a graphical display to demonstrate that specific weight selections would lead to a favorable balancing. Several statistics were also proposed to make quantitative comparisons.

15.5.4 Stated Preference Methods

A benefit–risk decision is essentially a judgment of the evidence assembled for benefits and risks. Data are observed for the treatment, but there are generally no data in the clinical trial for the judgment or preference. The stated preference methods rely on survey techniques to elicit the stakeholders' (e.g., patients and health-care providers) preferences on the value or choice for the alternatives and their attributes. A direct benefit–risk judgment can be formed. In addition, the survey data for such values or choices for the preference may provide insight into the relative importance of the benefit and risk criteria so as to inform other quantitative BRA methods.

Conjoint analysis (CA) is one of the most powerful techniques available for qualifying patient or physician preference and satisfaction (Ryan and Farrar 2000; Ryan et al. 2008). It provides a structured framework to elicit preferences that consists of the following steps:

1. *Defining product attributes*: Attributes and levels of each attribute include the most important health outcomes and product attributes associated with drug administration in a particular disease state. Once attributes and levels are determined, they can be combined into treatment profiles to be evaluated in CA. The experimental design includes profiles for additional hypothetical treatments to help identify relative preferences for each attribute level.

2. *Choice of CA formats*: Different CA formats are available. Discrete choice provides subjects with several different products or programs simultaneously and simply asks them to identify the most-preferred opinion in each choice set. DCE is based on random utility theory (Thurstone 1927; McFadden and Train 2000).

3. *Experimental design*: Statistical experimental design is needed to conduct the survey. Full-factorial design generates data based on all possible combination of attribute levels and may not be practical. An approximately orthogonal design can deduce the number of paired comparisons to the smallest number necessary for efficient estimation of utility weight (Johnson et al. 2010).

4. *Preference estimate*: Multiple regression techniques on the survey data are used to estimate the relative importance of each attribute level. CA allows estimation of the relative importance of each attribute such that the utility differences and values across all profiles can be evaluated. Several BR metrics can be derived such as minimum acceptable benefit for a given level of risk and maximum acceptable risk for a given level of benefit, net effective margin, and net safety margin, to be used in BRA.

CA can be directly used as part of BRA incorporating preference from stakeholders in addition to the data generated from the clinical studies. The results can also be used in assisting quantitative BRA, for example, providing ranking or weighting information for MCDA. See Ryan et al. (1997, 2001) for some examples using CA.

15.5.5 Health-Outcome Approaches

Health outcomes and modeling are often used in evaluating cost-effectiveness of a drug in comparison with alternative drugs. These health-outcome analyses consider combining benefits and risks and are usually conducted through the use of some health outcome indices such as quality-adjusted life years (QALYs), disability-adjusted life years (DALYs), and quality-adjusted time without symptoms and toxicity (Q-TWiST). Thus, these approaches can provide a viable method for BRA (Garrison et al. 2007). Among the indices, QALY is the most popular one, recommended by US Panel on Cost-effectiveness in Health and Medicine and NICE (Weinstein et al. 2009) and will be the focus here.

Providing a standardized approach to promote comparability in cost-effectiveness analyses of different interventions, QALY is a measure of the impact of a drug on the quality and length (quantity) of life of a patient. In general, QALY is defined as

$$QALY = \sum_{i=1}^{S} Q_i t_i$$

where
  S is the total number of health states
  Q_i represents the value or QoL associated with health state i
  t_i is the time or proportion of time spent in health state i
  Q is the utility value of a health state, ranging from 0 to 1 where 0 represents the worst possible condition or death and 1 the best possible condition or full health

When t_i is the proportion of time, the corresponding QALY is normalized to denote the quality-adjusted lifetime per life years.

To calculate QALY, we need to estimate the life years of the patient population based on clinical trial data and the health states and their durations during the life years. Health states are usually related to the study endpoints, for example, disease progression or deaths. Multistate analysis can be used to make these estimates. Once the health states are determined, the value or QoL (Q) associated with each state can be established using expert opinion, QoL data collected in the clinical studies, or direct or indirect research. The most frequently used method in practice is direct or indirect research by way of preference survey of health providers and patients. Time trade-off, standard gamble, and rating scale are among the ways to assess the preference of specific health state. As an alternative, indirect research uses questionnaires for health state in several health domains or attributes (e.g., EQ5D) and then to construct a multiattribute utility as a summary measure that reflects preferences both within and across health domains.

The QALY can be calculated separately related to benefits, QALY gain, and risks, QALY loss. The QALY gain and loss may be combined to derive incremental net health benefit:

$$INHB = QALY_g - QALY_l$$

The BRA may be made with a comparison of INHB between the intervention and alternatives.

In addition, there are several extensions of QALY, for example, DALYs, health-adjusted life expectancy (HALE) (Weinstein et al. 2009), and the Q-TWiST (Gelber et al. 1995) in cancer epidemiology.

In the case study, we considered health states of death and disease progression for QALY gain and only AE2 for QALY loss. The transition probabilities were estimated based on exponential distributions from randomization to progression, progression to death, and randomization to death without progression. Similarly, for QALY loss, we estimated the transition probability based on a Weibull distribution from randomization to AE2. The value was assumed to be 0.52 for progression and 0.81 for AE2 compared with 1 for state without progression and 0 for death. Based on 5000 simulations, the QALY gains were estimated at 0.17 and 1.11 months for the first 12 and 36 months, respectively, and the corresponding QALY losses were 0.01 and 0.33 months. Therefore, we have INHB = 0.16 and 0.78 months for the first 12 and 36 months, respectively.

15.6 Challenges and Conclusions

There are several inherent challenging issues with BRA as discussed in the following.

Weight selection: BRA entails integration or contrast of evidence for benefits and risks, for which the relative importance or weighting of benefits and risks is key. Weighting is implemented in any BRA either implicitly or explicitly. As weighing involves subjective judgment from the individual's knowledge, experiences, or behaviors, it will necessarily differ from person to person. Therefore, there does not exist unique weighting, and, whenever talking about weighting, one needs to clarify whose weighting. Weighting can be informal or formal. Informal weighting provides directional ranking and usually is implicitly used in qualitative assessment. Formal weighting assigns numerical values to each benefit and risk criterion, which typically is used in certain quantitative BR approaches (e.g., MCDA). As no data are collected in clinical trials on the weights, soliciting weights from the targeted stakeholders or patients through surveys is a useful and robust approach.

Uncertainties: There are many sources of uncertainties with any BRA, for example, strength of evidence for benefits or risks, weighting, and translation to the postmarketing setting. It is necessary to identify and document all major uncertainties. Consideration should be given to evaluate the impact of the uncertainties on the BR conclusion. Sensitivity analyses are needed to quantify the impact. Some actions may be taken to remedy uncertainties such as risk management plan (RMP) or risk evaluation and mitigation strategy (REMS).

Prespecification: Currently, BRA is primarily done after the fact in an ad hoc fashion, which may increase the chance of false-positive findings. What can or should be prespecified remains a challenge and is under discussion.

Subgroup analysis: The benefit–risk profile is usually not the same for all patients, especially for targeted therapies. Sometimes, the benefit–risk profile for the overall population may not be favorable. It is necessary to evaluate the benefit–risk profile for important subgroups. However, subgroup analysis is notoriously controversial, and it is important to plan and conduct robust subgroup analysis.

BRA using postmarketing observations: It is critical to continue monitoring the benefit–risk profile of the drugs postmarketing. The clinical trial experiences may be restricted in terms of patient population selected and concomitant care received. There may not be sufficient exposure or the length of follow-up to allow all important adverse reactions to be observed in clinical trials, particularly for those with rare incidences. In light of these issues, the PSUR will include new subsections for benefit evaluation and BRA. It is also known as the PBRER (EMA/CHMP/ICH/544553/1998). However, such assessment is challenging. The postmarketing data are mainly based on voluntary reporting and related to the drug safety. How to summarize and interpret the postmarketing data, which is typically underreported, inconsistent, and lacking of control data, is not an easy task. Together with potential missing data on benefits or efficacy, more work is needed to develop appropriate methodologies for the postmarketing BRA.

In conclusion, the BRA is an important and complex task, relevant across the life cycle of a medicinal product. Structured and systematic BR approaches can improve clarity and ensure consistency. While the final BR conclusion mainly relies on qualitative judgment, quantification and integration of the benefits and risks can provide useful support for the BR decision. Several quantitative methods have been discussed in this chapter, and which method to use depends upon the specific situation at hand. Further development and understanding in quantitative BRA are needed to address the challenges mentioned previously.

References

Aaron SD, Fergusson DA. Exaggeration of treatment benefits using the "event-based" number needed to treat. *CMAJ* September 23, 2008; 179(7): 669–671.

Altman DG. Confidence intervals for the number needed to treat. *BMJ* 1998; 317: 1309–1312.

Altman DG, Andersen PK. Calculating the number needed to treat for trials where the outcome is time to an event. *BMJ* 1999; 319: 1492–1495.

Centre for Innovation in Regulatory Science (CIRS). Unified Methodologies for Benefit–Risk Assessment (UMBRA) initiative. 2012. http://cirsci.org/UMBRA. Accessed August 5, 2014.

Chuang-Stein C. A new proposal for benefit–less-risk analysis in clinical trials. *Controlled Clinical Trial* 1994; 15: 30–43.

Coplan PM, Noel RA, Levitan BS, Ferguson J, Mussen F. Development of a framework for enhancing the transparency, reproducibility and communication of the benefit–risk balance of medicines. *Clinical Pharmacology and Therapeutics* 2010; 89(2): 312–315.

European Medicines Agency (EMA). Report of the CHMP working group on benefit–risk assessment models and methods, 2007.

European Medicines Agency. Benefit–risk methodology project work package 2 report: applicability of current tools and processes for regulatory benefit–risk assessment, 2010.

Garrison LP, Towse A, Bresnahan BW. Assessing a structured, quantitative health outcomes approach to drug risk-benefit analysis. *Health Affairs* 2007; 26(3): 684–695.

Gelber RD, Cole BF, Gelber S, Aron G. Comparing treatments using quality-adjusted survival: The Q-Twist method. *The American Statistician* May 1, 1995; 49(2): 161–169.

Guo JJ, Pandey S, Doyle J, Bian B, Lis Y, Raisch DW. A review of quantitative risk-benefit methodologies for assessing drug safety and efficacy—Report of the ISPOR risk-benefit management working group. *Value Health* August 2010; 13(5): 657–666.

Guyatt GH, Sinclair J, Cook DJ, Glasziou P. Users' guides to the medical literature: XVI. How to use a treatment recommendation. Evidence-Based Medicine Working Group and the Cochrane Applicability Methods Working Group. *JAMA* May 19, 1999; 281(19): 1836–1843.

Halpin DMG. Evaluating the effectiveness of combination therapy to prevent COPD exacerbations, the value of NNY analysis. *International Journal of Clinical Practice* 2005; 59: 1187–1194.

Hammond JS, Keeney RL, Raiffa H. *Smart Choices: A Practical Guide to Making Better Life Decisions.* New York: Broadway Books, 2002.

Holden WL, Juhaeri J, Dai W. Benefit–risk analysis: A proposal using quantitative methods. *Pharmacoepidemiology and Drug Safety* October 2003; 12(7): 611–616.

IMI-PROTECT. Benefit–risk group recommendations report: Recommendations for the methodology and visualization techniques to be used in the assessment of benefit and risk of medicines, 2013. Accessed May 8, 2014. Available at http://www.imi-protect.eu/benefitsRep.shtml.

Johnson FR, Lancsar E, Marshall D, Kilambi V, Mühlbacher A, Regier DA, et al. Constructing experimental designs for discrete-choice experiments: Report of the ISPOR conjoint analysis experimental design good research practices task force. *Value Health* 2013 January; 16(1): 3–13.

Laupacis A, Sackett DL, Roberts RS. An assessment of clinically useful measures of the consequences of treatment. *New England Journal of Medicine* June 30, 1988; 318(26): 1728–1733.

Lubsen J, Hoes A, Grobbee D. Implications of trial results: The potentially misleading notations of number needed to treat and average duration life gained. *Lancet* 2000; 356: 1757–1759.

Mayne TJ, Whalen E, Vu A. Annualized was found better than absolute risk reduction in the calculation of number needed to treat in chronic conditions. *Journal of Clinical Epidemiology* 2006; 59: 217–223.

McFadden D and Train K. Mixed MNL models for discrete response, *Journal of Applied Econometrics* 2000; 15(5): 447–470.

Mussen F, Salek S, Walker S. A quantitative approach to benefit–risk assessment of medicines—Part 1: The development of a new model using multi-criteria decision analysis. *Pharmacoepidemiology and Drug Safety* July 2007; 16 (Suppl. 1): S2–S15.

Ouellet D. Benefit–risk assessment: the use of clinical utility index. *Expert Opinion on Drug Safety* March 2010; 9(2): 289–300.

Ryan M, Bate A, Eastmond CJ, Ludbrook A. Use of discrete choice experiments to elicit preference. *Quality in Health Care* September 2001; 10: I55–I60.

Ryan M, Farrar S. Eliciting preference for healthcare using conjoint analysis. *BMJ* 2000; 320: 1530–1533.

Ryan M, Gerard K, Amaya-Amaya M. *Using Discrete Choice Experiments to Value Health and Health Care*. Dordrecht, the Netherlands: Springer, 2008.

Ryan M, Hughes J. Using Conjoint analysis to assess women's preference for miscarriage management. *Health Economics* 1997; 6(3): 261–273.

Schulzer M, Mancini GB. 'Unqualified success' and 'unmitigated failure': Number-needed-to-treat-related concepts for assessing treatment efficacy in the presence of treatment-induced adverse events. *International Journal of Epidemiology* August 1996; 25(4): 704–712.

Tervonen T, van Valkenhoef G, Buskens E, Hillege HL, Postmus D. A stochastic multi-criteria model for evidence-based decision making in drug benefit–risk analysis. *Statistics in Medicine* 2011; 30(12): 1419–1428.

Thurstone LL. A law of comparative judgment. *Psychological Review* 1927; 34: 273–286.

U.S. FDA. Enhancing benefit–risk assessment in regulatory decision-making. Draft PDUFA V implementation plan: Structured approach to benefit–risk assessment in drug regulatory decision-making, February 2013. http://www.fda.gov/downloads/ForIndustry/UserFees/PrescriptionDrugUserFee/UCM329758.pdf. Accessed August 5, 2014.

Weinstein MC, Torrance G, McGuire A. QALYs: The basics. *Value Health* March 2009; 12 Suppl. 1: S5–S9.

Zeng D, Chen M, Ibrahim JG, Wei G, Ding B, Ke C, Jiang Q. A counterfactual p-value approach for benefit–risk assessments in clinical trials. To appear in *Journal of Biopharmaceutical Statistics* 2013.

16

Detecting Safety Signals in Subgroups

Christy Chuang-Stein, Yoichi Ii, Norisuke Kawai,
Osamu Komiyama, and Kazuhiko Kuribayashi

CONTENTS

16.1 Introduction

On December 19, 2012, the U.S. Food and Drug Administration (FDA) informed health-care professionals and the public that Pradaxa (dabigatran etexilate mesylate, an anticoagulant) should not be used to prevent stroke or blood clots (major thromboembolic events) in patients with mechanical heart valves, also known as mechanical prosthetic heart valves. Results from a recently stopped trial in Europe found that Pradaxa users were more likely to experience strokes, heart attacks, and blood clots forming on the mechanical heart valves than users of another anticoagulant (warfarin). The observations led to a contraindication of Pradaxa in patients with mechanical heart valves (http://www.fda.gov/Drugs/DrugSafety/ucm332912.htm?source=govdelivery).

Direct-to-consumer advertisements of pharmaceutical products are allowed in the United States. It is not uncommon to hear a TV commercial about a medicinal product stating that the product is not to be used by certain patients. For example, advertisements for Cymbalta will typically include a statement that "patients using a MAOI (monoamine oxidase inhibitor) medication or patients with uncontrolled narrow-angle glaucoma should not take Cymbalta." Statements such as this reinforce the message that the risk of a product may outweigh its potential benefit for some patients.

Recognizing that not all users are equally affected by a pharmaceutical product, the U.S. FDA has required, for many years now, that sponsors assess the safety of their products in subgroups (Food and Drug Administration, 2005). Examples include subgroups defined by gender, race/ethnicity, age, geographic locations, baseline disease status, and certain concomitant drug usage. The U.S. government has passed legislation, requiring trials sponsored by the National Institutes of Health to be "designed and carried out in such a manner sufficient to provide for valid analysis of whether the variables being studied in the trial affect women or members of minority groups differently than other subjects in the trial" (Congressional Record-House, 1993). It has become a common practice to discuss the safety experience of patients in predefined subgroups in reports from clinical trials, often supplemented with tabular summary. This extends to publications from these trials although journal publications often include textual discussions without companion tables because of space limitation.

When a serious adverse reaction is observed with a product, substantial efforts and resources will be devoted to see if a subgroup uniquely predisposed to the adverse reaction could be identified. If such efforts are successful and strategies could be implemented to effectively manage the risk in the subgroup, the product could remain as a viable one. If the efforts are unsuccessful and the benefit/risk assessment is judged to be unfavorable for the target population, the product will be discontinued as a commercial candidate or will be withdrawn from the market place if it has already been marketed.

Much has been written about subgroup analysis for efficacy assessment (e.g., Yusuf et al., 1991; Assmann et al., 2000; Cui et al., 2002; Li et al., 2007; Wittes, 2009; Lipkovich et al., 2011). Many of the analytical issues related to efficacy subgroup analysis are applicable to subgroup analysis for safety. In this chapter, we will follow the definition adopted by Yusuf et al. (1991) who defined subgroup as a group of patients characterized by a common set of parameters prior to treatment initiation. In clinical trials, we refer to those characteristics as characteristics at baseline. Also, we focus on safety experience as reflected by adverse events, laboratory measurements, x-ray, ECG, etc. As a general rule, we do not include *lack of efficacy* as an adverse event in this chapter.

We structure this chapter as follows. In Section 16.2, we describe factors that could contribute to subgroup differences. In Section 16.3, we describe strategies commonly used to manage subgroups at a higher risk for a product. Section 16.4 describes statistical considerations in subgroup analysis. Section 16.5 discusses the role of meta-analysis in detecting safety signals in subgroups. Section 16.6 describes common methods to identify subgroups at a higher risk and to confirm the increase in risk. We conclude the chapter with some additional remarks in Section 16.7.

16.2 Factors That Could Contribute to Subgroup Differences

Common factors that have shown to contribute to subgroup differences in safety include demographics, comorbidity, concomitant medications, disease stage, culture, ethnic factor, genetic factor, and metabolism factor. We will give some examples on these factors in this section.

For some products, pediatric patients could be at a higher risk for serious reactions. Take the long-acting beta-agonists (LABAs) for asthma as an example. LABAs were approved as single-ingredient products or as an ingredient in combination products. This class of products was reviewed in a joint meeting among three FDA advisory committees (pediatric, pulmonary-allergy drugs, drug safety, and risk management) on December 11, 2008. A meta-analysis conducted and presented by the FDA showed an age–risk relationship with risk defined by a composite end point of asthma-related death, intubation, or hospitalization (Kramer, 2009; also see Section 16.5). Another highly visible debate involved the use of antidepressants. The debate took place several years ago. The center of the debate was whether children, adolescents, and even young adults (age 18–24) treated with antidepressant agents were more likely to complete suicide than older adults.

Elderly patients may also be at a higher risk for adverse reactions because of their generally decreased organ functions and frequent use of multiple medications. The *warnings and precautions* of Crestor states that the risks for skeletal muscle effects increase with the use of 40 mg dose, advanced age (≥65), hypothyroidism, renal impairment, and combination use with cyclosporine, lopinavir/ritonavir, or atazanavir/ritonavir. As the population ages and more people are taking multiple drugs, interactions between drugs have become an increasing concern. On January 17, 2013, FDA approved the revisions to the Kaletra (lopinavir/ritonavir, for HIV) label to provide additional information on interactions between Kaletra and several classes of drugs (http://www.accessdata.fda.gov/drugsatfda_docs/label/2013/021226s037lbl.pdf).

Ethnic could be an important factor for certain risks. Recent research has brought attention to a higher prevalence of drug-induced pneumonia in Japan, following the use of gefitinib and leflunomide (Azuma and Kudoh, 2007). According to the authors, drug-induced lung disease is more likely to occur when there is preexisting interstitial pneumonia. Since there is a higher prevalence of interstitial pneumonia in Japan, it is not surprising that comparisons between patients in different geographic regions suggest that drug-induced lung disease is more common in Japan than in many other countries.

Another noteworthy adverse drug reaction (ADR) with a higher incidence in Asian patients on the anticonvulsant drugs carbamazepine and phenytoin is the Stevens–Johnson syndrome (SJS) (Locharernkul et al., 2008). SJS is a serious and potentially life-threatening skin condition. There is a conjecture that this higher risk in SJS is associated with a higher presence of the HLA-B*1502 allele in the Asian population. Phillips et al. (2011) argued that carriers of this allele could also avoid other structurally similar anticonvulsants.

The association between the hypersensitivity reaction (HSR) and the presence of the major histocompatibility complex HLA-B*5701 in chromosome 6 in patients taking abacavir was conjectured and later confirmed in the trial PREDICT-1 (Hughes et al., 2008; see also Section 16.6). This association was added to the *warning* section of abacavir's label when the label was revised in 2009. The revised label states that patients who carry the HLA-B*5701 allele are at high risk for experiencing HSR to abacavir.

For a drug that acts through an active metabolite, patients' ability to metabolize the prodrug to its active metabolite has long been recognized as a contributing factor to the effect of the drug. For example, clopidogrel's label states that the effectiveness of clopidogrel is diminished in patients who are CYP2C19 poor metabolizers. Poor metabolizers with acute coronary syndrome or undergoing percutaneous coronary intervention treated with clopidogrel at recommended doses exhibit higher cardiovascular event rates than do patients with normal CYP2C19 function.

16.3 Strategies to Manage Subgroups at a Higher Risk for a Product

There are various strategies that could be used to manage the increased risk of a subgroup. They range from contraindicating the product for the subgroup to more closely monitoring patients in the identified subgroup.

For example, Pradaxa is contraindicated in patients with mechanical heart valves. With concerns over a suspected increase in the suicidality risk associated with antidepressant use in pediatric, adolescent, and young adult patients, close monitoring during the initial antidepressant treatment of these patients is recommended. This recommendation led the FDA to request a label change for all antidepressants on May 2, 2007 (http://www.fda.gov/NewsEvents/Newsroom/PressAnnouncements/2007/ucm108905.htm).

Occasionally, a reduced dose is recommended for patients considered to be at a higher risk for an ADR. For example, the starting Crestor dose for patients with severe renal impairment (but not on hemodialysis) is 5 mg and the dose should not exceed 10 mg. In contrast, 20 mg is an acceptable dose for patients with a normal renal function.

Following the joint advisory committee meeting on LABAs, the FDA issued LABA use guidelines (see Section 16.5) and requested LABA manufacturers to include additional statements in the product labels. For pediatric and adolescent patients who require LABA in addition to inhaled corticosteroid, the use guidelines state that these patients should use a combination product containing both an inhaled corticosteroid and a LABA to ensure adherence with both medications.

When a genetic factor is found to be associated with a serious adverse reaction to a product, patients could go through a genetic screening before taking the product. For example, screening for the HLA-B*5701 allele is recommended prior to initiating therapy with abacavir. Patients who are found to possess the HLA-B*5701 allele should not take abacavir. This approach was found to decrease the risk of clinically diagnosed HSR by more than 50% (see Section 16.6). Screening is also recommended prior to reinitiation of abacavir in patients of unknown HLA-B*5701 status who have previously tolerated abacavir.

16.4 Statistical Considerations in Subgroup Analysis

16.4.1 Overview

Statistical considerations surrounding subgroup analysis of clinical trial data have been well studied and widely documented (Assmann et al., 2000; Cui et al., 2002; Fleming, 1995; Wang et al., 2007; and references cited therein). Much of the considerations mentioned in these chapters are *issues* that arise from misalignment between the intended objectives of the subgroup analysis and the realistic limitations imposed by the study design and other specifications defined prior to the study. Therefore, it is important to clarify the objectives of the subgroup analyses cognizant of the various limitations, as the first step, followed by proper planning, analyses, and interpretation of the subgroup results. For example, Cui et al. (2002) classify common objectives of subgroup analysis into four categories: (1) to address particular concerns on efficacy and safety in some specific subgroups, (2) to provide supportive evidence for the main findings on an overall subject population, (3) to check the consistency or robustness of the main findings, and (4) to generate hypotheses for future studies. One must recognize that some of these objectives may not be tenable, given the limitations at hand. In most clinical trials, the objective of subgroup analysis in safety evaluation is exploratory in nature, thus fitting the third (check consistency) and the fourth categories (generate hypothesis) in the aforementioned list. On rare occasions, there may be a prespecified set of subgroups for which a predefined set of safety outcomes might be of interest, fitting the description of the 1st category.

The statistical issues in subgroup analysis, as raised in the aforementioned chapters, can be classified into three major categories: (1) multiplicity of statistical inference, (2) bias due to treatment group incomparability, and (3) precision of estimation for the subgroup (Cui et al., 2002). These issues apply to all subgroup analyses, whether the outcome of interest is efficacy or safety. When the outcome of interest is safety, however, there are additional points that need to be contemplated. The first is the relative importance of failure to identify safety signals, or *false negatives*. In contrast to efficacy end points where controlling *false positives* is important, in safety evaluation, false negatives that can lead to missing important safety signals are of primary concerns. The second is the inherent multiplicity in safety evaluation, even without regard to the subgroups. For example, numerous types of adverse events are typically examined in a clinical trial, compounding the multiplicity issue due to a potentially large number of subgroups of interest. Furthermore, extensive exploratory analyses are often carried out when prespecified analyses suggest some potential safety signals. Toward this end, a more balanced approach to false positive and false negative is required for safety evaluation. Prior medical and clinical knowledge and information from related drug class play a more prominent role in interpreting the subgroup results.

In this section, we focus on subgroup analysis for a single clinical trial, leaving that for multiple trials to Section 16.5 on meta-analysis.

16.4.2 False Positive and False Negative

16.4.2.1 False Positives and False Negatives due to Multiplicity

In subgroup analysis, it may become necessary to examine multiple, often numerous, subgroups based on baseline characteristics. This is in addition to a potentially large number of types of adverse events (e.g., MedDRA Preferred Term [PT] level, Standardized MedDRA Queries) that need to be evaluated for safety assessment.

TABLE 16.1

Probability of At Least One False Positive among K Independent Comparisons

	Number of Comparisons (K)										
α	1	2	3	4	5	10	20	30	40	50	100	1000
0.025	0.025	0.049	0.073	0.096	0.119	0.224	0.397	0.532	0.637	0.718	0.920	> 0.999
0.005	0.005	0.010	0.015	0.020	0.025	0.049	0.095	0.140	0.182	0.222	0.394	0.993

If there are K nonoverlapping subgroups, then the probability of at least one false positive in the evaluation of K subgroups is given by $1 - \prod_{i=1}^{K} (1 - \alpha_i)$, where α_i is the probability of false positive for the ith subgroup. For example, if a hypothesis test with significance level α is used to detect a safety signal, then under the null hypothesis of no signal in any subgroup, the probability of falsely indentifying at least one safety signal among the K subgroups is given by $1 - (1 - \alpha)^K$. This is shown for two values of α and a selected set of values of K in Table 16.1.

Suppose there are 20 nonoverlapping subgroups. A simple application of the aforementioned formula with $\alpha = 0.025$ for $K = 20$ gives a false-positive probability of approximately 40%. If we consider 50 adverse event (AE) terms that, for simplicity, are assumed to be nearly independent, then there would be 1000 comparisons resulting from possible combinations of 20 subgroup and 50 AE terms. Similar application of the formula would lead to a false-positive probability of nearly one. A more strict criterion of $\alpha = 0.005$ does not improve the situation by much.

False-negative rates can also be evaluated. For convenience, consider a continuous outcome where a two-sample t-test is applied to test the difference between two treatment arms. The null hypothesis states of no treatment difference and we will work with one-sided alternative hypothesis (positive difference indicates positive signal). Furthermore, consider a situation in which there is no treatment difference for all except for one subgroup (label this subgroup 1) among the K subgroups of equal size. Also, suppose Sidak's method (1967) for multiple comparisons is employed with an overall significance level α. The false-negative rate, then, is given by the following expression:

$$Pr(\text{false negative}) = Pr\left(T_{\gamma, 2Nn-2} \leq t_{2Nn_1 - 2, \alpha^*} \right)$$

In this expression, $T_{\gamma, df}$ is a random variable with noncentral t-distribution with noncentrality parameter γ and degrees of freedom df; $\alpha^* = 1 - (1 - \alpha)^{1/K}$; t_{df, α^*} is the upper $100\alpha^*$ percentile of central t-distribution with degrees of freedom df; and $\gamma = \sqrt{Nn_1/2} \, \delta/\sigma$, δ is the difference in mean between the treatment and control group in subgroup 1, σ is the common variance, N is the

TABLE 16.2

Probability of False Negative (2-Sample *t*-Test, One Sided)

	Adjustment	Number of Subgroups (K)									
α	Method	1	2	3	4	5	6	7	8	9	10
0.025	Sidak	0.49	0.80	0.89	0.93	0.96	0.97	0.97	0.98	0.98	0.99
	None	0.49	0.71	0.79	0.83	0.86	0.88	0.89	0.90	0.90	0.91
0.05	Sidak	0.36	0.71	0.84	0.89	0.93	0.94	0.96	0.96	0.97	0.98
	None	0.36	0.59	0.69	0.74	0.78	0.80	0.82	0.83	0.84	0.85

sample size for each treatment arm (assumed equal), and $r_1 = 1/K$. The false-negative rates as a function of K are shown in Table 16.2 in the case of $N = 200$, $K = 1$–10, $\delta = 1$, $\sigma = 5$, and $\alpha = 0.025, 0.05$. It can be seen from the rows with no multiple comparison adjustment that false-negative rate increases simply due to subdivisions into increasingly smaller subgroups as K increases. The false-negative rate is further increased by Sidak adjustment for multiple comparisons: nearly 80% with $K = 2$ and nearly 90% with $K = 3$. Analogous trend is seen for $\alpha = 0.05$.

The probability of finding at least one false negative when all subgroups are assumed to have a common positive signal can also be derived, following the development in Li et al. (2007). It can be shown that this probability also increases with increasing K.

These results highlight how the increase in the number of subgroups, and the corresponding decrease in sample size for each subgroup, leads to higher false-negative rates. This is an important observation, considering that false negative implies failure to detect safety signals. It should be noted that small sample size does not affect the false-positive rate.

What can be done to appropriately address the false-positive and false-negative rates? There is no established approach in this situation, but possible approaches are suggested in a report published by a PhRMA safety planning evaluation and reporting team (SPERT) (Crowe et al., 2009). The report describes a *3-tier approach* to address the multiplicity issue arising from numerous adverse event types. Under the 3-tier approach, AEs are classified into 1 of 3 tiers. Tier-1 events are events of clinical importance. They are often targeted medical events related to the mechanism of the product that are potentially serious or class effect of special interest. The list of tier-1 events should be kept to a manageable few and finalized prior to database unblinding. Tier-2 events are events that are not tier-1 but are *common*. SPERT suggests to use the *Rule of 4* to define tier-2 events. *Rule of 4* states that if a trial has 400 or fewer patients per group and there are at least 4 subjects with a given MedDRA PT in any treatment group, then that PT could qualify as a tier-2 event. Events that are neither tier-1 nor tier-2 are categorized as tier-3 events.

If an AE fits the description of *tier 1*, then one needs to make appropriate control of multiplicity only for the number of subgroups. Bonferroni or other alpha-spending approach may provide appropriate balance in such a case.

If system organ level frequencies are deemed relevant, it can be applied to this level. For studies whose primary objective is safety, such issues should be especially considered with care (see Section 16.4.4).

For AEs that fit the description of *tier 2*, it is easy to imagine that after dividing into subgroups, they may no longer be *tier 2* in their frequency. Thus, one may need to raise the cutoff from 4 *reports in at least one treatment group* (as in the Rule of 4) to a higher number to account for the subgroups. Even then, using statistical test to evaluate treatment group difference is not likely to be informative, since, on one hand, there might be too few subjects for each subgroup and on the other hand, increasing multiplicity can become intractable. Point and interval estimation for each subgroup together with a *descriptive use* of *p*-values from test of interaction might be a viable option. For meta-analysis of multiple studies, a closer adaptation of approaches described in SPERT report may be possible.

16.4.2.2 False Positive and Negative due to Bias

False positives and false negatives can arise not only from statistical multiplicity but also from bias, namely, the bias arising from treatment group incomparability. In the overall population, the balance in any important covariates (known or unknown) affecting safety outcomes is generally maintained on average by the randomization. If an important prognostic factor is known in advance, one may consider stratified randomization that would further increase the probability of balance. For some subgroups, however, such balance may not be achieved, especially for smaller subgroups. Cui et al. (2002) study degree of imbalance as a function of subgroup size, number of prognostic factors affecting the outcome, and the prevalence rate of subtypes for a prognostic factor. Unlike the case of statistical multiplicity in which the relationship between false-positive or false-negative rate and the number of subgroups is more predictable, the relationship is unpredictable in its direction in the case of bias. Therefore, false-positive or false-negative findings due to such imbalance are of particular concern (Cui et al., 2002).

For safety, especially for the AEs, it is likely that important factors are not measured or, more often, are unknown in advance for the AEs of interest. Unless information surrounding an AE is known in advance and is measured, it is difficult to make headways in terms of adjustments using models. Meta-analysis of multiple studies may alleviate the treatment imbalance within a subgroup by averaging the effect over multiple studies.

16.4.2.3 Estimation for a Subgroup

Due to the smaller sample size, an estimate of the effect for an individual subgroup is subject to larger variability than that for the overall population. A well-known approach to reducing the mean-squared error (MSE) in this type of situation is the empirical Bayes method (Davis and Leffingwell, 1990;

Lipsky et al., 2010). The empirical Bayes estimator for a given subgroup would be a weighted average between the individual subgroup estimate and the overall population estimate. The weight, sometimes called the *shrinkage* factor, depends on the relative size of the variance of the individual subgroup estimate and the variability of the subgroup estimates around the overall population estimate. The MSE-reducing property of empirical Bayes method is known to hold under very general conditions (Greenland, 2000). Alternatively, one may take a fully Bayesian approach. Hierarchical model that is relevant for modeling AEs is mentioned in the SPERT report (Crowe et al., 2009) and in others (Xia et al., 2011; Chuang-Stein and Xia, 2013).

The empirical Bayes estimator has a desirable property of reducing the MSE, but at the cost of introducing possible bias for some subgroups. For example, if there is a relatively small subgroup with a true positive signal among a majority of subgroups with no signal, then an estimate for such a subgroup would be shrunken toward the overall value of no signal. As such, it will be informative to show both the original (raw) estimate and the empirical Bayes estimate.

The parameter (estimand) of interest is usually a measure of divergence between treatment groups, such as difference in means (proportions) or odds ratios. To more directly evaluate the subgroup difference in these divergence measures, it might be informative to provide also the estimate of interaction effect (i.e., interaction between treatment and subgroups). This would mean, for example, displaying treatment group difference for each subgroup as well as the estimates of interaction (difference between the reference subgroup and each of the other subgroups). The latter will be covered in more detail in the next section.

16.4.3 Role of Interaction Test in Safety Analysis

A test of interaction between treatment and subgroup is sometimes recommended as a means of controlling false-positive findings (Cui et al., 2002). We examine possible roles of interaction test in safety analysis in this section.

It should be noted that the presence and the size of interaction depend on the choice of measure of divergence between the treatment groups. In the case of proportions, one may consider, for example, difference in proportions or ratio of the two proportions. An interaction in one measure may not be so in another. Interaction can be either qualitative or quantitative. In the case of difference in proportions, for example, a positive difference in one subgroup and a negative difference in another would be a qualitative interaction, whereas positive differences of differing magnitude between the subgroups would be a quantitative interaction. In addition to general tests of interactions, it is noted that tests of qualitative interaction are also available (Gail and Simon, 1985; Li and Mehrotra, 2008). Power of an interaction test is usually low, as a study is not normally powered to detect such interactions. Increasing the significance level of an interaction test to, say, 0.1 from the usual 0.05 would increase the power, but at the cost of a higher rate of false-positive findings.

In the case of safety end points, the aforementioned issues are further exacerbated. For example, unless the study is designed to look only for substantial treatment by subgroup interactions, the power is likely to be insufficient. Also, the number of subgroups and the number of end points being examined may become large. This can lead to increase in false-positive rates, which must be weighed against the decision to increase significance level in order to increase the power.

One approach that seems to strike a balance was proposed by Jackson et al. (2006). They display the point and interval estimates for the treatment group difference for each subgroup of the 13 subgrouping factors considered. Juxtaposed to the result for each subgroup is the p-value of the interaction test. This allows the interaction test p-value to serve as a flagging device to call attention to a selected set of subgroups, while not suppressing the individual subgroup estimates. Also, Jackson et al. report expected number of significant interactions under the null hypothesis of no interaction to provide a reference point that might be helpful during the interpretation of results.

16.4.4 Prespecification and Replication

In a study whose primary end point is safety, if there is a safety concern for specific subgroups, stratified randomization and preplanned subgroup analyses should be considered. Analysis results from such prespecified subgroup analyses are generally regarded as more credible.

In most situations, however, efficacy is the primary end points and most of the study design features are optimized for the primary objective. Even for this type of study, prespecification of safety end points and subgroups of interest would give more credibility to the analysis results than those based completely on post hoc assessments. For example, by specifying a limited number of safety end points and subgroups and by prioritizing the analyses, one can plan for a balanced approach to multiplicity that meets the objective. A common practice of enrolling a certain proportion of selected subgroup of subjects such as those defined by age range (e.g., age > 65 years), severity of target disease (e.g., NHYA class III), and organ functions (e.g., CrCl < 60 mL/min) can be a part of such planning. It should be noted, however, that prespecification does not guarantee reduction of false-positive and false-negative rates due to bias (i.e., comparability of treatment group). Stratification factors for randomization, typically based on considerations for the efficacy end points, do not necessarily match the prognostic factors for the safety end points of concern.

An appropriate level of exploratory subgroup analyses is warranted in any clinical trial. The objective of such analyses is typically hypothesis generating. Larger proportion of analyses tends to be exploratory in nature for an early phase study (e.g., a proof-of-concept study) in a relatively new therapeutic area where there is no approved drug in a similar class. However, even for later phase studies, such as a large phase 3 study or a long-term safety study, there are opportunities for explorations in rarer AEs or in AEs emerging after extended exposures. In many of these situations, a full specification of analyses in advance may be difficult. Still, one should make effort to keep track of the number and types of analyses conducted, so that multiplicity can be addressed appropriately during the interpretation of the results.

Study protocol or the associated statistical analysis plan offer the best venue for documenting prespecified subgroup analyses for a single study. Other documentation for subgroup analysis plan could be a program safety analysis plan (PSAP) across several studies of the same drug (see Section 16.5), which in turn may be a part of an analysis plan for the integrated summary of safety.

Replication can lend evidence to a finding observed in a subgroup since the latter may simply be a false-positive finding. There are several types of *replications*. For example, there may be other corroborating evidences such as laboratory measurements or a trend in signal across levels of subgroups (e.g., increasing level of baseline severity). When this happens, the credibility of the signal increases. Another type of replication might be similar pattern in other studies of the same investigational drug or that in products belonging to the same class.

16.4.5 Biological, Clinical, and Social Rationale

As mentioned in Section 16.4.1, the objective of the subgroup analysis in safety is primarily exploratory. Thus, it is particularly important that any findings be interpreted in the context of supporting information in order to consider the next step. For example, if there is a biological or medical rationale for more frequent occurrences of a particular AE in a specific subgroup (e.g., patients with diabetes), then the findings might be more credible. If a higher incidence of a particular laboratory abnormality in a particular subgroup is reported in other drugs of the same class, it would give corroborating evidence to the finding (e.g., less effect of beta blockers on blood pressure control in the black population). In a clinical trial that enrolls subjects from multiple countries or regions, one may find a geographical difference in safety outcomes. The latter may be a result of difference in the distribution of a biological factor (e.g., genetic factors), clinical practice (e.g., standard of care), or difference in social or cultural attitude. For example, some cultures may be reluctant to report AE on sexual function. In a Psychopharmacologic Drugs Advisory Committee on September 16, 2010, cultural differences in patients' thresholds for reporting subjective complaints such as nasal pharyngitis, insomnia, and headache were cited as possible contributing factor to the observed differences in the reported incidences of these common events between Russian and U.S. studies of Vivitrol for the treatment of opiate dependence (http://www.fda.gov/downloads/AdvisoryCommittees/CommitteesMeetingMaterials/Drugs/PsychopharmacologicDrugsAdvisoryCommittee/UCM232756.pdf).

While identifying possible attributable sources for the findings is important, one should not be excessively *creative* on the possible explanations, especially when numerous subgroups are evaluated without consideration to multiplicity—false positives can lead one astray. As mentioned in the previous section, replications of the finding in other studies of the test drug would give strong supportive evidence toward its authenticity.

16.4.6 Reporting and Interpreting Subgroup Results

16.4.6.1 Reporting

Prespecified analyses and post hoc analyses should be clearly distinguished when reporting the subgroup analysis results. In either case, the results should be reported in a transparent and consistent manner that facilitates proper interpretation of the findings. Graphical displays are particularly effective in presenting numerous subgroup results. Guidelines for reporting subgroup analysis for *New England Journal of Medicine* is given in Wang et al. (2007).

For the prespecified analyses, all planned subgroup analysis results should be included in the study report. If there is a predefined prioritization of the analyses, that should be described to aid the interpretation. If inferential statistical methods are used, approaches taken to account for multiplicity should be described.

For post hoc analyses, it is desirable to include in the study report (e.g., in an appendix) a listing of all subgroup analyses conducted in the interest of transparency. Description should include the end point, the subgroup or the subgrouping factor, and the description of the statistical method. Although a formal adjustment for multiplicity is often difficult for post hoc analyses, information that could aid the interpretation should be included. The latter includes the total number of analyses conducted and the number of positive results expected by pure chance.

Various techniques have been developed for effective presentation of safety data (Amit et al., 2008). Cuzick (2005) gives an advice on displaying subgroups results to aid proper interpretation.

16.4.6.2 Interpretation of Subgroup Results

A reasonable approach to interpreting subgroup results is to first assess the quality of subgroup analysis in terms of potential bias (Cui et al., 2002). Bias due to treatment group incomparability is particularly troublesome since the direction of bias (false positive or false negative) is unpredictable. For example, a well-planned, prespecified analysis in a study utilizing stratified randomization on key prognostic factor is likely to provide a high-quality analysis results with less chance of bias. In contrast, a finding from a large number of post hoc subgroup analyses would be of lower quality.

After the issue of bias is addressed, one can proceed to quantify the strength of the signal, taking statistical multiplicity into account. For safety end points, more concern is placed over failure to indentify safety signals, or false negatives, although an unrestrained inflation of false-positive rates can also severely hinder proper interpretation of the findings. Finally, one needs to interpret the finding in terms of biological, clinical, and social rationale, as described in Section 16.4.5.

Cui et al. (2002) organizes *common errors* in interpretation of subgroup results into a list of six categories, in a general setting that include both efficacy and safety end points. Many of the items on the list are mentioned earlier. Such list should serve as a useful reminder even for the case of safety end points. A proper interpretation allows one to take appropriate next steps, whether it is to plan for meta-analysis of the same drug prior to NDA, to plan for new clinical trial with target population, or to plan for postapproval safety activities.

16.5 Role of Meta-Analysis in Detecting Safety Signals in Subgroups

16.5.1 Introduction

Often, a subgroup analysis from a single trial is not adequate to detect safety signals because either the incidence of an adverse event of special interest is rare or the subgroup size is simply too small. Thus, meta-analysis is a powerful tool to combine results from multiple studies. In addition, when subgroup findings are displayed by study (e.g., using a forest plot), one can check the consistency of the finding across studies. The credibility of a subgroup finding is enhanced substantially when it is replicated over several studies.

A proper meta-analysis includes rigorous statistical methods and appropriate interpretation of results. To increase the validity of the results from a meta-analysis, the subgroup analysis should be specified a priori based on known biological mechanisms or in response to findings from some studies. To avoid selection bias, the inclusion/exclusion criteria for selecting studies and patients should be established and applied rigorously (Hahn et al., 2000). All statistical methods as well as data extraction criteria should be decided in advance. In a drug development program, the integration of information across studies is often facilitated by the use of consistent approaches in the collection, processing, and analysis of data across the entire development program. The latter is important to comparing like to like in a meta-analysis (Crowe et al., 2009).

This section presents the two most widely used statistical methods for meta-analysis, namely, the fixed effects model and the random effects model. In addition, we want to emphasize that an analysis based on crude pooling of adverse event numbers across different studies to compare treatment groups should be avoided, as the analysis is vulnerable to the mischief of Simpson's paradox (Chuang-Stein and Beltangady, 2011).

16.5.2 Fixed Effects Model

Suppose there are k independent studies each comparing a treatment group with a control group such as a placebo and we are interested in comparing the risk of a target adverse event between treatments in a subgroup of interest across the k studies. In the ith study, suppose there are n_{it} and n_{ic} subjects in the treatment and control groups, respectively, within the subgroup. Let x_{it} and x_{ic} represent the corresponding numbers of individuals with the target adverse event in the subgroup. The probabilities of experiencing the adverse event in the two treatment groups within the subgroup p_{it} and p_{ic} are estimated by $\hat{p}_{it} = x_{it}/n_{it}$ and $\hat{p}_{ic} = x_{ic}/n_{ic}$, respectively.

The fixed effects model assumes a common treatment effect across all studies. Denote this common effect by θ. Let $\hat{\theta}_i$ denote an estimate of θ from the ith study. An estimate for θ can be obtained by

$$\hat{\theta} = \frac{\sum_{i=1}^{k} \hat{\theta}_i w_i}{\sum_{i=1}^{k} w_i},$$

where w_i's are weights assigned to $\hat{\theta}_i$. There are several choices for w_i such as the Mantel–Haenszel weights and weights proportional to sample sizes in the studies (Mantel and Haenszel, 1959; Chuang-Stein and Beltangady, 2011).

The three most commonly used measures to compare two treatment groups are risk difference, risk ratio, and odds ratio. The estimate $\hat{\theta}_i$ and the Mantel–Haenszel estimate of w_i can be calculated for each measure as follows (Cochran, 1954; Mantel and Haenszel, 1959; Nurminen, 1981; Tanone, 1981):

1. Risk difference

$$\hat{\theta}_i = \hat{p}_{it} - \hat{p}_{ic}, \quad w_i = \frac{n_{it}n_{ic}}{n_i} \qquad (n_i = n_{it} + n_{ic}).$$

2. Relative risk (or risk ratio)

$$\hat{\theta}_i = \frac{\hat{p}_{it}}{\hat{p}_{ic}}, \quad w_i = \frac{n_{it}x_{ic}}{n_i}.$$

3. Odds ratio

$$\hat{\theta}_i = \frac{\hat{p}_{it}(1-\hat{p}_{ic})}{\hat{p}_{ic}(1-\hat{p}_{it})}, \quad w_i = \frac{(n_{it} - x_{it})x_{ic}}{n_i}.$$

In general, the choice of the most useful measure depends on the situation. Relative risk has the advantage of easy interpretation. For example, the risk of the treatment group for a particular adverse event is 25% higher than the control group, regardless of the adverse event rate in the control group. On the other hand, risk difference describes the number of additional patients who may experience the adverse event among, for example, 1000 patients exposed. The combined risk difference will, however, be applicable only to patients at levels of risk comparable to those included in the studies (Egger et al., 1997). Odds ratio has convenient mathematical properties and can be modeled in a logistic regression model (see Section 16.6). Also, odds ratio is approximately equal to relative risk when the adverse event rate is relatively rare, say less than 20% (Egger et al., 1997).

We usually provide not only point estimate for the overall fixed effect θ but also a confidence interval (CI) for θ. Assume for now that the Mantel–Haenszel weights are chosen. An approximate $100(1 - \alpha)\%$ CI for the overall risk difference θ_{RD} is given by

$$\hat{\theta}_{RD} \pm z_{\alpha/2} \sqrt{\frac{\hat{\theta}_{RD}\left(\sum_{i=1}^{k} U_i\right) + \left(\sum_{i=1}^{k} V_i\right)}{\left(\sum_{i=1}^{k} n_{it}n_{ic}/n_i\right)^2}},$$

where

$$U_i = \frac{n_{it}^2 x_{ic} - n_{ic}^2 x_{it} + n_{it}n_{ic}(n_{ic} - n_{it})/2}{n_i^2}$$

$$V_i = \frac{x_{it}(n_{ic} - x_{ic}) + x_{ic}(n_{it} - x_{it})}{2n_i}$$

$z_{\alpha/2}$ is the upper $100(\alpha/2)$th percentile of the standard normal distribution (Greenland and Robins, 1985; Sato, 1989).

When applying the logarithmic transformation for relative risk, an asymptotic $100(1 - \alpha)\%$ CI for overall effect θ_{RR} (relative risk) is given by (Greenland and Robins, 1985):

$$\exp\left\{\log \hat{\theta}_{RR} \pm z_{\alpha/2} \sqrt{\frac{\sum_{i=1}^{k} (x_{it} + x_{ic})n_{it}n_{ic}/n_i^2 - x_{it}x_{ic}/n_i}{\left(\sum_{i=1}^{k} x_{it}n_{ic}/n_i\right)\left(\sum_{i=1}^{k} x_{ic}n_{it}/n_i\right)}}\right\}.$$

Also, when applying the logarithmic transformation for odds ratio, an asymptotic $100(1 - \alpha)\%$ CI for the overall effect θ_{OR} (odds ratio) is given by (Robins et al., 1986):

$$\exp\left\{ \log \hat{\theta}_{OR} \pm z_{\alpha/2} \sqrt{ \frac{\sum_{i=1}^{k} P_i R_i}{2\left(\sum_{i=1}^{k} R_i\right)^2} + \frac{\sum_{i=1}^{k} (P_i S_i + Q_i R_i)}{2\left(\sum_{i=1}^{k} R_i\right)\left(\sum_{i=1}^{k} S_i\right)} + \frac{\sum_{i=1}^{k} Q_i S_i}{2\left(\sum_{i=1}^{k} S_i\right)^2} } \right\},$$

where
$P_i = (x_{it} + n_{ic} - x_{ic})/n_i$
$Q_i = (n_{it} - x_{it} + x_{ic})/n_i$
$R_i = x_{it}(n_{ic} - x_{ic})/n_i$
$S_i = (n_{it} - x_{it})x_{ic}/n_i$

We can test the null hypothesis that the treatment difference in all studies is equal to 0 by using the estimate of the standard error of $\hat{\theta}$ in each case. However, the magnitude of the summary statistics should also be examined. Because of the reduced overall variability and increased power resulting from pooling studies in the meta-analysis, a small treatment difference may yield a highly significant p-value. Therefore, it is necessary to examine the magnitude of the estimated treatment effect and statistical significance in tandem.

16.5.3 Precision of Estimate in a Fixed Effects Model

An advantage of meta-analysis is that after combining data from several studies, the precision of an estimate is improved due to the increase in the number of patients and the number of reported events. We conducted a simulation study to assess the improvement in precision by measuring the width of the 95% CI. For this simulation, we set $p_{ic} = 0.1$ for all i; $p_{it} = 0.2$, 0.3, 0.4 for all i; $k = 2, 4, 6, 8, 10$. The number of subjects in the control group was chosen by rounding random draws from a uniform distribution with min = 50 and max = 200. For simplicity, we set sample size in the treatment group to be the same as that in the control group, that is, $n_{it} = n_{ic}$. Next, we generated the responses x_{ic} for the control group based on a binomial distribution with the response probability of 0.1. We also generated the responses x_{it} for the treatment group based on a binomial distribution with different response probability of 0.2, 0.3, and 0.4. We have added 0.5 to all cells with a zero frequency in the calculation of the pooled estimate. We simulated 1000 replications for each combination of k, p_{ic}, and p_{it}.

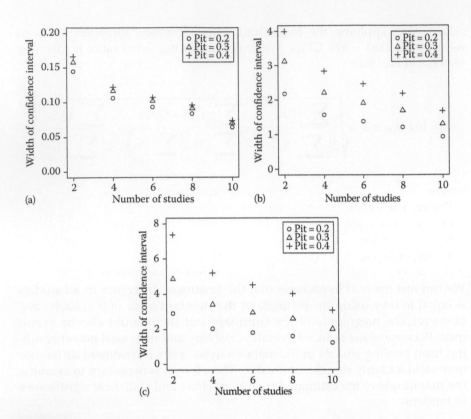

FIGURE 16.1
Plots of median width of 95% CI against the number of studies under the fixed effects model.
(a) Risk difference, (b) relative risk, and (c) odds ratio.

Figure 16.1 shows median width of a 95% CI for risk difference (Figure 16.1a), risk ratio (Figure 16.1b), and odds ratio (Figure 16.1c). We can see that the width of CI dramatically decreases as the number of studies increases regardless of the measure. Also, the plots of $p_{it} = 0.4$ are generally above those corresponding to $p_{it} = 0.2$ and 0.3, particularly, for relative risk and odds ratio. This is expected since the variance of a binomial distribution is larger as the probability is closer to 0.5.

16.5.4 Random Effects Model

A random effects model assumes that the treatment effect in the k studies $(\theta_1, \cdots, \theta_k)$ are a sample of independent observations from a normal distribution $N(\theta, \tau^2)$, where τ^2 represents the between-study variance. The estimate of the between-study variance $\hat{\tau}^2$ can be obtained by the DerSimonian–Laird approach (DerSimonian and Laird, 1986). By setting

$$w_i^* = (w_i + \hat{\tau}^2)^{-1},$$

the overall effect θ under the random effects model can be estimated by $\hat{\theta}^*$, where

$$\hat{\theta}^* = \frac{\sum_{i=1}^{k} \hat{\theta}_i w_i^*}{\sum_{i=1}^{k} w_i^*}.$$

In contrast to the fixed effects model, a random effects model explicitly assumes heterogeneity between studies and model it by τ^2. The implication of this is that under the random effects model, θ is an average effect. If $\hat{\tau}^2$ is small, then the overall estimate of treatment effect and the CI from the random effects model will be close to those from the fixed effects model. On the other hand, if $\hat{\tau}^2$ is large, then the CI may be much wider for the random effects model. Also, if $\hat{\tau}^2$ is large, the weight assigned to a study is less dependent on the sample size of the study. As a result, the random effects model may assign more weights to smaller studies compared with the fixed effect model. It should be noted that we may not have good precision in estimating τ^2 if we only have a small number of studies (Hardy and Thompson, 1996).

In choosing between fixed effects or random effects models, it is important to remember that the two models address different research questions. If the research question is concerned with an overall treatment effect in the existing studies, then a fixed effects model would be appropriate. If one wants to estimate the treatment effect that may be observed in a future study, then the heterogeneity of the treatment effect across studies should be accounted for and a random effects model would be appropriate. Also, sensitivity analyses could be conducted to assess the robustness of results despite the model chosen for the primary analysis.

16.5.5 Heterogeneity of Treatment Effects

The issue of heterogeneity across studies is critical in interpreting results from a meta-analysis. The Q statistic and I^2 statistic are commonly used to describe heterogeneity across studies. The Q statistic can be used to test whether there is evidence that the treatment effects are different across studies (Whitehead, 2002). As a function of the Q statistic, the I^2 statistic quantifies the extent of total variation in treatment effects across studies that is due to heterogeneity alone rather than chance (Higgins and Thompson, 2002). Values of the I^2 statistic can range from 0% to 100%. Higgins and Thompson (2002) suggested that any value of I^2 less than 30% suggests mild heterogeneity. A caveat about the I^2 statistic is that with a limited number of studies, it tends to have large uncertainty. A 95% CI for I^2 can be readily estimated and is useful to convey the extent of uncertainty (Ioannidis et al., 2007). One may also formally examine in sensitivity analyses the impact of specific studies on the extent of estimating between-study heterogeneity (Patsopoulos et al., 2008).

16.5.6 Examples

16.5.6.1 Example 1: Meta-Analysis of Studies of Inhaled Long-Acting Beta-Agonists

In 2008, the FDA convened a joint advisory committee meeting to review the risks and benefits of inhaled LABAs for the treatment of asthma in adults and children (Kramer, 2009). There were concerns over a potential increase in serious asthma exacerbations in some patients treated with these drugs (Chowdhury and Dal Pan, 2010). At the meeting, an FDA statistical reviewer presented results from a meta-analysis based on patient-level data from 110 randomized trials that studied LABAs for asthma control (Levenson, 2008).

The meta-analysis involved four products that contain LABA and were approved in the United States for the treatment of asthma. The analysis included 60,954 patients in 110 trials. To assess risk, a composite end point of asthma-related death, intubation, or hospitalization was used. The meta-analysis found 2.80 (95% CI: 1.11, 4.49) more events per 1000 patients in the group that received LABAs compared to the group that did not (Levenson, 2008; Kramer, 2009). The 4–11 age group had an estimated risk difference of 14.83 (95% CI: 3.24, 26.43) per 1000 subjects (Levenson, 2008). In addition to reviewing meta-analysis results, the advisory committees also discussed issues of the meta-analysis itself such as whether the data are adequate to address the important clinical and public health questions. For example, most patients in the FDA meta-analysis participated in studies years ago, when LABA monotherapy was common (Kramer, 2009).

Ultimately, the committees agreed unanimously that for single-agent LABAs, the benefits did not outweigh the risks in children 4–11 years of age. Most committee members also agreed that the benefits of single-agent LABAs did not outweigh the risks for adolescents and adults (Kramer, 2009).

Subsequently, FDA requested the labels for all LABAs to be updated with additional usage wording. The latter include statement such as "LABAs should only be used as additional therapy for patients with asthma who are currently taking but are not adequately controlled on a long-term asthma control medication, such as an inhaled corticosteroid" (Food and Drug Administration, 2010). As for pediatric and adolescent patients who require LABAs in addition to inhaled corticosteroid, FDA's use guidelines state that these patients should use a combination product containing both an inhaled corticosteroid and a LABA to ensure adherence with both medications (Food and Drug Administration, 2011).

This example tells us how a meta-analysis provides quantitative evidence to help make public health decisions, while recognizing the strengths and limitations of the evidence.

16.5.6.2 Example 2: Meta-Analysis of an Ethnic Factor in the Risk for an Adverse Reaction

McDowell et al. (2006) reviewed the evidence for ethnic differences in suscepti-bility to ADRs to cardiovascular drugs. They considered the incidence of angio-edema due to angiotensin converting enzyme (ACE) inhibitors in six studies. They excluded one small clinical study (out of the six) as it was classified as being at high risk for bias. They provided results from a meta-analysis compar-ing the risk of angioedema among black and nonblack patients in five studies using a fixed effects model. We reanalyzed the data to illustrate results from different meta-analysis approaches. Figure 16.2 shows results of meta-analyses using risk difference per 1000 subjects (Figure 16.2a), relative risk (Figure 16.2b), and odds ratio (Figure 16.2c) under both of the fixed and random effects mod-els. Heterogeneity between studies was assessed using the I^2 statistics.

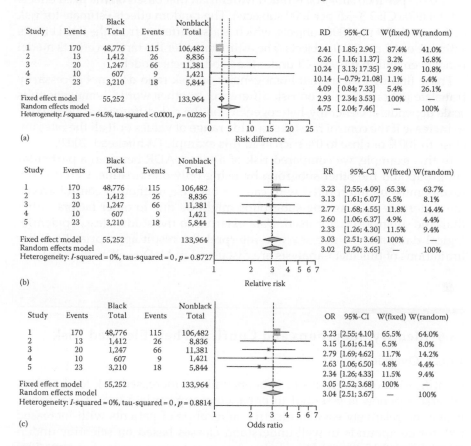

FIGURE 16.2

Meta-analyses of proportion of black and nonblack patients with angioedema associated with use of ACE inhibitors. (a) Risk difference (per 1000 subjects), (b) relative risk, and (c) odds ratio.

Pooled estimate of relative risk under the fixed effect model (Figure 16.2a) found that black patients had a relative risk of angioedema of 3.0 (95% CI: 2.5–3.7) compared with nonblack patients. The pooled estimate and the CI from the random effect model were almost equal to those from the fixed effects model because the I^2 statistics did not suggest noticeable heterogeneity among the studies. Meta-analysis using odds ratio provided similar results as risk ratio because the proportion of patients with angioedema was very low in all studies.

On the other hand, we see a clear evidence of heterogeneity among the studies for risk difference since the value of I^2 statistic is 64.5% (95% CI: 6.9%–86.5%), which is large enough to have a substantial impact on the weight. This has the effect of increasing the standard error and the width of the CI for the overall estimate of racial difference. The CI based on the random effects model (95% CI: 2.0–7.5 per 1000 subjects) is much wider than that based on the fixed effects model (95% CI: 2.3–3.5 per 1000 subjects). The random effects estimate for risk difference is 2.9 per 1000 subjects, which is also different from the fixed effects estimate of 4.7 per 1000 subjects. The estimate under the random effects model is not even within the 95% CI under the fixed effects model.

These findings tell us that homogeneity on risk ratio does not necessarily translate to homogeneity on risk difference. In other words, homogeneity is scale dependent. Generally, heterogeneity on the risk difference scale is likely to increase if the control rate takes a wide range of values or if all the rates are close to 100% or close to 0% such as in this example (Whitehead, 2002).

In this example, we compared risk of a target ADR between a particular subgroup and the other subgroup by using only one treatment group. The results must be interpreted with caution because, without a control group, the difference seen among subgroups might be due to other factors rather than the subgroup factor itself. Data external to the studies (e.g., epidemiological data) may be necessary to interpret the result appropriately. Other limitations of this meta-analysis are discussed in McDowell et al. (2006).

16.6 Identify Subgroups and Confirm Their Elevated Risk

16.6.1 Overview

Identification of subgroups of patients with an increased risk for a serious adverse reaction is essential to enhancing appropriate use of medicines. It may be relatively easy to identify subgroups of patients with increased risk for compounds in well-understood classes based on scientific understanding and historical knowledge. For novel compounds that have limited patient exposure, it is often not possible to identify subgroups with increased risk because there is a lack of knowledge about patient characteristics that may be related to risk. In recent years, the amount of patient characteristics available at baseline has rapidly increased. The latter include biomarker, genomic or phenotypic information, clinical observations, pharmacokinetic measures, environmental factors, lifestyle, and concomitant medications. This section describes a statistical approach to identify subgroups of patients with an increased risk for an adverse reaction of interest. Upon identifying meaningful subgroups, follow-up studies may be beneficial to confirm the findings.

In this section, we will describe in more detail the study PREDICT-1, which is the first adequately powered, randomized, double-blinded prospective study using pharmacogenetic screening strategy to reduce an ADR (Hughes et al., 2008).

16.6.2 Identifying Subgroups

To investigate a potentially different treatment effect on the incidence of an adverse outcome among subgroups in a multivariate manner, statisticians have traditionally relied on the inclusion of a treatment-by-subgroup interaction term in a model for risk. There are difficulties associated with this approach. First, statistical tests for interaction effects generally have low power. Next, subgroups are often assessed individually through repeated use of the same model with an additional main effect for the subgroup and an interaction effect between the subgroup and treatment. This process, applied to prespecified factors (and sometimes to post hoc factors also), may miss important interactions and could also generate false positives because of multiplicity. Statistical properties of this process can become intractable very quickly. With 100 possible factors, there are 4950 possible two-factor interactions. Under such circumstances, the traditional approach described earlier is not very useful in searching for important factors to define subgroups.

Recursive partitioning (Breiman et al., 1984), proposed for such problems, has received much attention recently (Ruberg et al., 2010; Foster et al., 2011; Lipkovich et al., 2011). This is a predictive approach and models the relationship in terms of trees or dendrograms. It has minimal statistical or model assumptions and has the ability to find complex interactions and nonlinear relations among predictors through its branching process. It is particularly appropriate for initial exploration of large data sets and may either validate other approaches with strong model assumptions or lead to a final analysis in its own right. It can be used for end points measured on either a nominal scale or an interval scale.

A recursive partitioning algorithm consists of three major steps: (1) growing a large initial tree, (2) a pruning algorithm, and (3) a method for determining the best tree size. The first step constructs the classification tree by recursively splitting each parent node into two child nodes. For each node, all available covariates are evaluated for every possible split based on selecting a numerical cutoff for continuous covariates or a subset of categories for nominal covariates, and the best split is searched. The splitting criterion is set as information gain based on reduction in entropy due to the split. The splitting is continued until the minimal number of patients required to form a child node is reached (Therneau and Atkinson, 2011). Second, the constructed tree is further pruned using a cost-complexity algorithm, which results in a sequence of optimally pruned subtree by iteratively truncating the *weakest link* (Breiman et al., 1984). Third, the best tree is selected from the nested subtree sequence based on the cost-complexity measure. Breiman et al. offer details on how Monte Carlo techniques can be used to choose the best value of the cost-complexity measure.

When identifying subgroups based on the optimal tree, it is important to assess if the performance can be reproduced in an independent data set. For example, when focusing on the incidence of a specific adverse event, the positive predictive value and negative predictive value can be used for evaluation. To do this, the original data set may be divided into training and validation subsets beforehand. The training subset is used to construct the subgroups, and the validation subset is used to examine the performance. The proportion of data going into the training set varies across situations and will largely depend on incidence of adverse outcomes and/or distributions covariates of interest. Large data sets are typically needed for this exercise.

While recursive partitioning is highly interpretable, it also has a few undesirable characteristics. This method can produce predictive models that may not be scientifically interpretable, especially if the meaningful predictors are correlated with less meaningful predictors. Tree-based routines may select the less meaningful predictors for the partition if it meets the optimization criteria. Hence, the user should preprocess the data to identify and remove highly correlated predictors, keep scientifically relevant predictors in the model, and remove correlated, but less relevant predictors.

Recursive partitioning methods are a simple and powerful mechanism for searching large groups of potential predictors while also addressing potential interactions/relationships between predictors. Additionally, clinicians could find the resulting decision trees easy to understand and in many circumstances, easy to apply in clinical practice.

16.6.3 Managing an Elevated Risk

Once a risk factor is identified and a risk management strategy is proposed, a sponsor may consider conducting a study to confirm the effectiveness of the strategy. In this section, we describe the study PREDICT-1, which investigated a biomarker strategy for abacavir. PREDICT-1 is the first adequately powered, randomized, double-blinded prospective study using pharmacogenetic screening to reduce the adverse effect of abacavir (Hughes et al., 2008).

Abacavir is a nucleoside reverse-transcriptase inhibitor with activity against the human immunodeficiency virus (HIV). The most important adverse effect of abacavir that limits its use in therapy and mandates a high degree of clinical vigilance is an immunologically mediated HSR affecting between 5% and 8% of patients during the first 6 weeks of treatment (Mallal et al., 2008). Retrospective analyses have identified several risk factors for HSR to abacavir, with carriage of the major histocompatibility complex class I allele HLA-B*5701 as having the strongest association with the incidence of HSR (Hetherington et al., 2002; Mallal et al., 2002). Studies of cohorts with HIV infection have shown that avoiding abacavir in HLA-B*5701-positive patients significantly reduced the incidence of HSR in Australia (Rauch et al., 2006) and in France (Zucman et al., 2007). Although observational cohorts and case-controlled studies provided valuable information on the role of genetics and variation in response to medicines, the manufacturer of abacavir decided to study the effectiveness of HLA-B*5701 screening in reducing HSR in a randomized clinical trial.

The objective of the study was to test the hypothesis that prospective pharmacogenetic screening for HLA-B*5701 and excluding patients carrying the allele from abacavir treatment will reduce the incidence of HSR to abacavir. The primary end points were the proportion of patients with clinically diagnosed HSR to abacavir during the 6-week observation period and the proportion of patients with immunologically confirmed HSR. The study consisted of a screening period, a randomized 6-week observation period, and an epicutaneous patch testing period for patients with a clinically suspected HSR as well as a subset of abacavir-tolerant patients. The key entry criteria included no prior abacavir treatment, no information on a patient's HLA-B*5701 status, and a clinical need for abacavir treatment. Eligible patients were randomized 1:1 to either going through the screening procedure or receiving abacavir in the usual manner. Patients identified as HLA-B*5701-positive in the screening arm did not receive abacavir and were excluded from the study. Patients in both arms with clinically diagnosed HSR to abacavir during the 6-week observation period were withdrawn from abacavir and received the epicutaneous patch testing.

A total of 1956 patients were enrolled and randomized. Of the 980 patients in the prospective-screening group, 55 (5.6%) were excluded from the study because they were carriers of the HLA-B*5701 allele. The populations that could be evaluated for clinically diagnosed HSR consisted of 803 patients in the prospective-screening group and 847 patients in the control group. The incidences of clinically diagnosed and immunologically confirmed HSR to abacavir were significantly lower in the prospective-screening group than in the control group (clinically diagnosed: odds ratio, 0.40; 95% CI, 0.25–0.62; p-value < 0.001; immunologically confirmed: odds ratio, 0.03; 95% CI, 0.00–0.18; p-value < 0.001) (Mallal et al., 2008). The statistical analysis was based on a logistic regression model adjusted for self-reported race (white versus nonwhite), history of receipt of antiretroviral therapy (none versus any), introduction of a new nonnucleoside reverse-transcriptase inhibitor (yes or no), and concurrent use or nonuse of a protease inhibitor.

This study confirmed the value of a pharmacogenetic test in reducing HSR to abacavir.

16.7 Discussion

All pharmaceutical products have side effects. Many factors affect an individual's ability to remain on a product. Depending on the safety profile of the products, the proportion of patients who will eventually discontinue the products for safety reason will vary. At times, alternative products were developed for patients who cannot tolerate a commonly prescribed medication. For example, a medication patch may be developed to avoid gastrointestinal-related side effects. When an adverse reaction is serious, researchers will typically investigate patient characteristics to help identify patients who are at a higher risk for the serious adverse reaction. Such investigations, if successful, will lead to useful risk-mitigation strategies.

By and large, subgroup analysis has become a routine practice when analyzing results from a clinical trial. Children and elderly patients always receive special attention, if they are intended users. If a product is cleared through liver or kidney, patients with impaired hepatic or renal functions will receive special attention during the development of the product. Similarly, products that are metabolized via a particular metabolic pathway (e.g., the cytochrome P450 system) will be investigated in multiple drug–drug interaction studies. Risk factors that have been identified in the past for similar products will usually be incorporated into the development plan of the new product in the same class.

Interest in subgroup analysis has intensified in recent years due to the development of targeted therapies under the broad theme of personalized medicine. While most of the attention on targeted therapies is currently focused on efficacy, safety will play an important role as the ultimate decision on a therapy is the balance between benefit and risk. While increasing benefit is one way to increase the benefit/risk balance, reducing adverse reaction could also lead to a more favorable benefit/risk profile for a subgroup of patients.

Even though we did not state it explicitly, we have so far treated factors defining subgroups as categorical. When a subgroup is defined by cutoffs on a continuous variable at baseline, Wittes (2009) argued that nearby subgroups matter. If the subgroup effect is real, one should see a reasonably smooth (monotone) relationship between the subgroups and the size of the treatment effect. A smooth relationship will give credibility to the factor as a risk factor. To do this, we will need enough data to support the division of patients into multiple groups.

We did not use terms such as *prognostic* or *predictive* to describe factors in this chapter. Subgroups may be identified by factors that are prognostic or predictive. If the average treatment effect (e.g., the difference in the proportions of patients with an adverse event of interest between a new treatment and a comparator) varies between subgroups, the factor defining the subgroups is predictive. This is what an interaction test tries to detect. On the other hand, if the treatment effect is constant across subgroups but the absolute response to the new (control) treatment differs across subgroups, the factor is considered as prognostic. For example, even though patients with a more severe condition typically report more adverse events, disease severity will not be a predictive factor if the comparative risk measure for a new treatment (versus a comparator) is similar between the more severe patients and their less severe peers. Of course, the decision on *prognostic* versus *predictive* may depend on the scale chosen to measure the comparative risk as illustrated in Section 16.5.

We have discussed subgroups in randomized trials. Increasingly, researchers are looking for new data sources to carry out safety assessment. For example, claims database and electronic medical records are being used in the Observational Medical Outcome Partnership initiative (www.omop. gov) to explore their utility in detecting safety signals. Olfson et al. (2006) reported results from observational studies where data from 1999 to 2000 in the national Medicaid Analytic Extract Files were examined. The authors concluded that their findings suggest a possible association between anti-depressant drug treatment and suicide attempts and completed suicide in severely depressed children and adolescents in the Medicaid program after hospital discharge. The use of claims database or electronic health-care records are sometimes used to conduct postauthorization safety studies that are as part of postapproval commitments required by regulators.

The ultimate goal of risk assessment is to be able to predict risk for a clinically important event at the individual subject level. This is akin to the construction of the Framingham risk scores for individuals 20 years and older who do not have heart disease or diabetes. The Framingham risk score is based on an individual's age, gender, total cholesterol, low-density lipoprotein (LDL) cholesterol, smoking status, systolic blood pressure, and whether the individual is taking medications to control for blood pressure. The risk score is used to predict the individual's chance of having a heart attack in the next 10 years (http://cvdrisk.nhlbi.nih.gov/calculator.asp). Data continue to be collected in the Framingham study and the prediction model continues to be refined.

One can regard the concept of individual risk scores similar to letting the number of subgroup go to infinity so that each individual forms their own subgroup. To develop a risk calculator for a clinically important event after exposure to a pharmaceutical product will require a large amount of data. In our opinion, the latter may become possible with advancements in computing technology and our ability to access data in claims database and electronic medical records. Statisticians should participate in such efforts to ensure that appropriate statistical methods are used to address possible sources of bias, especially selection bias, that are often associated with data from nonrandomized sources.

References

Amit O, Heiberger RM, Lane PW. Graphical approaches to the analysis of safety data from clinical trials. *Pharmaceutical Statistics,* 7:20–35, 2008.

Assmann, SF, Pocock SJ, Enos LE, Kasten LE. Subgroup analysis and other (mis) uses of baseline data in clinical trials. *Lancet,* 355:1064–1069, 2000.

Azuma A, Kudoh S, High prevalence of drug-induced pneumonia in Japan. *Japan Medical Association Journal,* 50(5):405–411, 2007.

Breiman L, Freidman JH, Olshen RA, Stone CJ. *Classification and Regression Trees,* Wadsworth, Belmont, CA, 1984, Chapters 10–11, pp. 279–317.

Chowdhury BA, Dal Pan G. The FDA and safe use of long-acting beta-agonists in the treatment of asthma. *The New England Journal of Medicine,* 362(13):1169–1171, 2010.

Chuang-Stein C, Beltangady M. Reporting cumulative proportion of subjects with an adverse event based on data from multiple studies. *Pharmaceutical Statistics,* 10(1):3–7, 2011.

Chuang-Stein C, Xia A. The practice of pre-marketing safety assessment in drug development. *Journal of Biopharmaceutical Statistics,* 23(1):3–25, 2013.

Cochran WG. Some methods for strengthening the common chi-square tests. *Biometrics,* 10:417–451, 1954.

Congressional Record-House. NIH Revitalization Act. Sec. 131, Inclusion of Women and Minorities in Clinical Research, 20 May, page H2623, 1993.

Crowe BJ, Xia HA, Berlin JA, Watson DJ, Shi H, Lin SL, Kuebler J et al. Recommendations for safety planning, data collection, evaluation and reporting during drug, biologic and vaccine development: A report of the safety planning, evaluation, and reporting team. *Clinical Trials,* 6(5):430–440, 2009.

Cui L, Hung HM, Wang SJ, Tsong Y. Issues related to subgroup analysis in clinical trials. *Journal of Biopharmaceutical Statistics,* 12:347–358, 2002.

Cuzick J. Forest plots and the interpretation of subgroups. *Lancet,* 365:1308, 2005.

Davis CE, Leffingwell DP. Empirical Bayes estimates of subgroup effects in clinical trials. *Controlled Clinical Trials,* 11:37–42, 1990.

DerSimonian R, Laird N. Meta-analysis in clinical trials. *Control Clinical Trials,* 7(3):177–188, 1986.

Egger M, Smith GD, Phillips AN. Meta-analysis: Principles and procedures. *British Medical Journal,* 315(7121):1533–1537, 1997.

Fleming TR. Interpretation of subgroup analyses in clinical trials. *Drug Information Journal,* 29:1681S–1687S, 1995.

Food and Drug Administration. Guidance for industry—Premarketing risk assessment. http://www.fda.gov/downloads/Drugs/GuidanceComplianceRegulatory Information/Guidances/UCM072002.pdf, 2005. Accessed January 10, 2014.

Food and Drug Administration. FDA Drug safety communication: Drug labels now contain updated recommendations on the appropriate use of long-acting inhaled asthma medications called Long-Acting Beta-Agonists (LABAs). http:// www.fda.gov/Drugs/DrugSafety/PostmarketDrugSafetyInformationfor PatientsandProviders/ucm213836.htm, 2010. Accessed January 10, 2014.

Food and Drug Administration. FDA Drug safety communication: FDA requires post-market safety trials for Long-Acting Beta-Agonists (LABAs). http://www.fda.gov/Drugs/DrugSafety/ucm251512.htm, 2011. Accessed January 10, 2014.

Foster JC, Taylor JMG, Ruberg SJ. Subgroup identification from randomized clinical trial data. *Statistics in Medicine*, 30:2867–2880, 2011.

Gail M, Simon R. Testing for qualitative interactions between treatment effects and patient subsets. *Biometrics*, 41(2):361–372, 1985.

Greenland S. Principles of multilevel modelling. *International Journal of Epidemiology*, 29:158–167, 2000.

Greenland S, Robins JM. Estimation of a common effect parameter from sparse follow-up data. *Biometrics*, 41:55–68, 1985.

Hahn S, Williamson PR, Hutton JL, Garner P, Flynn EV. Assessing the potential for bias in meta-analysis due to selective reporting of subgroup analyses within studies. *Statistics in Medicine*, 19:3325–3336, 2000.

Hardy RJ, Thompson SG. A likelihood approach to meta-analysis with random effects. *Statistics in Medicine*, 15:619–629, 1996.

Hetherington S, Hughes AR, Mosteller M, Shortino D, Baker KL, Spreen W, Lai E et al. Genetic variations in HLA-B region and hypersensitivity reactions to abacavir. *Lancet*, 359:1121–1122, 2002.

Higgins JPT, Thompson SG. Quantifying heterogeneity in a meta-analysis. *Statistics in Medicine*, 21:1539–1558, 2002.

Hughes S, Hughes A, Brothers C, Spreen W, Thorborn D. PREDICT-1 (CNA106030): The first powered, prospective trial of pharmacogenetic screening to reduce drug adverse events. *Pharmaceutical Statistics*, 7:121–129, 2008.

Ioannidis JP, Patsopoulos NA. Evangelou E. Uncertainty in heterogeneity estimates in meta-analyses. *British Medical Journal*, 335:914–916, 2007.

Jackson RD, LaCroix AZ, Gass M, Wallace RB, Robbins J, Lewis CE, Bassford T et al. Calcium plus vitamin D supplementation and the risk of fractures. *The New England Journal of Medicine*, 354:669–683, 2006.

Kramer JM. Balancing the benefits and risks of inhaled long-acting beta-agonists—The influence of values. *The New England Journal of Medicine*, 360(16):1592–1595, 2009.

Levenson M. Long-acting beta-agonists and adverse asthma events meta-analysis: Statistical briefing package for joint meeting of the Pulmonary–Allergy Drugs Advisory Committee, Drug Safety and Risk Management Advisory Committee, and Pediatric Advisory Committee on December 10–11, 2008.

Li J, Mehrotra DV, An efficient method for accommodating potentially underpowered primary endpoints. *Statistics in Medicine*, 27:5377–5391, 2008.

Li Z, Chuang-Stein C, Hoseyni C. The probability of observing negative subgroup results when the treatment effect is positive and homogeneous across all subgroups. *Drug Information Journal*, 41(1):47–56, 2007.

Lipkovich I, Dmitrienko A, Denne J, Enas G. Subgroup identification based on differential effect search: A recursive partitioning method for establishing response to treatment in patient subpopulations. *Statistics in Medicine*, 30:2601–2621, 2011.

Lipskly AM, Gaushe-Hill M, Vienna M, Lewis RJ. The importance of "Shrinkage" in subgroup analyses. *Annals of Emergency Medicine*, 55(6):544–552, 2010.

Locharernkul C, Loplumlert J, Limotai C, Korkij W, Desudchit T, Tongkobpetch S, Kangwanshiratada O et al. Carbamazepine and phenytoin induced Stevens–Johnson syndrome is associated with HLA-B*1502 allele in Thai population. *Epilepsia*, 49:2087–2091, 2008.

Mallal S, Nolan D, Witt C, Masel G, Martin AM, Moore C, Sayer D et al. Association between presence of HLA-B*5701, HLA-DR7, and HLA-DQ3 and hypersensitivity to HIV-1 reverse-transcriptase inhibitor abacavir. *Lancet*, 359:727–732, 2002.

Mallal S, Phillips E, Carosi G, Molina JM, Workman C, Tomazic J, Jagel-Guedes E et al. HLA-B*5701 screening for hypersensitivity to abacavir. *The New England Journal of Medicine*, 358:568–579, 2008.

Mantel N, Haenszel W. Statistical aspects of the analysis of data from retrospective studies of disease. *Journal of the National Cancer Institute*, 22:719–748, 1959.

McDowell SE, Coleman JJ, Ferner RE. Systematic review and meta-analysis of ethnic differences in risks of adverse reactions to drugs used in cardiovascular medicine. *British Medical Journal*, 332(7551):1177–1181, 2006.

Nurminen M. Asymptotic efficiency of general noniterative estimators of common relative risk. *Biometrika*, 68:525–530, 1981.

Olfson M, Marcus SC, Shaffer D. Antidepressant drug therapy and suicide in severely depressed children and adults: A case-control study. *Archives of General Psychiatry*, 63(8):865–872, 2006.

Patsopoulos NA, Evangelou E, Ioannidis JP. Sensitivity of between-study heterogeneity in meta-analysis: Proposed metrics and empirical evaluation. *International Journal of Epidemiology*, 37:1148–1157, 2008.

Phillips EJ, Simon A. Mallal SA, HLA-B*1502 screening and toxic effects of carbamazepine. *The New England Journal of Medicine*, 365(7):672–673, 2011.

Rauch A, Nolan D, Martin A, McKinnon E, Almeida C, Mallal S. Prospective genetic screening decreases the incidence of abacavir hypersensitivity reactions in the western Australian HIV cohort study. *Clinical Infectious Diseases*, 43:99–102, 2006.

Robins J, Greenland S, Breslow NE. A general estimator for the variance of the Mantel–Haenszel odds ratio. *American Journal of Epidemiology*, 124(5):719–723, 1986.

Ruberg SJ, Chen L, Wang Y. The mean does not mean as much anymore: Finding subgroups for tailored therapeutics. *Clinical Trials*, 7:574–583, 2010.

Sato T. On the variance estimator for the Mantel–Haenszel risk difference (letter). *Biometrics*, 45:1323–1324, 1989.

Sidak Z. Rectangular confidence regions for the means of multivariate normal distributions. *Journal of the American Statistical Association*, 62:626–633, 1967.

Tarone RE. On summary estimators of relative risk. *Journal of Chronic Diseases*, 34:463–468, 1981.

Therneau TM, Atkinson EJ. An introduction to recursive partitioning using the RPART routines. Technical report, Mayo Clinic, Section of Biostatistics, 2011.

Wang R, Lagakos SW, Ware JH, Hunter DJ, Drazen JM. Statistics in medicine—Reporting of subgroup analyses in clinical trials. *The New England Journal of Medicine*, 357:2189–2194, 2007.

Whitehead A. *Meta-Analysis of Controlled Clinical Trials*, John Wiley & Sons, Ltd., Hoboken, NJ, 2002.

Wittes J. On looking at subgroups. *Circulation*, 119:912–915, 2009.

Xia HA, Ma H, Carlin BP. Bayesian hierarchical modeling for detecting safety signals in clinical trials. *Journal of Biopharmaceutical Statistics*, 21(5):1006–1029, 2011.

Yusuf S, Wittes J, Probstfield J, Tyroler HA. Analysis and interpretation of treatment effects in subgroups of patients in randomised clinical trials. *JAMA*, 266:93–98, 1991.

Zucman D, Truchis P, Majerholc C, Stegman S, Caillat-Zucman S. Prospective screening for human leukocyte antigen-B*5701 avoids abacavir hypersensitivity reaction in the ethnically mixed French HIV population. *Journal of Acquired Immune Deficiency Syndromes*, 45:1–3, 2007.

17

Overview of Safety Evaluation and Quantitative Approaches during Preclinical and Early Phases of Drug Development

John Sullivan and Hisham Hamadeh

CONTENTS

The drug discovery and development process is long and often filled with unforeseeable hurdles. The major factor in selecting a promising drug candidate is the potential benefit it provides in treating a given disease or syndrome. All beneficial effects of drug, however, are counterbalanced by safety considerations since all therapeutics have adverse effects. This chapter discusses quantitative approaches to early drug development safety. These approaches differ from those in later drug development and in the postmarket phase in which greater numbers of subjects are treated. In early drug development, there is a strong reliance on biomarkers for both efficacy and safety evaluations on account of small numbers of subjects. Regulatory approval standards require larger numbers of subjects to demonstrate differences from control groups and rely on evidence of patients being made to feel better or live longer and occasionally on the presence of a surrogate marker for efficacy.

The quantitative approaches to safety can be summarized by stating that safety assessments are based initially on theoretical and later emerging concerns related to our understanding of biology, on genetics, on generalizations around similar compounds, on observational considerations, and on careful evaluations of small numbers of animal and human subjects. A strong reliance on dose–response or concentration–response is needed, along with specific quantitative approaches to safety biomarkers or pharmacodynamic (PD) endpoints coupled with a good understanding of human physiology and pathophysiology. If the target is inside the cell, a small molecule is the selected modality. On the other hand, if the target is outside the cell, a biologic such as a monoclonal antibody may be preferred (since biologics typically do not enter cells). Some targets may be amenable to either approach. The modality (small molecule or biologic) automatically determines some of the safety considerations a drug developer may need to address. In general, biologics have more specificity for a given target and have less "off-target" effects (though not necessarily fewer side effects). Biologics are not well absorbed orally and so must be administered parenterally, in contrast to small molecules that are often administered orally in pill form. Small molecules generally have more pharmacokinetic (PK) drug interactions compared to biologics and these interactions not uncommonly have safety considerations.

A discussion of issues around drug discovery where theoretical safety concerns often dominate, will be followed by approaches to animal studies, and followed by early human studies. This chapter includes analyses of data related to the following topics:

- Mechanism of action
- Profiling techniques
- Safety pharmacology studies
- Toxicology studies in animals
- Potential for neutralizing antibodies
- Potential for drug interactions
- Early clinical development safety

Animal studies evaluating pharmacological, toxicological, and histopathologic outcomes provide a means by which potential safety and toxicity issues that could arise during the development process are identified. Human studies commence once acceptable outcomes in animal studies are observed. Similar to animals, the first studies in humans (first in human [FIH] studies) involve a small number of subjects (usually healthy volunteers) where the drug is studied at a range of doses. For studies in life-threatening diseases where the first doses will be administered to patients (usually with advanced cancer where usual therapeutic options have been exhausted), more toxicity can be tolerated and initial doses are potentially therapeutic. Identification of a maximum tolerated dose (MTD) is usually a specific goal. This chapter describes various approaches used for designing and analyzing safety-related issues with specific examples provided as appropriate.

17.1 Analyses of Safety Based on a Drug's Mechanism of Action

Understanding the biological mechanism of action of a drug allows one to predict both its therapeutic and toxic effects without necessarily needing a particular quantitative approach. For example, a therapeutic that targets bone marrow stem cells and raises platelet levels can have both therapeutic and toxic effects and the likelihood of adverse reactions are increased with increasing doses. Romiplostim (a biologic) and eltrombopag (a small molecule) both target the thrombopoietin (TPO) receptor c-Mpl and are used to correct low platelet counts in patients with idiopathic or immune thrombocytopenic purpura in order to prevent bleeding (Amgen 2013; GlaxoSmithKline 2013). Both may increase platelet counts to prevent bleeding but may also cause adverse effects such as thrombosis. Off-target effects such as hepatotoxicity have been observed with the small molecule but not the biologic. Similarly, drug interactions based on cytochrome P450 (CYP) enzymes or transporter function have also been observed with the small molecule but not the biologic (package insert eltrombopag, package insert romiplostim). These safety issues were predicted based on the mechanism of action of the therapeutic—in this case, a biologic (peptibody or Fc construct protein) with more specificity for the target versus a small molecule with some off-target effects and liability for drug interactions.

Another example at the drug discovery level would relate to our understanding of structure–function relationships based on chemistry. Quinolones utilized as antibiotic therapies, for example, have been studied extensively in humans to the extent that various toxicities have been related to the various parts of the active molecule. One arrangement of the quinolone pharmacophore can cause phototoxicity, another CNS disturbances, another QT prolongation, and another disordered glucose homeostasis (Tillotson 1996; Lipsky and Baker 1999). The intricacy in the design of quinolones is to find a structure that maximizes benefit while minimizing risk. It takes many years to understand the relationships between structure and function and requires data from both animals and humans treated with a variety of different molecules from a given class of compound.

17.2 Analyses of Drug Safety Using Profiling Techniques

"Toxicogenomics is defined as the application of genomic technologies (for example, genetics, genome sequence analysis, gene expression profiling, proteomics, metabolomics, and related approaches) to study the adverse effects of environmental and pharmaceutical chemicals on human health and the environment" (National Research Council 2007). Toxicogenomics evaluates changes in gene, protein, and metabolite expression patterns associated with chemical exposure, in conjunction with phenotypic responses in organisms, tissues, and cells. These techniques have allowed us to understand how structurally unrelated chemicals can produce similar gene expression profiles. For example, the expression profiles of chemically distinct compounds belonging to the peroxisome proliferator class produce similar patterns of gene expression but distinct from those of phenobarbital that produces similar histopathologic changes in the liver (Hamadeh et al. 2002). Quantitative approaches to this discipline involve novel methods using small numbers of animals.

In general, systems-level understanding of molecular perturbations is crucial for evaluating chemical-induced toxicity risks. Microarray data provide comprehensive gene expression responses against chemical exposure, and therefore the toxicogenomics approach is highly advantageous for understanding systems-level biological perturbations. To solve the difficulty in handling huge amounts of toxicogenomics data sets, preparation of toxicogenomics biomarker gene sets and implementation of gene-set-level data analysis are effective.

The first step is to identify biological pathways that were affected by the chemical exposure, and detailed expression analysis of the individual genes will be needed to focus on the affected biological pathways. When available, large-scale toxicogenomics reference databases will be utilized for comparative analysis to appropriately evaluate the toxicological significance. To comprehend the systems-level molecular dynamics, relationships among pathways should also be taken into consideration. Such an analytical flow can be automated if appropriate computational skills are available. Refining toxicogenomic biomarker gene sets, scoring algorithm, and development of user-friendly integrative software will substantially help the utilization of the toxicogenomics data set to evaluate biological response by which hazardous effects of exposed chemicals could be appropriately managed.

Another quantitative approach examines genetic variations that influence individual response to a drug and use this knowledge to predict treatment outcome. Screening for key genomic biomarkers in early development would establish the patient population that should (or should not) be treated with the drug from both an efficacy and a safety perspective. For example, pharmacogenomic studies indicate there are at least 30 hERG gene polymorphisms that correlate with arrhythmias or prolonged QT intervals (He et al. 2013). Genome association studies with iloperidone, an antipsychotic drug, indicate six single nucleotide polymorphism (SNPs) that are significantly associated with change in QT in iloperidone-treated patients (Volpi et al. 2009). In another example, genome-wide associated studies show that genetic variations for certain human leukocyte antigens (HLAs) are associated with drug-induced hepatotoxicity with flucloxacillin (Daly 2012). In another example, lumiracoxib, a selective cyclooxygenase-2 inhibitor used for the treatment of symptomatic osteoarthritis and acute pain, is strongly associated with the HLA-DQAI*0102 allele (Singer et al. 2010). Carriers of this gene have been identified as those at risk for lumiracoxib-induced hepatotoxicity. Lumiracoxib was not approved in the United States on account of hepatic safety signals during clinical development but was approved and subsequently withdrawn from the European Union (EU) market because of cases of hepatotoxicity prior to the discovery of the genetic predisposition to hepatic toxicity. A further example of identification of safety issues in the postmarketing phase that relies on genetic or genomic information is the subgrouping of the population of cancer patients that respond to antibodies to epidermal growth factor receptor (EGFR) (cetuximab and panitumumab). Analyses of tumor samples from patients with metastatic colorectal cancer have revealed that patients with mutant KRAS tumors (about 40% of the population) have zero chance of responding to the drug and will only experience toxicity when administered these drugs (Amado et al. 2008). Only patients with wild-type KRAS have an opportunity to respond to this class of therapeutics. Accordingly, the genetic testing of the tumor and identification of those not able to respond to the therapeutic becomes both a safety and an efficacy issue.

17.3 Analyses of Safety Pharmacology Studies

Prior to the conduct of human trials, safety pharmacology studies are typically conducted especially with small molecules. Pharmacological studies evaluate various organ systems to understand or rule out toxic effects. The studies conducted depend on the modality and the target. For example, a compound for the treatment of cardiovascular disease could have a battery of assessments using both *in vitro*, *ex vivo*, and *in vivo* methods. The proarrhythmic potential of new drugs can be predicted from a variety of nonclinical methods including testing the binding properties of the drug candidate against a panel of crucial cardiac ion channel targets, such as the hERG K+ channel (Kv11.1), testing its effect on the action potential on mammalian heart cells, and *in vivo* testing its effects on QT interval in dogs (Redfern et al. 2003). Such tests can inform the need for formal clinical studies designed to assess the cardiac activities of drugs (ICH 2005). Untoward modulation of ion channels may cause potentially fatal arrhythmias such as torsades des pointes (a specific type of ventricular tachycardia usually related to prolonged repolarization of the heart muscle as reflected in measurement of the QT interval). Cardiovascular evaluations in FIH studies would include measuring blood pressure, heart rate, and assessing QT, PR, and QRS intervals on the electrocardiogram (ECG). Nonclinical safety studies are conducted with small cohorts for practical reasons and because there is typically a clear dose- and concentration-related relationship between the small molecule drug and the safety PD endpoint, such as a QT interval.

The analyses are typically descriptive but may fulfill criteria for statistical significance depending on the outcome measure and strength of the relationship between drug response and dependent variable. Other safety pharmacology studies may include effects of high single doses on the respiratory system, the central nervous system, endocrine system, renal system, or gastrointestinal system depending on the compound or disease. In general, high single doses are studied, which would far exceed the doses intended to be administered to people but which may indicate potential toxicity in humans. The dose- and concentration-related relationships between drug and receptor usually conform to a log concentration–response relationship as described in the Hill equation (Hill 1910).

17.4 Analyses of Toxicology Studies in Animals

These studies are required prior to the administration to humans. The various International Conference on Harmonization (ICH) guidelines (ICH M3 (R2) 2009; ICH S9 2010; ICH S6 (R1) 2011) describe the studies required by regulatory authorities prior to administration to humans. S6 deals with biologic products, while M3 is prescriptive for all products and addresses harmonization of nonclinical studies to support various stages of human drug development. S9 addresses the abbreviated package required for life-threatening diseases typically first administration of a therapeutic to a patient with advanced cancer. For S6 and M3, the general idea is to identify and describe toxicity in animals at doses and concentrations that far exceed those that would be administered to humans, thus identifying potential on-target toxicities for humans. These toxicology studies involve small numbers of healthy animals (initially studied in rodents such as mice and rats, then progressing to larger animals, such as monkeys and dogs) that are treated at a range of doses. Some biologics are not cross-reactive with rodents or dogs and toxicology studies are often conducted in cynomolgus monkeys. Occasionally, a target for a biologic is only expressed in humans or higher primates. In such cases, selection of a FIH dose would be made based on other criteria such as minimally active biological effect level (MABEL) (Duff 2006; Tibbits et al. 2010). Toxicology studies carefully evaluate for relatively common effects (usually extension of pharmacological effect) and dose–response or concentration–response relationships. Such effects may include detection of biomarkers, PK markers, clinical chemistry, clinical signs, and qualitative evaluations, such as histopathology. Most analyses are descriptive. Animal studies may identify potential toxicities in humans and typically a no adverse effect level (NOAEL) is calculated. This is the dose studied in animals below that in which findings in animals are considered to have effects considered "adverse." Findings are considered adverse when the general health of the animal is thought to be impacted (Dorato and Englehardt 2005). Statistical analyses are descriptive and much of the data are subject to qualitative analyses; for example, the histological results are generally descriptive interpretations. Some molecular biology and molecular pathology endpoints are starting to be assessed from preclinical toxicology studies and are used to help elucidate the mechanisms of disease or treatment related conditions. The toxicology findings provide a basis for calculating a safe starting dose in humans. The FDA guideline (FDA Guidance for Industry 2005) outlines a process (algorithm) and vocabulary for deriving the maximum recommended starting dose (MRSD) for FIH clinical trials of new molecular entities in adult healthy volunteers. It starts with the determination of a NOAEL as described earlier, and then conversion of a NOAEL to a human equivalent dose (HED), and application of a safety factor. The NOAEL used is typically that of the most sensitive species since the HED would generate the most conservative starting dose. The conversion of NOAEL to HED is based on normalization to body surface area. The body surface area normalization and extrapolation of the animal dose to the human dose is done in one step by dividing the NOAEL in a given species by the appropriate body surface area conversion factor (BSA-CF). This conversion factor is a unitless number that converts mg/kg dose for each animal species to the mg/kg dose in humans, which is equivalent to the animal's NOAEL on a mg/m^2 basis. Once the HED has been derived, a "safety factor" is added to attempt to ensure that the first dose in humans will not cause adverse effects. "The safety factor" is typically at least 10, which translates to the MRSD as one-tenth of the HED. How much of a safety factor is selected will depend on a variety of factors including knowledge of the drug and class, the disease, and the nature of the toxicology findings as well as the perceived risks to humans. Considerations that would tend to increase the safety factor include severe or not easily monitorable toxicity or a steep dose–response relationship for the drug. An alternative approach to this algorithm can also be proposed in which primary emphasis is on animal PKs and modeling rather than dose (Mahmood et al. 2003). Whether this approach is taken depends on the amount of PK data and the robustness of the model used to predict human PK. A safety factor is still usually added to the projected starting dose in humans. In the study by Mahmood, four methods of allometric scaling were compared for small molecules administered either orally or intravenously for 10 marketed drugs. In the first approach, animal drug clearance was calculated against body weight on a log–log scale and an allometric equation used to predict clearance in humans. The equation was $CL = a(W)^b$, where W is the body weight and a and b are the coefficient and exponent of the allometric equation, respectively. In the second approach, maximum life span potential was used in addition to body weight to calculate human clearance. In the third approach, brain weight and body weight were used to calculate human clearance, and in the fourth approach, a correction factor was added to the species tested that was closest to human clearance on a mg/kg basis.

All four approaches appeared to produce a safe starting dose. For biologics, it was demonstrated that when PK was linear in cynomolgus monkeys, it was linear in humans and clearance could be estimated by within 2.3-fold with allometric scaling using body weight (Dong et al. 2011). For nonlinear PK, a population modeling approach using nonlinear mixed-effects modeling software, NONMEM®, was applied to describe monkey data by a two-compartment PK model with parallel linear and nonlinear elimination from the central compartment. The best results were obtained when the doses achieved target saturating concentrations as might have been anticipated.

For drugs intended to be administered to patients with life-threatening diseases such as advanced cancer, ICH S9 provides guidance on toxicology studies and how FIH doses are selected. The concept is that the first dose administered to humans should be administered on a continual basis (rather than as a single dose) and should be a potentially therapeutic dose, even though data indicate that there is a low chance (about 5%) that a given subject will benefit from a potential therapeutic in a traditional FIH trial (Horstmann et al. 2005). "A common approach for small molecules is to set a start dose at 1/10 the severely toxic dose in 10% of the animals (STD10) in rodents. If the nonrodent is the most appropriate species then one-sixth of the highest non-severely toxic dose (HNSTD) is considered an appropriate starting dose. The HNSTD is defined as the highest dose level that does not produce evidence of lethality, life-threatening toxicities or irreversible findings." (ICH S9)

17.5 Analyses of Data Evaluating Potential for Neutralizing Antibodies

Nonclinical studies with biologics often result in the development of both binding and neutralizing antibodies. Rates of antibody formation can be problematic from a toxicological viewpoint since the antibody may interfere with drug action thus not allowing potential toxicities to manifest. It is not surprising that an animal will develop an immune response to a human protein. Antibody effects on PK have been documented to both reduce and increase concentrations of the drug of interest and to not alter PK. Occasionally, antibody formation can lead to significant toxicity in the animal, which is not predictive for what will happen in humans.

In humans, the greatest concern for biologic products relates to those products that are most similar to endogenous proteins. For example, manufactured erythropoietin administered exogenously can rarely induce an immune response in which the antibody neutralizes the endogenous protein resulting in pure red cell aplasia (PRCA). Thus, instead of raising the hemoglobin and treating anemia, the result is a falling hemoglobin and production of anemia. Some years back, an outbreak of PRCA occurred in the European Union, Canada, and Australia, which was attributed to a form of erythropoietin (Eprex) (Boven et al. 2005). The most probable cause related to organic compounds leached from uncoated rubber stoppers in prefilled syringes containing polysorbate 80. A change to fluororesin stoppers along with an EU directive to stop subcutaneous administration resulted in a marked decrease in the problem. Analyses of epidemiological data indicated that the rate ratio fell by a factor of 17 ($p < 0.0001$) when coated stoppers were introduced along with intravenous administration only. Similarly, a therapeutic designed to treat thrombocytopenia (megakaryocytic growth and differentiating factor [MGDF]) produced neutralizing antibodies. Because of the sequence homology to the endogenous protein (TPO), the clinical consequence of the neutralizing antibody was a drop in platelets instead of an elevation (Li et al. 2001). Analyses of these data were descriptive, observational, and correlative and resulted in discontinuation of development of this product. Robust assay methodology is required for antibody detection and confirmation of whether the screening result is neutralizing. Most biologics currently in development are monoclonal antibodies. The potential consequence of neutralizing antibody formation to a monoclonal antibody is an absence of therapeutic effect and rarely a change in PKs. While this is still a concern, it is less of a concern in comparison to a neutralizing antibody to an endogenous protein. Low rates of neutralizing antibodies to monoclonal antibodies may be a manageable issue for drug development depending on the product and indication.

17.6 Analyses of Data Evaluating Potential for Drug Interactions

Nonclinical studies for small molecules using a variety of techniques such as *in vitro* human CYP inhibitor profiles will identify a given compound's potential for drug interaction as an inhibitor of other drugs that are metabolized by CYP enzymes (e.g., CYP 1A2, 2B6, 2C8, 2C9, 2C19, 2D6, 3A). Similarly, evaluation of metabolic pathways for a given compound will identify whether phase 1 metabolism is involved and what potential CYPs act as substrates for the new drug of interest. Confirmation of potential for drug interactions will rely on studies during human clinical development. These particular studies are known as extended phase 1 studies and use small cohorts of healthy subjects and usually evaluate the single-dose PK profile of the drug being studied in terms of a "worst-case scenario." A drug such as ketoconazole or itraconazole will provide maximum inhibition of CYP 3A metabolism, which is the most common CYP enzyme that metabolizes small molecules. If the new drug utilizes CYP 3A in its metabolism, a single-dose study with a strong inhibitor will demonstrate how much of a change in concentration is likely with the novel compound (where the new drug is a victim).

Conversely, a drug such as midazolam happens to be an almost pure CYP 3A substrate and can be used as a probe "victim" to reliably detect whether the new drug is likely to cause an interaction by inhibiting CYP3A (where the new drug is a perpetrator). The selection of the number of subjects for a given drug–drug interaction study will depend on how small an effect is clinically important to detect or rule out, and the inter- and intrasubject variability in PK measurements. The analysis is based on a ratio of the geometric means and 90% confidence intervals for the PK variables area under the curve (AUC) and observed maximum concentration (C_{max}) with and without the interacting drug. Interpretation of the results will depend on the clinical importance. For example, a less than twofold difference would not likely be clinically important. In contrast, a more than threefold increase in PK exposure would be potentially important depending on the usual concentrations of the drug and the therapeutic index. Conversely, a ratio of the geometric means of close to one with a confidence interval within the 0.8–1.25 boundary would allow one to confidently exclude an interaction.

In recent years, new interactions have been discovered with drug transporters at various tissue locations (e.g., GI tract, liver, brain, and kidney). Depending on nonclinical findings, studies in humans may be required during drug development to inform labeling and safe use of the drug. The general approach to these interaction studies is similar to those of CYPs described earlier. Various regulatory and industry guidelines describe the approach (Bjornsson et al. 2003; Giacomini et al. 2010; FDA Draft Guidance for Industry 2012). Important drug interactions have resulted in the withdrawal of several drugs from the market. Examples include terfenadine (Seldane), which was an antihistamine used by millions, and mibefradil (Posicor), a calcium channel blocker for angina and hypertension. Terfenadine was a prodrug metabolized by CYP 3A, which caused QT prolongation at high concentrations. When strong CYP 3A inhibitors were administered such as antifungals, increases in plasma concentrations resulted in QT prolongation and a sometimes fatal arrhythmia ("torsades de pointes" a type of ventricular tachycardia). Mibefradil was an inhibitor of CYP 3A4 and 2D6 and caused a multiplicity of drug interactions, some of which were fatal. The drug was withdrawn since there were alternative therapies available that did not have these liabilities.

17.7 Quantitative Approaches in Early Clinical Development Safety

FIH studies are typically conducted in healthy subjects beginning with a single ascending dose study with the primary objective being safety assessment. Secondary objectives include PK and PD assessments if possible. Statistical analyses are typically descriptive. Dose escalation continues until potential drug-related toxicities are identified if this is reasonable. Sometimes, an upper dose limit is prespecified if a particular toxicity could be anticipated and it would not be considered reasonable to have healthy subjects subjected to this experience. PK measures confirm that the drug is getting into the systemic circulation at biologically important levels. Noncompartmental measures such as C_{max}, time after drug administration that this occurs (T_{max}), and AUC are determined. Regulatory authorities typically prefer these measures since they are less susceptible to judgment calls or subjectivity (in comparison to compartmental parameters used for modeling approaches or even calculations of half-life). PK measures also will confirm that drug concentrations are "linear" or dose proportional as opposed to nonlinear or nondose proportional. FIH protocols typically require a review of data after each cohort has completed dosing. This is usually conducted by the investigator along with representatives from the sponsor such as the medical monitor and the safety officer. Others such as the statistician or PK scientist may also participate. In recent years, the first subject to receive the drug is sometimes treated as one of a "sentinel pair" with the other subject receiving a placebo. This approach has been more frequently adopted following the TeGenero incident (Duff 2006) where a cohort of subjects were dosed within a short period of time and all subjects receiving the active treatment developed a severe form of cytokine release syndrome within about an hour of drug administration. Protocols are also written with stopping rules for both individuals and cohorts. The common terminology criteria for adverse events (CTCAE) scale is typically used to describe and quantitate these toxicities. The latest version 4.0 was published in 2009. The National Cancer Institute (NCI) CTCAE is a descriptive terminology that can be utilized for adverse event (AE) reporting (CTCAE, U.S. Department of Health and Human Services 2009). A grading (severity) scale is provided for each AE term typically ranging from 1 (mild) to 4 (life threatening). For example, a grade 3 toxicity in a healthy subject that was considered drug related (i.e., an adverse drug reaction [ADR]) may trigger the stopping rule. The team (investigator and sponsor representatives) would then evaluate all data and make a decision on next steps, which could range from stopping the study to no change in plans. Intermediate decisions could consist of amending the protocol to repeat a cohort or expand the number at a given dose or to study a smaller dose increment than planned. Sometimes, if there are particular concerns around potential toxicities, an external committee or data monitoring committee (DMC) will rule on these issues. It is uncommon to use a DMC for phase 1 studies and typically only on request from a regulator or ethics committee. When the single-dose study has been completed, a multiple ascending dose (MAD) study is conducted, which may study fewer doses but obviously for a longer period of time usually ranging from 14 to 30 days for small molecules where the drug is usually administered once or twice daily depending on the duration of activity. For biologics with long durations of activity, it may not be necessary to study multiple doses in healthy subjects and a single dose may suffice before proceeding to multiple-dose studies in patients. Evaluations of toxicities in these early human trials are typically descriptive and observational. Occasionally, dose- or concentration-related toxicities may be identified. This is more usually the case where the toxicity is an extension of pharmacological effect. An example would be a drug to reduce glucose levels where therapeutic doses ensured normoglycemia (as opposed to hyperglycemia) and higher doses caused hypoglycemia. This is an example of a drug in which the PD measure (also sometimes called a biomarker) is both an efficacy and a safety endpoint.

The main goal in FIH studies is to identify safety issues. Typically, only common adverse effects will be detected because of the small sample size. Uncommon or rare adverse effects may be detected but this is less likely with small cohort sizes. The occurrence of a rare serious drug-related AE early in drug development is problematic in that it is not possible to assess the frequency of the event given the small number of subjects studied. The ability to study a range of doses and look for dose–response relationships adds considerable statistical power to the detection of safety issues for common on-target effects. The emergence of nonmanageable safety concerns may result in discontinuation of a given drug candidate. Once successful safety-related outcomes are observed in single- and multiple-dose cohorts, studies in humans are expanded to larger trials to assess the efficacy and safety of the drug in disease populations. These studies are variously described as phase 1b or 2a or phase 1/2. The studies are designed to have several dose levels studied and are potentially analyzed as formal dose–response studies, which adds statistical power (FDA Guidance for Industry 2003).

For FIH advanced disease oncology trials using the ICH S9 approach, initial dosing is usually based on a 28-day cycle where the therapeutic is administered either daily or a certain number of days in a cycle for a small molecule (depending on putative mode of action and expected toxicities). For biologics with longer durations of activity, the therapeutic may be administered weekly or at longer intervals depending on the half-life or clearance. Assessment of toxicity during the first cycle is made by the investigator (with or without a data review team) and according to what is specified in the protocol. Traditionally, a modified Fibonacci design has been used where three subjects are dosed, and if one develops a dose limiting toxicity (DLT), three more subjects are added to the cohort. If no further subjects develop DLTs, dose escalation may proceed. If however another subject develops a DLT, that dose will be declared a nontolerated dose, and the prior lower dose will be declared the MTD. There are variations on this approach with some investigators using just one subject per dose for traditional cytotoxics (Eisenhauer et al. 2000), others using continual reassessment methods, and more recently, investigators using studies in targeted patient populations. Assessment of human toxicity is made using CTCAE version 4.0. Typically, a grade 4 event would be considered a DLT.

Early clinical trials also formally analyze PK and PD using modeling approaches. For example, erythropoiesis-stimulating proteins have both efficacy endpoints (reduction in transfusions) and safety issues (increase in cardiovascular morbidity at high target hemoglobin concentrations). Modeling approaches (Agoram et al. 2007) can act as a suitable tool to predict PK responses in untested doses and schedules and thereby improve potential efficacy and safety. PK in chronic kidney disease patients is not different from that in healthy subjects. In this example, the quantitative approach relied on obtaining PK profiles from 140 healthy subjects receiving single and multiple intravenous and subcutaneous doses of darbepoetin alfa over a range of fixed and weight-based dosing. Data were analyzed by a nonlinear mixed effect modeling approach using NONMEM software. Covariates were identified and the model was evaluated by comparing simulated profiles to observed profiles in a test data set. The population PK model, which included first-order absorption, two-compartment disposition, and first-order elimination, adequately described the data. Modeling indicated that there was a nearly twofold disproportionate dose–exposure relationship at the highest compared to the lowest dose, which appeared to reflect changing bioavailability. The covariate analysis showed that increasing body weight appeared to relate to increasing clearance and central compartment volume. The absorption rate constant appeared to decrease with increasing age (as might be anticipated given age related changes in lean body mass). The full covariate model performed adequately in a fixed effect prediction test against the external data set. A further modeling example allowed a dose and schedule in pediatrics (from 0.4 mg/kg subcutaneously twice weekly to 0.8 mg/kg subcutaneously once weekly) for etanercept in polyarticular juvenile idiopathic arthritis (JIA) to be approved without a study. A trial with a twice weekly regimen had demonstrated efficacy and safety in this indication. An adult trial had demonstrated equivalence between a twice-weekly and once-weekly regimen in rheumatoid arthritis (Lee et al. 2003). A logistic regression analysis with NONMEM was applied to describe the exposure response relationship and the 95% confidence intervals constructed by bootstrapping. A PK model had demonstrated similar PK on a weight-based regimen for children and adults.

Quantitative approaches in early development safety are well illustrated by abuse liability studies that are conducted to determine whether drugs should be federally scheduled in the United States and if so, what schedule they should be given. These studies select subjects that have histories of drug addiction but are not currently using drugs or addicted. The volunteers are reliably able to detect euphoric or reinforcing effects of new chemical entities using validated questionnaires in human experimental pharmacology studies. The studies are designed as a Latin square crossover trial in which all subjects receive all treatments that include at least three doses of the drug of interest, several doses of an active control (to confirm assay validity), and a negative control (placebo). In the prototypical study of diazepam abuse liability (Sullivan et al. 1993), treatments were administered orally under double-blind conditions to 12 subjects in an inpatient setting with a

3-day washout according to two 6 × 6 balanced Latin squares. Drug conditions tested were placebo, 120 and 240 mg of pentobarbital (positive controls), and 10, 20, and 40 mg diazepam. Effects were assessed on measures of subjective and behavioral responses. Validated questionnaires derived from subscales of the Addiction Research Center Inventory (ARCI) were administered at regular intervals, a "euphoria" scale derived from the morphine–benzedrine group subscale of the ARCI, a "sedation" scale derived from the pentobarbital-chlorpromazine-alcohol group subscale of the ARCI, and a subject liking categorical scale; symptoms and observer-rated scales were also measured. Changes from baseline scores were calculated for each measure of drug response. These were graphed and areas under the curve (AUC) for 8 h after drug administration (typical duration of drug effect) calculated by the method of trapezoids. The AUC scores were analyzed by the use of a Latin square analysis of variance (ANOVA). Pooled replicate Latin square analysis was performed (Cochran and Cox 1957). This analysis estimated mean squares for treatments subjects and occasions as well as residuals. The between treatment sums of squares were partitioned into orthogonal comparisons of interest. These were (1) comparison of mean placebo response, with diazepam response and with pentobarbital response, and (2) the mean squares for the validity measures of a five-point parallel-line bioassay, which are regression, preparations, nonparallelism, and deviation from linearity (Finney 1964). Statistical significance for F ratios was accepted at $p < 0.05$. The results showed that for a variety of measures, pentobarbital was distinguished from placebo in a dose-responsive manner, which replicated previous studies of pentobarbital thus validating the experiment for testing the psychoactive effects of diazepam. The time course of effects was similar for both drugs. Pentobarbital showed a dose–response relationship for euphoria, sedation, subject liking, and symptoms. Diazepam was also distinguished from placebo and showed dose–response relationships on these measures. Observer-rated measures showed similar results and relative potency measures were reliably calculated for pentobarbital and diazepam. Diazepam (Valium) is a benzodiazepine schedule IV drug under the Controlled Substances Act of the United States. This means that it has a low abuse potential compared to drugs in schedules III and lower (e.g., opioids) and has a currently accepted medical use, and abuse of the drug may lead to physical or psychological dependence to a lesser degree than drugs scheduled as III or lower, and the prescribing physician must have a current DEA license. Hypnotics such as zolpidem (Ambien) and eszopiclone (Lunesta) have provided similar abuse liability assessments and are also schedule IV. Similar abuse liability assessments of other psychoactive drugs and analgesics have resulted in lack of scheduling for sumatriptan, transnasal butorphanol, and tramadol (Sullivan et al. 1992; Jasinski et al. 1993; Preston et al. 1994). These analyses demonstrate that studies with small numbers of subjects and the use of powerful crossover designs can provide valid safety assessments of subjective PD responses.

17.8 Conclusions

This chapter has outlined a variety of approaches to analysis of early drug development safety. There is reliance on a variety of techniques ranging from assessment of genetic characteristics to modeling and traditional statistical approaches. Because of small numbers in both the nonclinical and clinical settings, there is also a strong reliance on qualitative and descriptive data along with clinical judgment as it pertains to safety evaluation.

Acknowledgments

The authors sincerely thank Jing Huang PhD and Yeshi Mikyas PhD for assistance with editing and review, and Linda Godfrey for administrative and other assistance.

References

Agoram B, Sutjandra L, Sullivan JT. 2007. Population pharmacokinetics of darbepo-etin alfa in healthy subjects. *British Journal of Clinical Pharmacology* 63(1): 41–52.

Amado RG, Wolf M, Peeters M et al. 2008. Wild-type KRAS is required for panitu-mumab efficacy in patients with metastatic colorectal cancer. *Journal of Clinical Oncology* 26: 1626–1634.

Amgen. 2013. Romiplostim package insert. Available from: http://pi.amgen.com/united_states/nplate/nplate_pi_hcp_english.pdf. Accessed May 2013.

Bjornsson TD, Callaghan JT, Einolf HJ et al. 2003. The conduct of in vitro and in vivo drug-drug interaction studies: A pharmaceutical research and manufacturers of America (PhRMA) perspective. *Drug Metabolism and Disposition: The Biological Fate of Chemicals* 31(7): 815–832.

Boven K, Stryker S, Knight J et al. 2005. The increased incidence of pure red cell apla-sia with an Eprex formulation in uncoated rubber stopper syringes. *Kidney International* 67(6): 2346–2353.

Cochran WG, Cox GM. 1957. *Experimental Designs*, 2nd edn. New York: John Wiley.

Daly AK. 2012. Using genome-wide association studies to identify genes important in serious adverse drug reactions. *Annual Review of Pharmacology and Toxicology* 52: 21–35.

Dong JQ, Salinger DH, Endres CJ et al. 2011. Quantitative prediction of human pharma-cokinetics for monoclonal antibodies: Retrospective analysis of monkey as a sin-gle species for first-in-human prediction. *Clinical Pharmacokinetics* 50(2): 131–142.

Dorato MA, Engelhardt JA. 2005. The no-observed-adverse-effect-level in drug safety evaluations: Use, issues, and definition(s). *Regulatory Toxicology and Pharmacology* 42(3): 265–274.

Duff G. 2006. *Expert Scientific Group on Phase One Clinical Trials: Final Report*. London, U.K.: The Stationary Office.

European Medicines Agency. 2011. ICH guideline S6 (R1)—Preclinical safety evaluation of biotechnology—Derived pharmaceuticals. Available from: http://www.ema.europa.eu/docs/en_GB/document_library/Scientific_guideline/2009/09/WC500002828.pdf. Accessed June 2011.

FDA Draft Guidance for Industry. 2012. Drug interaction studies—Study design, data analysis, implications for dosing, and labeling recommendations. Washington, DC: U.S. Department of Health and Human Services, Food and Drug Administration, Center for Drug Evaluation and Research (CDER).

FDA Guidance for Industry. 2003. Exposure-response relationships—Study design, data analysis, and regulatory applications. Washington, DC: U.S. Department of Health and Human Services, Food and Drug Administration, Center for Drug Evaluation and Research (CDER), Center for Biologics Evaluation and Research (CBER).

FDA Guidance for Industry. 2005. Estimating the maximum safe starting dose in initial clinical trials for therapeutics in adult healthy volunteers. Washington, DC: U.S. Department of Health and Human Services, Food and Drug Administration, Center for Drug Evaluation and Research (CDER).

Finney DJ. 1964. *Statistical Method in Biological Assay*, 2nd edn. London, U.K.: Charles Griffin, pp. 99–136.

Food and Drug Administration. 2010. ICH guidelines S9—Guidance for industry S9 nonclinical evaluation for anticancer pharmaceuticals. Available from: http://www.fda.gov/downloads/Drugs/.../Guidances/ucm085389.pdf. Accessed November 10, 2013 (ICH Topic S9 March 2010).

Giacomini KM, Huang SM, Tweedie DJ et al. 2010. Membrane transporters in drug development. *Nature Reviews, Drug Discovery* 9(3): 215–236.

GlaxoSmithKline. 2013. Eltrombopag package insert. Available from: https://www.gsksource.com/gskprm/htdocs/documents/PROMACTA-PI-MG-COMBINED.PDF. Accessed November 2012.

Hamadeh HK, Bushel PR, Jayadev S et al. 2002. Gene expression analysis reveals chemical-specific profiles. *Toxicological Sciences* 67(2): 219–231.

He FZ, McLeod HL, Zhang W. 2013. Current pharmacogenomic studies on hERG potassium channels. *Trends in Molecular Medicine* 19(4): 227–238.

Hill AV. 1910. The possible effects of the aggregation of the molecules of haemoglobin on its dissociation curves. *Journal of Physiology* 40: iv–vii.

Horstmann E, McCabe MS, Grochow L et al. 2005. Risks and benefits of phase 1 oncology trials, 1991 through 2002. *New England Journal of Medicine* 352(9): 895–904.

ICH. 2005. ICH topic E14 CHMP/IC/2/04. The clinical evaluation of QT/QTc interval prolongation and proarrhythmic potential for non-antiarrhythmic drugs. Available from: http://www.ema.europa.eu/docs/en_GB/document_library/Scientific_guideline/2009/09/WC500002879.pdf. Accessed March 5, 2013.

ICH. 2009. ICH topic. ICH guideline M3(R2) on non-clinical safety studies for the conduct of human clinical trials and marketing authorisation for pharmaceuticals. Available from: http://www.ema.europa.eu/docs/en_GB/document_library/Scientific_guideline/2009/09/WC500002720.pdf. Accessed November 11, 2013.

ICH. 2011. S6 addendum to the preclinical safety evaluation of biotechnology-derived pharmaceuticals. Available from: http://www.fda.gov/downloads/Drugs/GuidanceComplianceRegulatoryInformation/Guidances/UCM194490.pdf. Accessed February 9, 2014.

Jasinski DR, Preston KL, Sullivan JT et al. 1993. Abuse potential of oral tramadol. *NIDA Research Monograph* 132: 103.

Lee H, Kimko HC, Rogge M, Wang D, Nestorov I, Peck CC. 2003. Population pharmacokinetic and pharmacodynamic modeling of etanercept using logistic regression analysis. *Clinical Pharmacology and Therapeutics* 73(4): 348–365.

Li J, Yang C, Xia Y et al. 2001. Thrombocytopenia caused by the development of antibodies to thrombopoietin. *Blood* 98(12): 3241–3248.

Lipsky BA, Baker CA. 1999. Fluoroquinolone toxicity profiles: A review focusing on newer agents. *Clinical Infectious Diseases* 28(2): 352–364.

Mahmood I, Green MD, Fisher JE. 2003. Selection of the first-time dose in humans: Comparison of different approaches based on interspecies scaling of clearance. *Journal of Clinical Pharmacology* 43: 692–697.

National Research Council United States Committee on Applications of Toxicogenomic Technologies to Predictive Toxicology. 2007. *Applications of Toxicogenomic Technologies to Predictive Toxicology and Risk Assessment.* Washington, DC: National Academies Press. Available from: http://www.ncbi.nlm.nih.gov/books/NBK10219/.

Preston KL, Sullivan JT, Testa M, Jasinski DR. 1994. Psychopharmacology and abuse potential of transnasal butorphanol. *Drug and Alcohol Dependence* 35(2): 159–167.

Redfern WS, Carlsson L, Davis AS et al. 2003. Relationships between preclinical cardiac electrophysiology, clinical QT interval prolongation and torsade de pointes for a broad range of drugs: Evidence for a provisional safety margin in drug development. *Cardiovascular Research* 58(1): 32–45.

Singer JB, Lewitzky S, Leroy E et al. 2010. A genome-wide study identifies HLA alleles associated with lumiracoxib-related liver injury. *Nature Genetics* 42(8): 711–714.

Sullivan JT, Jasinski DR, Johnson RE. 1993. Single-dose pharmacodynamics of diazepam and pentobarbital in substance abusers. *Clinical Pharmacology and Therapeutics* 54: 645–653.

Sullivan JT, Preston KL, Testa MP, Busch M, Jasinski DR. 1992. Psychoactivity and abuse potential of sumatriptan. *Clinical Pharmacology and Therapeutics* 52(6): 635–642.

Tibbitts J, Cavagnaro JA, Haller CA, Marafino B, Andrews PA, Sullivan JT. 2010. Practical approaches to dose selection for first-in-human clinical trials with novel biopharmaceuticals. *Regulatory Toxicology and Pharmacology* 58: 243–251.

Tillotson GS. 1996. Quinolones: Structure-activity relationships and future predictions. *Journal of Medical Microbiology* 44(5): 320–324.

U.S. Department of Health and Human Services. May 28, 2009. Common terminology criteria for adverse events (CTCAE). Version 4.02: September 15, 2009. National Institutes of Health, National Cancer Institute. Washington, DC.

Volpi S, Heaton C, Mack K et al. 2009. Whole genome association study identifies polymorphisms associated with QT prolongation during iloperidone treatment of schizophrenia. *Molecular Psychiatry* 14(11): 1024–1031.

Index

Printed and bound by CPI Group (UK) Ltd, Croydon, CR0 4YY

24/10/2024

01778302-0011